Spatial Statistics and
Spatio-Temporal Data

WILEY SERIES IN PROBABILITY AND STATISTICS

Established by WALTER A. SHEWHART and SAMUEL S. WILKS

A complete list of the titles in this series appears at the end of this volume

Spatial Statistics and Spatio-Temporal Data

Covariance Functions and Directional Properties

Michael Sherman
Texas A&M University, USA

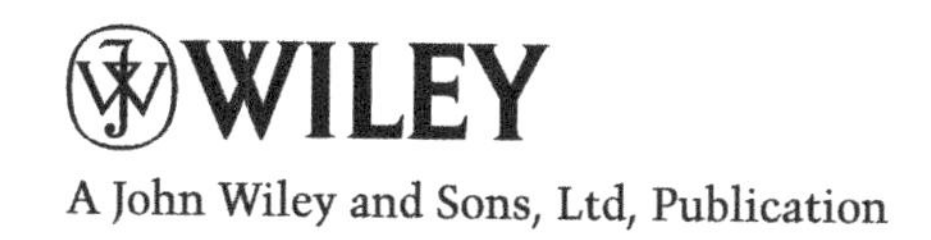

A John Wiley and Sons, Ltd, Publication

This edition first published 2011

Registered office
John Wiley & Sons Ltd, The Atrium, Southern Gate, Chichester, West Sussex, PO19 8SQ, United Kingdom

For details of our global editorial offices, for customer services and for information about how to apply for permission to reuse the copyright material in this book please see our website at www.wiley.com.

Library of Congress Cataloging-in-Publication Data

Sherman, Michael, 1963–
Spatial statistics and spatio-temporal data : covariance functions and directional properties / Michael Sherman.
p. cm.
Includes bibliographical references and index.
ISBN 978-0-470-69958-4 (cloth)
1. Spatial analysis (Statistics) 2. Analysis of covariance. I. Title.
QA278.2.S497 2010
519.5–dc22
2010029551

A catalogue record for this book is available from the British Library.

Print ISBN: 978-0-470-69958-4
ePDF ISBN: 978-0-470-97440-7
oBook ISBN: 978-0-470-97439-1
ePub ISBN: 978-0-470-97492-6

Typeset in 11/13pt Times by SPi Publishers Services, Pondicherry, India

Contents

Preface

The fields of spatial data and spatio-temporal data have expanded greatly over the previous 20 years. This has occurred as the amount of spatial and spatio-temporal data has increased. One main tool in spatial prediction is the covariance function or the variogram. Given these functions, we know how to make optimal predictions of quantities of interest at unsampled locations. In practice, these covariance functions are unknown and need to be estimated from sample data. Covariance functions and their estimation is a subset of the field of geostatistics. Several previous texts in geostatistics consider these topics in great detail, specifically those of Cressie (1993), Chiles and Delfiner (1999), Diggle and Ribeiro (2007), Isaaks and Srivastava (1989), and Stein (1999).

A common assumption on variogram or covariance functions is that they are isotropic, that is, not direction dependent. For spatio-temporal covariance functions, a common assumption is that the spatial and temporal covariances are separable. For multivariate spatial observations, a common assumption is intrinsic correlation; that is, that the variable correlations and spatial correlations are separable. All these types of assumptions make models simpler, and thus aid in effective parameter estimation in these covariance models. Much of this book details the effects of these assumptions, and addresses methods to assess the appropriateness of such assumptions for these various data structures.

Chapters 1–3 are an introduction to the topics of stationarity, spatial prediction, variogram and covariance models, and estimation for these models. Chapter 4 gives a brief survey of spatial models, highlighting the Gaussian case and the binary data setting and the different methodologies for these two data structures. Chapter 5 discusses the assumption of isotropy for spatial covariances, and methods to assess and correct for anisotropies; while Chapter 6 discusses models for spatio-temporal covariances and assessment of symmetry and separability assumptions. Chapter 7 serves as an introduction to spatial point patterns. In this chapter we discuss testing for spatial randomness and models for both regular and clustered point patterns. These and further topics in the analysis of point patterns can be found in, for example, Diggle (2003) or Illian *et al.* (2008). The isotropy assumption for point pattern models has not been as often addressed as in the geostatistical setting. Chapter 8 details methods for testing for isotropy based on spatial point pattern observations. Chapter 9 considers models for multivariate spatial and spatio-temporal observations and covariance functions for these data. Due to spatial correlations and unwieldy likelihoods in the spatial setting, many statistics are complicated. In particular, this means that variances and other distributional properties are difficult to derive analytically. Resampling methodology can greatly aid in estimating these quantities. For this reason, Chapter 10 gives some background and details on resampling methodology for independent, time series, and spatial observations.

The first four chapters and Chapters 7 and 10 of this book are relatively non-technical, and any necessary technical items should be accessible on the way. Chapters 5, 6, 8, and 9 often make reference to large sample theory, but the basic methodology can be followed without reference to these large sample results. The chapters that address the testing of various assumptions of covariance functions, Chapters 5, 6, 8, and 9, often rely on a common testing approach. This approach is repeated separately, to some extent, within each of these chapters. Hopefully, this will aid the data analyst who may be interested in only one or two of the data structures addressed in these chapters. There are no exercises given at the end of chapters. It is hoped that some of the details within the chapters will lend themselves to further exploration, if desired, for an instructor. All data analyses have been carried out using the R language, and various R packages. I have not listed any specific packages, as the continual growth and improvement of these packages would make this inappropriate. Furthermore, as R is freeware, users can experiment, and find the software they are most comfortable with.

This book introduces spatial covariance models and discusses their importance in making predictions. Whenever building models based on data, a key component is to assess the validity of any model assumptions. It is hoped that this book shows how this can be done, and hopefully suggests further methodology to expand the applicability of such assessments.

The content of this book could never have come into being without the benefits of associations with mentors, colleagues, collaborators, and students. Specifically, I greatly appreciate Ed Carlstein and Martin Tanner for their imparting of wisdom and experience to me when I was a student.

I greatly thank colleagues, with whom many of the results in this book have been obtained. Specifically, I thank Tanya Apanosovich, Jim Calvin, Ray Carroll, Marc Genton, Yongtao Guan, Bo Li, Johan Lim, Arnab Maity, Dimitris Politis, Gad Ritvo, and Michael Speed for their collaborative efforts over the years. I also thank Professor Christopher K. Wikle for the use of the Pacific Ocean wind-speed data in Chapters 5 and 6, and Professor Sue Carrozza for use of the leukemia data in Chapter 8. Lastly, I sincerely appreciate the loving aid of my wife, Aviva Sherman, who provided the utmost emotional and technical support in the writing of this book.

1

Introduction

Spatial statistics, like all branches of statistics, is the process of learning from data. Many of the questions that arise in spatial analyses are common to all areas of statistics. Namely,

i. What are the phenomena under study.
ii. What are the relevant data and how should it be collected.
iii. How should we analyze the data after it is collected.
iv. How can we draw inferences from the data collected to the phenomena under study.

The way these questions are answered depends on the type of phenomena under study. In the spatial or spatio-temporal setting, these issues are typically addressed in certain ways. We illustrate this from the following study of phosphorus measurements in shrimp ponds.

Figure 1.1 gives the locations of phosphorus measurements in a 300 m × 100 m pond in a Texas shrimp farm.

i. The phenomena under study are:
   a. Are the observed measurements sufficient to measure total phosphorus in the pond? What can be gained in precision by further sampling?
   b. What are the levels of phosphorus at unsampled locations in the pond, and how can we predict them?

Spatial Statistics and Spatio-Temporal Data: Covariance Functions and Directional Properties Michael Sherman

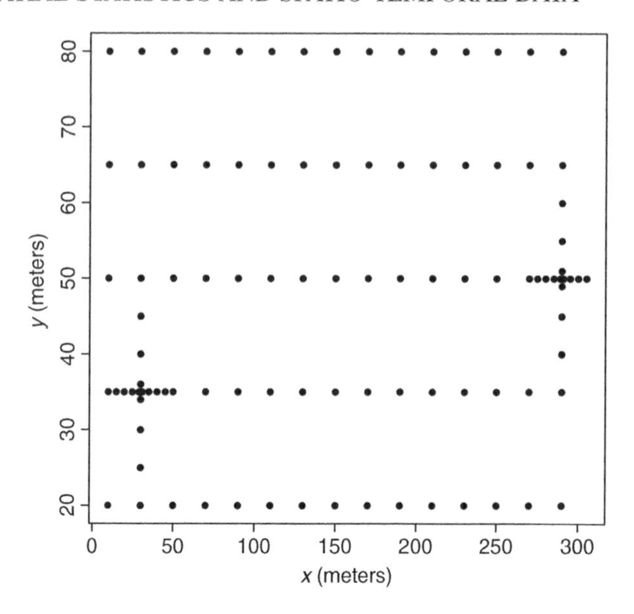

Figure 1.1 Sampling locations of phosphorus measurements.

c. How does the phosphorus level at one location relate to the amount at another location?

d. Does this relationship depend only on distance or also on direction?

ii. The relevant data that are collected are as follows: a total of $n = 103$ samples were collected from the top 10 cm of the soil from each pond by a core sampler with a 2.5 cm diameter. We see 15 equidistant samples on the long edge (300 m), and 5 equidistant samples from the short edge (100 m). Additionally, 14 samples were taken from each of the shallow and deep edges of each pond. The 14 samples were distributed in a cross shape. Two of the sides of the cross consist of samples at distances of 1, 5, 10, and 15 m from the center while the remaining two have samples at 1, 5, and 10 m from the center.

iii. The analysis of the data shows that the 14 samples in each of the two cross patterns turn out to be very important for both the analysis, (iii), and inferences, (iv), drawn from these data. This will be discussed further in Section 3.5.

iv. Inferences show that the answer to (d) helps greatly in answering question (c), which in turn helps in answering question (b) in an informative and efficient manner. Further, the answers to (b), (c), and (d) determine how well we can answer question (a). Also, we will see that increased sampling will not give much better answers to (a); while addressing (c), it is found that phosphorus levels are related but only up to a distance of about 15–20 m. The exact meaning of 'related,' and how these conclusions are reached, are discussed in the next paragraph and in Chapter 2.

We consider all observed values to be the outcome of random variables observed at the given locations. Let $\{Z(\mathbf{s}_i),\ i = 1, \ldots, n\}$ denote the random quantity Z of interest observed at locations $\mathbf{s} \in D \subset \mathbb{R}^d$, where D is the domain where observations are taken, and d is the dimension of the domain. In the phosphorus study, $Z(\mathbf{s}_i)$ denotes the log(phosphorus) measurement at the ith sampling location, $i = 1, \ldots, 103$. The dimension d is 2, and the domain D is the 300 m × 100 m pond. For usual spatial data, the dimension, d, is 2.

Sometimes the locations themselves will be considered random, but for now we consider them to be fixed by the experimenter (as they are, e.g., in the phosphorus study). A fundamental concept for addressing question (iii) in the first paragraph of the introduction is the covariance function.

For any two variables $Z(\mathbf{s})$ and $Z(\mathbf{t})$ with means $\mu(\mathbf{s})$ and $\mu(\mathbf{t})$, respectively, we define the covariance to be

$$Cov[Z(\mathbf{s}), Z(\mathbf{t})] = E[(Z(\mathbf{s}) - \mu(\mathbf{s}))(Z(\mathbf{t}) - \mu(\mathbf{t}))].$$

The correlation function is then $Cov[Z(\mathbf{s}), Z(\mathbf{t})]/(\sigma_{\mathbf{s}}\sigma_{\mathbf{t}})$, where $\sigma_{\mathbf{s}}$ and $\sigma_{\mathbf{t}}$ denote the standard deviations of the two variables. We see, for example, that if all random observations are *independent*, then the covariance and the correlation are identically zero, for all locations $\mathbf{s}$ and $\mathbf{t}$, such that $\mathbf{s} \neq \mathbf{t}$. In the special case where the mean and variances are constant, that is, $\mu(\mathbf{s}) = \mu$ and $\sigma_{\mathbf{s}} = \sigma$ for all locations $\mathbf{s}$, we have

$$Corr[Z(\mathbf{s}), Z(\mathbf{t})] = Cov[Z(\mathbf{s}), Z(\mathbf{t})]/\sigma^2.$$

The covariance function, which is very important for prediction and inference, typically needs to be estimated. Without any replication this is usually not feasible. We next give a common assumption made in order to obtain replicates.

1.1 Stationarity

A standard method of obtaining replication is through the assumption of second-order stationarity (SOS). This assumption holds that:

i. $E[Z(\mathbf{s})] = \mu$;

ii. $Cov[Z(\mathbf{s}), Z(\mathbf{t})] = Cov[Z(\mathbf{s}+\mathbf{h}), Z(\mathbf{t}+\mathbf{h})]$ for all shifts $\mathbf{h}$.

Figure 1.2 shows the locations for a particular shift vector $\mathbf{h}$. In this case we can write

$$Cov[Z(\mathbf{s}), Z(\mathbf{t})] = Cov[Z(\mathbf{0}), Z(\mathbf{t}-\mathbf{s})] =: C(\mathbf{t}-\mathbf{s}),$$

so that the covariance depends only on the spatial lag between the locations, $\mathbf{t}-\mathbf{s}$, and not on the two locations themselves. Second-order stationarity is often known as 'weak stationarity.' Strong (or strict) stationarity assumes that, for any collection of k variables, $Z(\mathbf{s}_i)$, $i = 1, \ldots, k$, and constants a_i, $i = 1, \ldots, k$, we have

$$\begin{aligned} &P[Z(\mathbf{s}_1) \leq a_1, \ldots, Z(\mathbf{s}_k) \leq a_k] \\ &\quad = P[Z(\mathbf{s}_1+\mathbf{h}) \leq a_1, \ldots, Z(\mathbf{s}_k+\mathbf{h}) \leq a_k], \end{aligned}$$

for all shift vectors $\mathbf{h}$.

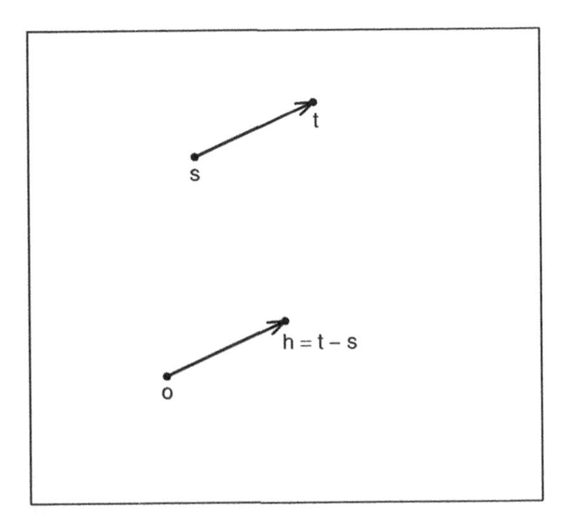

Figure 1.2 A depiction of stationarity: two identical lag vectors.

This says that the entire joint distribution of k variables is invariant under shifts. Taking $k = 1$ and $k = 2$, and observing that covariances are determined by the joint distribution, it is seen that strong stationarity implies SOS. Generally, to answer the phenomenon of interest in the phosphorus study (and many others) only the assumption of weak stationarity is necessary. Still, we will have occasions to use both concepts in what follows.

It turns out that the effects of covariance and correlation in estimation and prediction are entirely different. To illustrate this, the role of covariance in estimation and prediction is considered in the times series setting ($d = 1$). The lessons learned here are more simply derived, but are largely analogous to the situation for spatial observations, and spatio-temporal observations.

1.2 The effect of correlation in estimation and prediction

1.2.1 Estimation

Consider equally spaced observations, Z_i, representing the response variable of interest at time i. Assume that the observations come from an autoregressive time series of order one. This AR(1) model is given by

$$Z_i = \mu + \rho(Z_{i-1} - \mu) + \epsilon_i,$$

where the independent errors, ϵ_i, are such that $E(\epsilon_i) = 0$ and $Var(\epsilon_i) = \eta^2$. For the sake of simplicity, take $\mu = 0$ and $\eta^2 = 1$, and then the AR(1) model simplifies to

$$Z_i = \rho Z_{i-1} + \epsilon_i,$$

with $Var(\epsilon_i) = 1$.

For $-1 < \rho < 1$, assume that $Var(Z_i)$ is constant. Then we have $Var(Z_i) = (1 - \rho^2)^{-1}$, and thus direct calculations show that $Cov(Z_{i+1}, Z_i) = \rho/(1 - \rho^2)$. Iteration then shows that, for any time lag k, we have:

$$Cov(Z_{i+k}, Z_i) = \rho^{|k|}/(1 - \rho^2).$$

Noting that the right hand side does not depend on i, it is seen that SOS holds, and we can define $C(k) := \rho^{|k|}/(1 - \rho^2)$. Further, note that the distribution of Z_i conditional on the entire past is the same as the

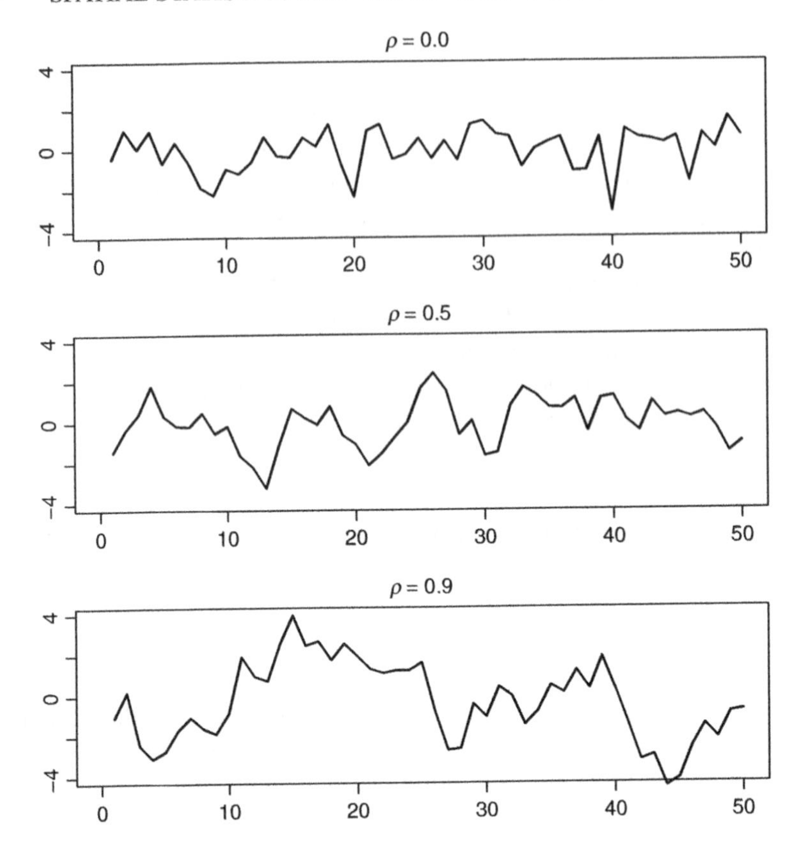

Figure 1.3 Outcomes of three AR(1) time series.

distribution of Z_i given only the immediate past, Z_{i-1}. Any such process is an example of a *Markov process*. We say that the AR(1) process is a Markov process of order one, as the present depends only on the one, immediately previous observation in time.

Figure 1.3 shows the outcomes of three AR(1) time series, the first an uncorrelated series ($\rho = 0.0$), the second with moderate correlation ($\rho = 0.5$), and the third with strong correlation ($\rho = 0.9$). Each time series consists of $n = 50$ temporal observations. Note that as ρ increases the oscillations of the time plots decrease. For example, the number of crossings of the mean ($\mu = 0$) decreases from 22 ($\rho = 0$), to 17 ($\rho = 0.5$), to 8 ($\rho = 0.9$). In other words, the 'smoothness' of the time plots increases. This notion of 'smoothness' and its importance in spatial prediction is discussed in Section 3.3.

To examine the effect of correlation on estimation, assume that SOS holds. From observations $Z_i, i = 1, \ldots, n$, we seek to estimate and draw inferences concerning the mean of the process, μ. To do this we desire a confidence interval for the unknown μ. Under SOS, it holds that each observation has the same variability, $\sigma^2 = Var(Z_i)$, and to simplify we assume that this value is known. The usual large sample 95 percent confidence interval for μ is then given by

$$\left(\overline{Z}_n - 1.96\frac{\sigma}{n^{1/2}}, \overline{Z}_n + 1.96\frac{\sigma}{n^{1/2}}\right),$$

where

$$\overline{Z}_n = \sum_{i=1}^{n} Z_i/n$$

denotes the sample mean of the observations.

We hope that the true coverage of this interval is equal to the nominal coverage of 95%. To see the true coverage of this interval, continue to assume that the data come from a (SOS) time series. Using the fact that, for any constants $a_i, i = 1, \ldots, n$, we have

$$Var\left(\sum_{i=1}^{n} a_i Z_i\right) = \sum_{i=1}^{n}\sum_{j=1}^{n} a_i a_j Cov(Z_i, Z_j),$$

and setting $a_i = 1/n$, for all i, gives:

$$Var(\overline{Z}_n) = \frac{1}{n^2}\sum_{i=1}^{n}\sum_{j=1}^{n} Cov(Z_i, Z_j)$$

$$= \frac{1}{n^2}\left[nC(0) + 2\sum_{j=1}^{n-1}(n-j)C(j)\right],$$

where the second equality uses SOS and counting. To evaluate this for large n, we need the following result named after the 19th century mathematician, Leopold Kronecker:

Lemma 1.1 *Kronecker's lemma*

For a sequence of numbers $a_i, i = 1, \ldots,$ such that

$$\sum_{i=1}^{\infty} |a_i| < \infty,$$

we have that

$$\frac{\sum_{i=1}^{n} i\ a_i}{n} \to 0, \quad \textit{as } n \to \infty.$$

In a direct application taking $a_i = C(i)$ in Kronecker's lemma, it is seen that:

$$(1/n)\sum_{j=1}^{n-1} jC(j) \to 0, \text{ as } n \to \infty \text{ whenever } \sum_{j=-\infty}^{\infty} |C(j)| < \infty,$$

and thus that

$$nVar(\overline{Z}_n) \to \tilde{\sigma}^2 := \sum_{j=-\infty}^{\infty} C(j),$$

whenever $\sum_{j=-\infty}^{\infty} |C(j)| < \infty$. This last condition, known as 'summable covariances' says that the variance of the mean tends to 0 at the same rate as in the case of independent observations. In particular, it holds for the AR(1) process with $-1 < \rho < 1$. The same rate of convergence does not mean, however, that the correlation has no effect in estimation.

To see the effects of correlation, note that in the case of independent observations, it holds that the variance of the standardized mean, $nVar(\overline{Z}_n) = \sigma^2 = C(0)$. It is seen that in the presence of correlation, the true variance of the standardized mean, $\tilde{\sigma}^2$, is quite different from σ^2. In particular, for the stationary AR(1) process (with $\eta = 1$), $\sigma^2 = C(0) = (1-\rho^2)^{-1}$, while arithmetic shows that $\tilde{\sigma}^2 = (1-\rho)^{-2}$, so that the ratio of the large sample variance of the mean under independence to the true variance of the mean is $R = \sigma^2/\tilde{\sigma}^2 = (1-\rho)/(1+\rho)$. In the common situation where correlation is positive, $0 < \rho < 1$, we see that ignoring correlation leads to underestimation of the correct variance.

To determine the practical effect of this, let $\Phi(\cdot)$ denote the cumulative distribution function of a standard normal variable. The coverage of the interval that ignores the correlation is given by

$$P\left[\overline{Z}_n - 1.96\frac{\sigma}{n^{1/2}} \le \mu \le \overline{Z}_n + 1.96\frac{\sigma}{n^{1/2}}\right]$$

$$= P\left[-1.96(\sigma/\tilde{\sigma}) \le \frac{n^{1/2}(\overline{Z}_n - \mu)}{\tilde{\sigma}} \le 1.96(\sigma/\tilde{\sigma})\right]$$

$$\to \Phi(1.96R^{1/2}) - \Phi(-1.96R^{1/2}).$$

We have assumed that the Central Limit theorem holds for temporary stationary observations. It does under mild moment conditions on the Z_is and on the strength of correlation. In particular, it holds for the stationary AR(1) model. Some details are given in Chapter 10.

Evaluating the approximate coverage from the last expression, we see that when $\rho = 0.2$, the ratio $R = 0.667$ and the approximate coverage of the usual nominal 95% confidence interval is 89%. When $\rho = 0.5$, $R = 0.333$ and the approximate coverage is 74%. The true coverage has begun to differ from the nominal of 95% so much that the interval is not performing at all as advertised. When $\rho = 0.9$, $R = 0.053$, and the true coverage is approximately 35%. This interval is completely unreliable. It is seen that the undercoverage becomes more severe as temporal correlation increases.

Using the correct interval, with $\tilde{\sigma}$ replacing σ, makes the interval wider, but we now obtain approximately the correct coverage. Note, however, that the estimator, $\overline{Z}_n$, is still (mean square) consistent for its target, μ, as we still have $Var(\overline{Z}_n) \to 0$, as $n \to \infty$, whenever $\sum_{j=-\infty}^{\infty} |C(j)| < \infty$.

To generalize this to the spatial setting, first note that we can write the conditional mean and the conditional variance for the temporal AR(1) model as:

$$E[Z_i | Z_j : j < i] = \mu + \rho(Z_{i-1} - \mu)$$

and

$$Var[Z_i | Z_j : j < i] = \eta^2.$$

A spatial first-order autoregressive model is a direct generalization of these two conditional moments. Specifically, conditioning on the past is replaced by conditioning on all other observations. In the temporal AR(1) case, it is assumed that the conditional distribution of the present given the past depends only on the immediate past. The spatial analogue assumes that the conditional distribution of $Z(\mathbf{s})$ depends only on the nearest neighbors of $\mathbf{s}$. Specifically, with equally spaced observations in two dimensions, assume that:

$$E[Z(\mathbf{s}) | Z(\mathbf{t}), \mathbf{t} \neq \mathbf{s}] = \mu + \gamma \sum_{d(\mathbf{s},\mathbf{t})=1} [Z(\mathbf{t}) - \mu]$$

and

$$Var[Z(\mathbf{s}) | Z(\mathbf{t}), \mathbf{t} \neq \mathbf{s}] = \eta^2.$$

Note how these two conditional moments are a natural spatial analogue to the conditional moments in the temporal AR(1) model. If the observations follow a normal distribution, then we call this spatial model a Gaussian first-order autoregressive model. The first-order Gaussian autoregressive model is an example of a (spatial) Markov process. Figure 1.4 shows sample observations from a first-order Gaussian model on a 100×100 grid with $\gamma = 0.0$ and $\gamma = 0.2$. Note how high values (and low values) tend to accumulate near each other for the $\gamma = 0.2$ data set. In particular, we find that when $\gamma = 0.0$, 2428 of the observations with a positive neighbor sum (of which there are 4891) are also positive (49.6 percent), while when $\gamma = 0.2$, we have that 3166 of the observations with a positive neighbor sum (of which there are 4813) are also positive (65.8 percent). To see the effects of spatial correlation on inference for the mean, we again compare the true variances of the mean with the variances that ignore correlation.

First we need to find the variance of the mean as a function of the strength of correlation. Analogously to the temporal case, we have that

$$nVar(\overline{Z}_n) \to \tilde{\sigma}^2 := \sum_{\mathbf{s} \in \mathbf{Z}^2} Cov[Z(\mathbf{0}), Z(\mathbf{s})]$$

as $n \to \infty$. Unfortunately, unlike in the temporal case of an AR(1), it is not a simple matter to evaluate this sum for the conditionally specified spatial model. Instead, we compute the actual finite sample variance for any given sample size n and correlation parameter γ.

Towards this end, let $\mathbf{Z}$ denote the vector of n spatial observations (in some order). Then $Var(\mathbf{Z})$ is an $n \times n$ matrix, and it can be shown using a factorization theorem of Besag (1974), that $Var(\mathbf{Z}) := \mathbf{\Sigma} = \eta^2(\mathbf{I} - \mathbf{\Gamma})^{-1}$, where $\mathbf{\Gamma}$ is an $n \times n$ matrix with elements $\gamma_{st} = \gamma$ whenever locations $\mathbf{s}$ and $\mathbf{t}$ are neighbors, that is, $d(\mathbf{s}, \mathbf{t}) = 1$. This model is discussed further in Chapter 4.

Using this and the fact that $nVar(\overline{Z}_n) = (1/n)\mathbf{1}^{\mathrm{T}}\mathbf{\Sigma}\mathbf{1}$, for any sample size n, we can compute the variance for any value of γ (this is simply the sum of all elements in $\mathbf{\Sigma}$ divided by the sample size n). Take this value to be $\tilde{\sigma}^2$. In the time series AR(1) setting, we were able to find the stationary variance explicitly. In this spatial model this is not simply done. Nevertheless, observations from the center of the spatial field are close to the stationary distribution. From the diagonal elements of $Var(\mathbf{Z})$, for observations near the center of the field we can see the (unconditional) variance of a single observation, σ^2, for various values of γ.

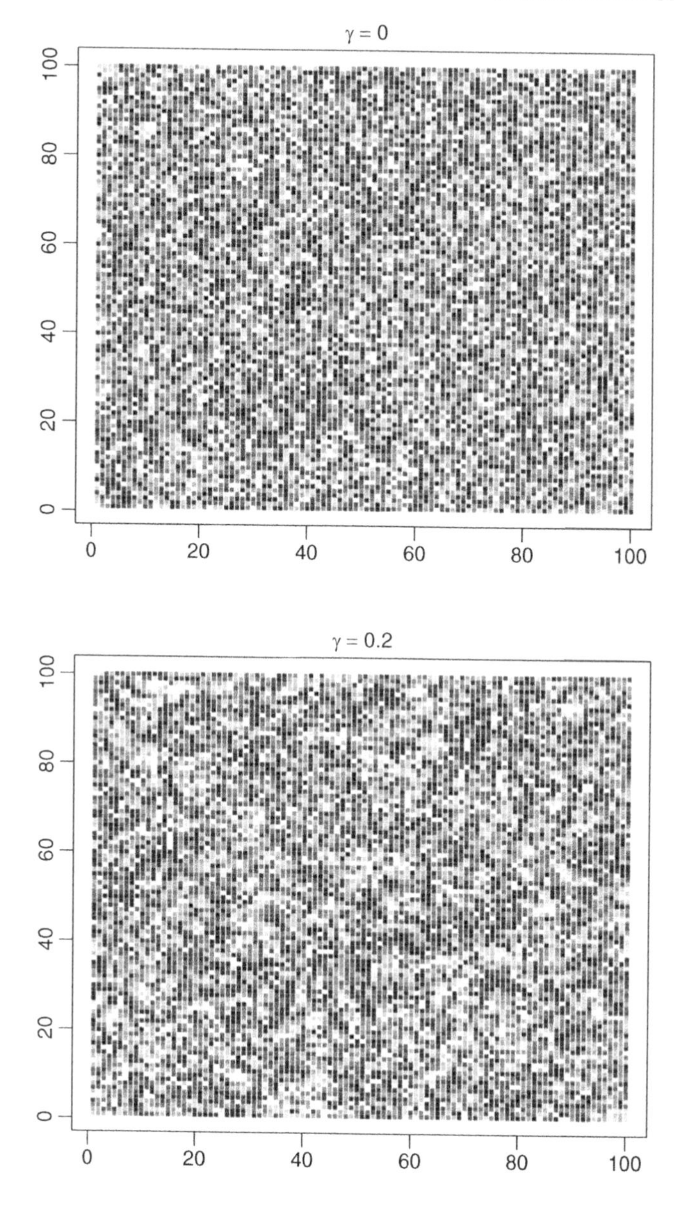

Figure 1.4 *Output from two* 100×100 *first-order spatial Gaussian models.*

For a 30×30 grid of observations (with $\eta^2 = 1.0$), direct calculation shows that $\tilde{\sigma}^2 = 1.24$ for $\gamma = 0.05$ ($\sigma^2 = 1.01$), $\tilde{\sigma}^2 = 1.63$ for $\gamma = 0.10$ ($\sigma^2 = 1.03$), and $\tilde{\sigma}^2 = 4.60$ for $\gamma = 0.20$ ($\sigma^2 = 1.17$). It is seen that, as in the temporal setting, the variance of the mean increases as spatial correlation increases, and that the ratio $R = \sigma^2/\tilde{\sigma}^2 = 0.813, 0.632, 0.254$ for $\gamma = 0.05, 0.10, 0.20$, respectively. This leads to approximate coverages of the usual 95% nominal confidence interval for μ of 92%, 88%, and 68%, respectively. We have seen that, as in the temporal setting, accounting for spatial correlation is necessary to obtain accurate inferences. Further, it is seen that when correlations are positive, ignoring the correlation leads to undercoverage of the incorrect intervals. This corresponds to an increased type-I error in hypothesis testing, and thus the errors are often of the most serious kind. Further, to obtain accurate inferences, we need to account for the spatial correlation and use the correct $\tilde{\sigma}^2$, or a good estimate of it, in place of the incorrect σ^2.

1.2.2 Prediction

To see the effects of correlation on prediction, consider again the temporal AR(1) process. In this situation, we observe the first n observations in time, Z_i, $i = 1, \ldots, n$, and seek to predict the unobserved Z_{n+1}. If we entirely ignore the temporal correlation, then each observation is an equally good predictor, and this leads to the predictor $\widehat{Z}_{n+1} := \overline{Z}_n$. Direct calculation shows that the true expected square prediction error for this estimator, $E[(\widehat{Z}_{n+1} - Z_{n+1})^2]$, is approximately given by

$$MSE(\widehat{Z}_{n+1}) \simeq \sigma^2 \left[1 + \frac{1}{n}\left(1 - \frac{2\rho}{n(1-\rho)} - \frac{2\rho^2}{n(1-\rho)^2}\right)\right],$$

where the error is in terms of order smaller than $1/n^2$. From this equation we see that, as $n \to \infty$, $MSE(\widehat{Z}_{n+1}) \to \sigma^2 \neq 0$. This is in stark contrast to the situation in Section 1.2.1, where the sample mean estimator has asymptotic MSE equal to 0. This is generally true in prediction. No amount of data will make the prediction error approach zero. The reason is that the future observation Z_{n+1} is random for any sample size n. Additionally, unlike in Section 1.2.1, we see that as ρ increases, $MSE(\widehat{Z}_{n+1})$ decreases. So, although strong correlation is hurtful when the goal is estimation (i.e., estimation becomes more difficult), strong correlation is helpful when the goal is prediction (prediction becomes easier).

Consider the unbiased linear estimator, $\widetilde{Z}_{n+1} = \sum_{i=1}^{n} a_i Z_i$, with $\sum_{i=1}^{n} a_i = 1$, that minimizes MSE over $a_1, \ldots, a_n$. Then, it can be shown using the methods in Chapter 2, Section 2.2, that

$$\widetilde{Z}_{n+1} = \rho Z_n + \frac{(1-\rho)^2 \sum_{i=2}^{n-1} Z_i}{n(1-\rho)+2\rho} + (1-\rho)\frac{Z_1 + Z_n}{n(1-\rho)+2\rho}.$$

Note that this predictor is approximately the weighted average of Z_n and the average of all previous time points. Also, the weight on Z_n increases as correlation increases. The methods in Section 2.2 further show that, for this predictor,

$$MSE(\widetilde{Z}_{n+1}) \simeq \sigma^2 \left[1 - \rho^2 + \frac{(1-\rho)(1+\rho)}{n}\right],$$

where the error is in terms of order smaller than $1/n$.

Imagine that we ignore the correlation and use the predictor $\widehat{Z}_{n+1}$ (i.e., we assume $\rho = 0$). Then we would approximately report

$$MSE^*(\widehat{Z}_{n+1}) = \sigma^2 \left[1 + \frac{1}{n}\right],$$

which is approximately equal to $MSE(\widehat{Z}_{n+1})$ (for large n and/or moderate ρ, the error is of order $1/n^2$), and thus $\widehat{Z}_{n+1}$ is approximately *accurate*. Accurate means that the inferences drawn using this predictor and the assumed MSE would be approximately right *for this predictor*. In particular, prediction intervals will have approximately the correct coverage for the predictand Z_{n+1}. This is in stark contrast to the estimation setting in Section 1.2.1, where ignoring the correlation led to completely inaccurate inferences (confidence intervals with coverage far from nominal). It seems that, in prediction, ignoring the correlation is not as serious as in estimation. It holds on the other hand, that

$$\frac{MSE(\widetilde{Z}_{n+1})}{MSE(\widehat{Z}_{n+1})} \simeq 1 - \rho^2.$$

This shows that $\widehat{Z}_{n+1}$ is not the *correct* predictor under correlation. Correct means that the inferences drawn using this predictor are the 'best' possible. Here 'best' means the linear unbiased predictor with minimal variance. The predictor $\widetilde{Z}_{n+1}$ is both accurate and correct for the AR(1) model with known AR(1) parameter ρ. In estimation, the estimator

$\widehat{Z}_{n+1} := \overline{Z}_n$ is approximately correct for all but extremely large $|\rho|$, but is only approximately accurate when we use the correct variance, $\tilde{\sigma}^2$.

The conclusions just drawn concerning prediction in the temporal setting are qualitatively similar in the spatial setting. Ignoring spatial correlation leads to predictions which are approximately accurate, but are not correct. The correct predictor is formed by accounting for the spatial correlation that is present. This is done using the kriging methodology discussed in Chapter 2.

In summary, it is seen that, when estimation is the goal, we need to account for correlation to draw accurate inferences. Specifically, when positive correlation is present, ignoring the correlation leads to confidence intervals which are too narrow. In other words, in hypothesis testing there is an inflated type-I error. When prediction is the goal, we can obtain approximately accurate inferences when ignoring correlations, but we need to account for the temporal or spatial correlation in order to obtain correct (i.e., efficient) predictions of unobserved variables.

We now discuss in the temporal setting a situation where ignoring correlation leads to inaccurate and surprising conclusions in the estimation setting.

1.3 Texas tidal data

A court case tried to decide a very fundamental question: where is the coastline, that is, the division between land and water. In many places of the world, most people would agree to within a few meters as to where the coastline is. However, near Port Mansfield, TX (south of Corpus Christi, TX), there is an area of approximately six miles between the intercoastal canal and a place where almost all people would agree land begins. Within this six-mile gap, it could be water or land depending on the season of the year and on the observer. To help determine a coastline it is informative to consider the history of this question.

In the 1300s, the Spanish 'Las Siete Partidas,' Law 4 of Title 28, stated that the '...sea shore is that space of ground ... covered by water in their ... highest annual swells.' This suggests that the furthest reach of the water in a typical year determines the coastline. This coastline is approximately six miles away from the intercoastal canal. In 1935, the US Supreme Court, in *Borax* v. *Los Angeles*, established MHW – 'Mean High Water' as the definition of coastal boundary. This states that the coastline is the average of the daily high-tide reaches of the water. In 1956, in

Rudder s. *Ponder*, Texas adopted MHW as the definition of coastal boundaries. This coastline is relatively close to the intercoastal canal. The two definitions of coastline do not agree in this case and we seek to understand which is more appropriate. The development here follows that in Sherman *et al.* (1997).

The hourly data in a typical year are given by Y_t, $t = 1, \ldots, 8760$, where Y_t denotes the height of the water at hour t at a station at the intercoastal canal. The horizontal projection from this height determines the coastal boundary. The regression model dictated by NOAA (National Oceanographic and Atmospheric Administration) is

$$Y_t = a_0 + \sum_{i=1}^{37} a_i \cos(\pi t S_i)/180) + \sum_{i=1}^{37} b_i \sin(\pi t S_i)/180) + \epsilon_t,$$

where a_i and b_i are amplitudes associated with S_i, the speed of the ith constituent, $i = 1, \ldots, 37$, and ϵ_ts are random errors. The speeds are assumed to be known, while the amplitudes are unknown and need to be estimated. This model is similar to that in classical harmonic analysis and periodogram analysis as discussed in, for example, Hartley (1949).

The basic question in the coastal controversy is: which constituents best explain the variability in water levels? If annual or semiannual constituents explain a large proportion of the overall variability in tidal levels, this suggests that the flooded regions between the intercoastal canal and land are an important feature in the data, and suggests that the contested area cannot be called land. If, however, daily and twice-daily constituents explain most of the variability in tidal levels, then the contested area should be considered land. Note that the regression model is an example of a general linear model, and the amplitudes can be estimated using least squares estimation. In an effort to assess goodness of fit, consider the residuals from this fitted model. Figure 1.5 shows (a) the first 200 residuals, e_t, $t = 1, \ldots, 200$, and (b) residuals e_t, $t = 1001, \ldots, 1200$, from the least squares fit. One typical assumption in multiple regression is one of independent errors, that is, $Cov[\epsilon_s, \epsilon_t] = 0$ whenever $s \neq t$.

Notice that the plot of the first 200 residuals shows a stretch of approximately 60 consecutive negative residuals. This suggests that the errors are (strongly) positively correlated. The second residual plot similarly suggests a clear lack of independence in the errors, as do most stretches of residuals. From the results in estimation in Section 1.2.1, we know that ignoring the correlation would likely be a serious error if our

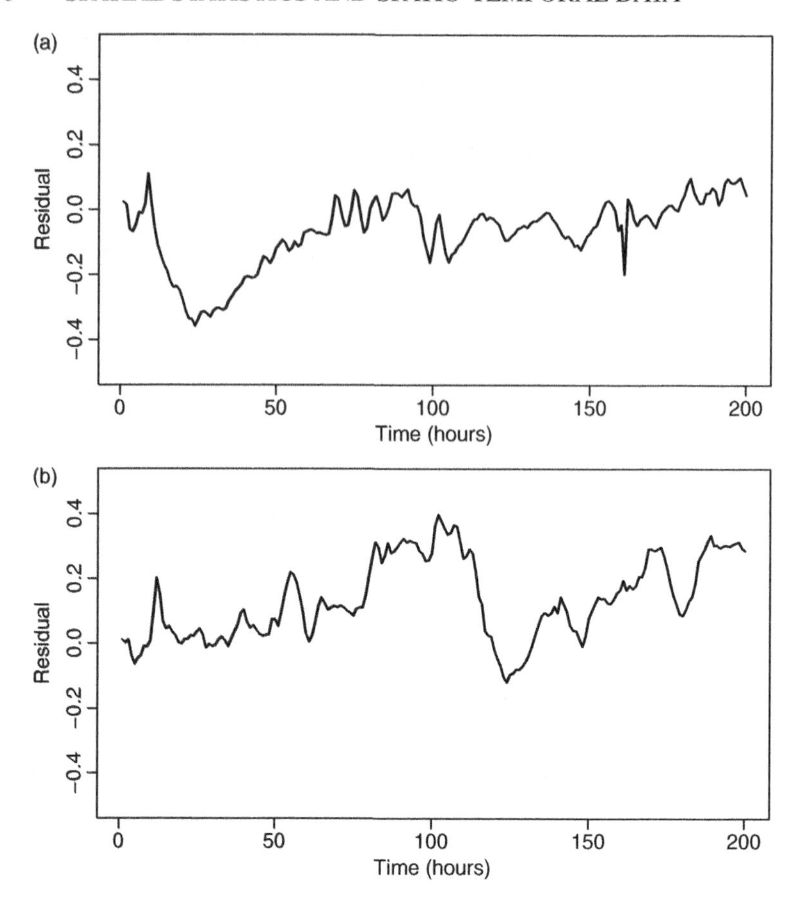

Figure 1.5 Two sets of residuals from OLS in the tidal data. (a) Residuals 1–200; (b) residuals 1001–1200.

goal is to estimate the mean of the process. The goal here, however, is to estimate the regression parameters in the harmonic analysis, and it is not clear in the regression setting what the effect of ignoring the correlation would be. To explore this, consider the setting of a simple linear regression model:

$$Y_t = \alpha + \beta x_t + \epsilon_t, \quad t = 1, \ldots, T,$$

where Y_t is the response, x_t denotes a covariate, and ϵ_t are stationary errors. The ordinary least squares (OLS) estimator of β is

$$\widehat{\beta} := \frac{\sum_{t=1}^{T}(x_t - \overline{x})Y_t}{\sum_{t=1}^{T}(x_t - \overline{x})^2},$$

with an associated variance of

$$Var(\widehat{\beta}) = \frac{\sum_{t=1}^{T}\sum_{u=1}^{T}(x_t - \overline{x})(x_u - \overline{x})E(\epsilon_t\epsilon_u)}{\left[\sum_{t=1}^{T}(x_t - \overline{x})^2\right]^2}.$$

It is seen that the variance of the estimated slope depends on the correlations between errors *and* on the structure of the design. To see the effect of the latter, consider AR(1) errors, $\epsilon_t = \rho\epsilon_{t-1} + \eta_t$ [with $Var(\eta_t) = 1.0$], under the following two designs:

Design 1: Monotone

$$x_i = i, \quad i = 1, \ldots, T$$

Design 2: Alternating

$$x_i = \begin{cases} (i+1)/2, & \text{if } i \text{ is odd;} \\ T - (i/2) + 1, & \text{if } i \text{ is even.} \end{cases}$$

To numerically compare the variance under these two scenarios, consider $T = 10$ and $\rho = 0.5$. In this case we have

$$Var(\widehat{\beta}) = \begin{cases} 0.0306 \text{ in Design 1;} \\ 0.0031 \text{ in Design 2.} \end{cases}$$

If we ignore the correlation, we then report the variance to be:

$$Var_{ind}(\widehat{\beta}) := \frac{Var(\epsilon_i)}{\sum_{t=1}^{T}(x_t - \overline{x})^2} = 0.0162$$

for *both* designs.

The conclusion is that, in contrast to the stationary case in Section 1.2.1, OLS variance estimates that ignore the positive correlation can under *or* over estimate the correct variance, depending on the structure of the design. In the tidal data, constituents of fast speed (small periods) correspond to the alternating design, while constituents of low speed (long periods) correspond to the monotone design.

Table 1.1 P-value comparison between ignoring and accounting for correlation in tidal data.

Period	Correlation	OLS
8765 (annual)	0.550	$\leq$0.001
4382 (semiannual)	$\leq$0.001	$\leq$0.001
327	0.690	0.002
25.8 (day)	$\leq$0.001	$\leq$0.001
12	$\leq$0.001	0.145

OLS are the p-values for parameter estimates that ignore correlation. Correlation p-values account for the correlation.

Table 1.1 gives the p-values for the test that the given constituent is not present in the model, based on the usual t-statistic for a few selected constituents for data from 1993. One set of p-values is computed under an assumption of independent errors, while the second set of p-values is based on standard errors which account for correlation. Variances in these cases are constructed using the block bootstrap in the regression setting.

We discuss the block bootstrap in more detail in Chapter 10. The block bootstrap appears to be a reliable tool in this case, as the residual process is well approximated by a low-level autoregressive moving-average (ARMA) process.

From the table we see that the OLS p-values that ignore correlation cannot be trusted. Further, the errors in ignoring the correlation are as predicted from the simple linear regression example. Namely that, for long periods, OLS variance estimates underestimate the correct variance, and thus lead to large t-statistics and hence p-values which are too small. For short periods, however, this is reversed and the OLS variances are too large, leading to small t-statistics and overly large p-values. The block bootstrap accounts for the temporal correlation and gives reliable variance estimates and thus reliable p-values. A careful parametric approach that estimates the correlation from within the ARMA(p, q) class of models gives results similar to those using the block bootstrap.

Finally, the semiannual period (Period = 4382) is very significant. This suggests that the flooding of the contested area is a significant feature of the data and thus this area cannot reasonably be considered as land. This outcome is qualitatively similar for other years of data as well. Although the Mean High Water criterion may be reasonable for tides in Los Angeles,

CA (on which the original Supreme Court decision was based), it does not appear to be reasonable for tides in Port Manfield, TX.

Much of the discussion in this chapter has focused on the role of correlation and how the effects of correlation are similar in the time series and spatial settings. There are, however, several fundamental differences between time series and spatial observations. Some of these will become clear as we develop spatial methodology. For now, note that in time there is a natural ordering, while this is not the case in the spatial setting. One effect of this became clear when considering the marginal variance of time series and spatial fields in Section 1.2.1. A second major difference between the time series and spatial settings is the effect of edge sites, observations on the domain boundary. For a time series of length n, there are only two observations on the boundary. For spatial observations on a $n \times n$ grid, there are approximately $4n$ observations on the boundary. The effects of this, and methods to account for a large proportion of edge sites are discussed in Section 4.2.1. A third fundamental difference is that, in time series, observations are typically equally spaced and predictions are typically made for future observations. In the spatial setting, observations are often not equally spaced and predictions are typically made for unobserved variables 'between' existing observations. The effects of this will be discussed throughout the text, but especially in Chapters 3 and 5. Other differences in increased computational burden, complicated parameter estimation, and unwieldy likelihoods in the spatial setting will be discussed, particularly in Chapters 4 and 9.

2

Geostatistics

The name geostatistics comes from the common applications in geology. The term is somewhat unfortunate, however, because the general methodology is quite general, and appropriate to the effort to understand the type and extent of spatial correlation, and to use this knowledge to make efficient predictions. The basic situation is as follows:

A process of interest is observed at n locations $\mathbf{s}_1, \ldots, \mathbf{s}_n$. These n observations are denoted by $Z(\mathbf{s}_1), \ldots, Z(\mathbf{s}_n)$, although sometimes this will be denoted as $Z_1, \ldots, Z_n$. This is simply shorthand, as we typically need to know the specific locations $\mathbf{s}_1, \ldots, \mathbf{s}_n$ to carry out any estimation and inferences. In the phosphorus example, described in Chapter 1, the $n = 103$ locations are displayed in Figure 1.1. The main goals are:

i. Predict the process at unobserved locations. For example, let $\mathbf{s}_0$ denote a specific location of interest, and let $Z(\mathbf{s}_0)$ denote the amount of log(Phosphorus) at the specific location of interest. We desire the unobserved value $Z(\mathbf{s}_0)$. If, however, for example, we desire the total amount of phosphorus in a region B of the pond, then we seek to predict $\int_B Z(\mathbf{s})d\mathbf{s}$.

ii. Understand the connection between process values at 'neighboring' locations. In the phosphorus study this helps to understand the additional information gained by more intensive sampling.

Spatial Statistics and Spatio-Temporal Data: Covariance Functions and Directional Properties Michael Sherman

In order to describe the relationship between nearby observations, we (re)introduce the following models of correlation:

Second-order stationarity (SOS)

i. $E[Z(\mathbf{s})] - \mu = 0$,
ii. $Cov[Z(\mathbf{s}+\mathbf{h}), Z(\mathbf{s})] = Cov[Z(\mathbf{h}), Z(\mathbf{0})] = C(\mathbf{h})$ for all shifts $\mathbf{h}$.

This is identical to the notion introduced in Section 1.1.

Intrinsic stationarity (IS)

i. $E[Z(\mathbf{s}) - \mu] = 0$,
ii. $Var[Z(\mathbf{s}+\mathbf{h}) - Z(\mathbf{s})] =: 2\gamma(\mathbf{h})$ for all shifts $\mathbf{h}$,

which is a closely related notion.

Although it is stated in terms of a difference, assumption (i) for IS is identical to assumption (i) in SOS. Under IS, from (i), we can write $2\gamma(\mathbf{h}) = E[Z(\mathbf{s}+\mathbf{h}) - Z(\mathbf{s})]^2$.

Typically, $2\gamma(\mathbf{h})$ is called the variogram function, and $\gamma(\mathbf{h})$ is called the semivariogram, although some also call $\gamma(\mathbf{h})$ the variogram function.

A reasonable question is: why consider the assumption of IS when the assumption of SOS is apparently more immediate and interpretable. There are four main reasons to consider the variogram and the assumption of IS.

i. IS is a more general assumption than SOS.

   Assume that SOS holds, then direct calculation shows that $2\gamma(\mathbf{h}) = 2[C(\mathbf{0}) - C(\mathbf{h})]$. Thus, we can define the variogram as the right hand side of the previous equality and IS then holds.

   On the other hand, for temporal data consider an AR(1) process with $\rho = 1$, that is, $Z_i = Z_{i-1} + \epsilon_i$. Iterating, we have

$$Z_{i+j} = Z_i + \delta_j \quad \text{with } Var(\delta_j) = j\sigma^2. \tag{2.1}$$

   Then $Cov[Z_{i+j}, Z_i] = i\sigma^2$, which is a function of i, so SOS does not hold. We can, however, simply write $2\gamma(j) = j\sigma^2$ so IS holds. This example is called an unstable AR(1) in the time series literature. It is also known as a random walk in one dimension, or one-dimensional Brownian motion.

ii. The variogram adapts more easily to nonstationary observations.

   One such example is when the mean function is not constant. This is explored in Section 3.5.1.

iii. To estimate the variogram, no estimate of μ is required.

Using increments filters out the mean. This is not the case when estimating the covariance function. For this, an estimate of μ is required. This is shown in Section 3.1.

iv. Estimation of the variogram is easier than estimation of the covariance function. This is shown, for example, in Cressie (1993).

Although, from (i), IS is a more general assumption, (iv) states that estimation of the variogram is easier than estimation of the covariance function. Easier means that the common moment estimator is more accurate in most cases. In particular, the natural moment estimator of the variogram is unbiased under IS, while the covariance estimator is biased. This is surprising given reason (i): the more general estimator is more easily estimated.

2.1 A model for optimal prediction and error assessment

To begin, start with the simplest structure for the underlying mean function, namely that it is constant. Specifically, the initial model is

$$Z(\mathbf{s}) = \mu + \delta(\mathbf{s}), \tag{2.2}$$

where μ is an unknown constant, and $\delta(\mathbf{s})$ is intrinsically stationary, IS. Notice this is exactly the assumption that $Z(\mathbf{s})$ is IS. Although simple, this is a model that is often practically useful. When sampling is relatively dense, and the mean function changes smoothly, there is often little harm in assuming that the mean function is constant.

Consider an unsampled location, $\mathbf{s}_0$, where we desire to know the unobserved process value $Z(\mathbf{s}_0)$. The goal is to find the 'best' prediction of $Z(\mathbf{s}_0)$ based on the observations $Z(\mathbf{s}_1), \ldots, Z(\mathbf{s}_n)$.

First we need to define what is meant by best. As in the time series setting, described in Chapter 1, this is taken to mean having the lowest squared prediction error, that is, the desire is to minimize $MSE[\widehat{Z}(\mathbf{s}_0)] = E\{[\widehat{Z}(\mathbf{s}_0) - Z(\mathbf{s}_0)]^2\}$.

We now show that the solution is:

$$\widehat{Z}(\mathbf{s}_0) = E[Z(\mathbf{s}_0)|Z(\mathbf{s}_1), \ldots, Z(\mathbf{s}_n)].$$

To see this, let $\mathbf{Z}_n := [Z(\mathbf{s}_1), \ldots, Z(\mathbf{s}_n)]$ denote the vector of n observations. Then, adding and subtracting the solution inside the square, we have

$$\begin{aligned}
&E\{[\widehat{Z}(\mathbf{s}_0) - Z(\mathbf{s}_0)]^2\} \\
&\quad = E\left(\left\{\widehat{Z}(\mathbf{s}_0) - E[Z(\mathbf{s}_0)|\mathbf{Z}_n] + E[Z(\mathbf{s}_0)|\mathbf{Z}_n] - Z(\mathbf{s}_0)\right\}^2\right) \\
&\quad = E\left(\left\{\widehat{Z}(\mathbf{s}_0) - E[Z(\mathbf{s}_0)|\mathbf{Z}_n]\right\}^2\right) + E\left(\{E[Z(\mathbf{s}_0)|\mathbf{Z}_n] - Z(\mathbf{s}_0)\}^2\right) \\
&\qquad + 2E\left(\left\{\widehat{Z}(\mathbf{s}_0) - E[Z(\mathbf{s}_0)|\mathbf{Z}_n]\right\}\{E[Z(\mathbf{s}_0)|\mathbf{Z}_n] - Z(\mathbf{s}_0)\}\right).
\end{aligned}$$

The third term in the last expression is 0. This can be seen by conditioning on $\mathbf{Z}_n$, and observing that both $\widehat{Z}(\mathbf{s}_0)$ and $E[Z(\mathbf{s}_0)|\mathbf{Z}_n]$ are constant given $\mathbf{Z}_n$. Finally, $E\{E[Z(\mathbf{s}_0)|\mathbf{Z}_n] - Z(\mathbf{s}_0)\}$ given $\mathbf{Z}_n$ is identically 0 for any value of $\mathbf{Z}_n$. Thus,

$$\begin{aligned}
&MSE[\widehat{Z}(\mathbf{s}_0)] \\
&\quad = E\left(\left\{\widehat{Z}(\mathbf{s}_0) - E[Z(\mathbf{s}_0)|\mathbf{Z}_n]\right\}^2\right) + E\left(\{E[Z(\mathbf{s}_0)|\mathbf{Z}_n] - Z(\mathbf{s}_0)\}^2\right),
\end{aligned}$$

and this is minimized when $\widehat{Z}(\mathbf{s}_0) = E[Z(\mathbf{s}_0)|\mathbf{Z}_n]$. The *prediction variance* for this predictor is then

$$E\left(\{E[Z(\mathbf{s}_0)|\mathbf{Z}_n] - Z(\mathbf{s}_0)\}^2\right).$$

The predictor, $\widehat{Z}(\mathbf{s}_0) = E[Z(\mathbf{s}_0)|\mathbf{Z}_n]$, makes great intuitive sense. The best guess of the response at any unsampled location based on observed values is its expectation given all the observed values. The fact that we can easily derive this optimal predictor is the good news. The difficulty occurs in attempting to calculate this predictor. When $Z(\mathbf{s})$ has a continuous distribution (i.e., a density function), this conditional expectation depends on the *joint* distribution of the observations $Z(\mathbf{s}_0), Z(\mathbf{s}_1), \ldots, Z(\mathbf{s}_n)$. Thus, computation of this conditional mean requires knowledge of this joint density, and calculation of a $(n+1)$-dimensional integral. Given that we have only n available data points, there is no hope of estimating this joint distribution, and thus little hope of estimating the conditional expectation.

A more modest goal is to seek the predictor, $\widehat{Z}(\mathbf{s}_0)$, that is the best *linear* function of the observed values, that is,

$$\widehat{Z}(\mathbf{s}_0) = \sum_{i=1}^{n} \lambda_i Z(\mathbf{s}_i)$$

for constants λ_i, $i = 1, \ldots, n$. This has reduced the problem from estimating an $(n+1)$-dimensional *distribution* to estimating n *constants*.

Certainly we could take all $\lambda_i = 1/n$, which gives $\widehat{Z}(\mathbf{s}_0) = \overline{Z}_n$, the sample mean. The question is: can we do better? Before launching into finding the best linear estimator, an immediate question arises: are there any constraints on the coefficients λ_i? Given that the mean is assumed to be constant in Equation 2.2, it seems natural to require that $E[\widehat{Z}(\mathbf{s}_0)] = \mu$. This would imply that the expectation of our predictor is the expectation of the predictand, that is, the predictor is unbiased. For this to hold, it is clear that $\sum_{i=1}^{n} \lambda_i = 1$ is required. Thus the goal has now become: minimize

$$E\left\{\left[\sum_{i=1}^{n} \lambda_i Z(\mathbf{s}_i) - Z(\mathbf{s}_0)\right]^2\right\} \quad \text{subject to} \sum_{i=1}^{n} \lambda_i = 1.$$

Can any of the λ_is be negative? At first glance, allowing negative weights may seem inappropriate. Negative weights can lead to inappropriate predictions for variables known to be positive. On the other hand, allowing negative weights is necessary to allow for predictions to be larger than any observed values (or smaller than any observed values), and this is often desirable. For this reason, the development here allows for negative weights in prediction. Further, negative weights will be seen in examples, for instance in Section 2.2.2 and in Section 2.5.1.

2.2 Optimal prediction (kriging)

Now that we have defined what is meant by best, we can address obtaining the best linear predictor. First expand the square error of prediction:

$$\left[Z(\mathbf{s}_0) - \widehat{Z}(\mathbf{s}_0)\right]^2 = Z^2(\mathbf{s}_0) - 2\sum_{i=1}^{n} \lambda_i Z(\mathbf{s}_0) Z(\mathbf{s}_i) + \sum_{i=1}^{n}\sum_{j=1}^{n} \lambda_i \lambda_j Z(\mathbf{s}_i) Z(\mathbf{s}_j).$$

Now, completing the square on the first two terms on the right hand side of the previous equation by adding $\sum_{i=1}^{n} \lambda_i Z^2(\mathbf{s}_i)$, subtracting the same quantity from the third term, and using the fact that $\sum_{i=1}^{n} \lambda_i = 1$, we have that

$$\left[Z(\mathbf{s}_0) - \widehat{Z}(\mathbf{s}_0)\right]^2 = \sum_{i=1}^{n} \lambda_i [(Z(\mathbf{s}_i) - Z(\mathbf{s}_0)]^2 - \sum_{i=1}^{n}\sum_{j=1}^{n} \lambda_i \lambda_j \frac{[Z(\mathbf{s}_i) - Z(\mathbf{s}_j)]^2}{2}.$$

Now, taking the expectation of both sides gives:

$$E\{[Z(\mathbf{s}_0) - \widehat{Z}(\mathbf{s}_0)]^2\} = 2\sum_{i=1}^{n} \lambda_i \gamma(\mathbf{s}_0 - \mathbf{s}_i) - \sum_{i=1}^{n}\sum_{j=1}^{n} \lambda_i \lambda_j \gamma(\mathbf{s}_i - \mathbf{s}_j), \tag{2.3}$$

using the definition of the variogram function and the fact that the semivariogram, $\gamma(\mathbf{h}) = \frac{1}{2}E[Z(\mathbf{s}+\mathbf{h}) - Z(\mathbf{s})]^2$ for all $\mathbf{h}$. Note that this last expression in Equation 2.3 is the prediction variance for a linear predictor using *any* weights λ_i, $i = 1, \ldots, n$.

Writing the right hand side of the previous equation as $F(\lambda_1, \ldots, \lambda_n)$, say, we seek to minimize $F(\lambda_1, \ldots, \lambda_n)$ subject to $G(\lambda_1, \ldots, \lambda_n) := \sum_{i=1}^{n} \lambda_i - 1 = 0$. To solve this, let m denote a Lagrange multiplier, and set $H(\lambda_1, \ldots, \lambda_n) = F - mG$. Differentiating H with respect to λ_i for $i = 1, \ldots, n$, and m, gives

$$\frac{dH}{d\lambda_i} = 2\gamma(\mathbf{s}_0 - \mathbf{s}_i) - 2\sum_{j=1}^{n} \lambda_j \gamma(\mathbf{s}_i - \mathbf{s}_j) - m,$$

and

$$\frac{dH}{dm} = -\sum_{i=1}^{n} \lambda_i + 1.$$

Setting all $n+1$ derivatives equal to 0 gives $n+1$ equations. There are $n+1$ unknowns (the λ_is and m), so there is a unique solution given by:

$$\sum_{j=1}^{n} \lambda_j \gamma(\mathbf{s}_i - \mathbf{s}_j) + \frac{m}{2} = \gamma(\mathbf{s}_0 - \mathbf{s}_i), \quad i = 1, \ldots, n,$$

and

$$\sum_{i=1}^{n} \lambda_i = 1.$$

This is a linear system, and writing $\gamma_{ij} = \gamma(\mathbf{s}_i - \mathbf{s}_j)$, we have that:

$$\begin{bmatrix} \gamma_{11} & \gamma_{12} & \dots & \gamma_{1n} & 1 \\ \gamma_{21} & \gamma_{22} & \dots & \gamma_{2n} & 1 \\ \vdots & & & & \vdots \\ \gamma_{n1} & \gamma_{n2} & \dots & \gamma_{nn} & 1 \\ 1 & 1 & \dots & 1 & 0 \end{bmatrix} \begin{bmatrix} \lambda_1 \\ \lambda_2 \\ \vdots \\ \lambda_n \\ m/2 \end{bmatrix} = \begin{bmatrix} \gamma_{01} \\ \gamma_{02} \\ \vdots \\ \gamma_{0n} \\ 1 \end{bmatrix}.$$

This can be simply written as $\boldsymbol{\Gamma\lambda} = \boldsymbol{\gamma}$. If, as is typically the case, the matrix $\boldsymbol{\Gamma}$ is invertible, then we have:

$$\boldsymbol{\lambda} = \boldsymbol{\Gamma}^{-1}\boldsymbol{\gamma}.$$

The first n elements of the vector $\boldsymbol{\lambda}$ give the weights for the best linear predictor of $Z(\mathbf{s}_0)$. The predictor, $\widehat{Z}(\mathbf{s}_0) = \sum_{i=1}^{n} \lambda_i Z(\mathbf{s}_i)$, with these optimal weights, is called the 'kriging' predictor of $Z(\mathbf{s}_0)$ in honor of D. G. Krige, a South African engineer. Using this predictor is commonly known as 'kriging' or, more precisely, as 'ordinary kriging.' The case where μ is assumed to be known is known as 'simple kriging.' Notice that, by using the variogram function, knowledge of μ was not required to obtain the kriging predictor. This is not the case if we use the covariance function, which explicitly requires μ. More general mean functions allowing for a linear dependence on covariates come under the name of 'universal kriging.' This is discussed in Section 2.4.

The errors associated with the kriging predictor are given by the *kriging variance*:

$$\begin{aligned}
\sigma^2_{z_0} &:= E\{[Z(\mathbf{s}_0) - \widehat{Z}(\mathbf{s}_0)]^2\} \\
&= 2\sum_{i=1}^{n} \lambda_i \gamma(\mathbf{s}_0 - \mathbf{s}_i) - \sum_{i=1}^{n}\sum_{j=1}^{n} \lambda_i \lambda_j \gamma(\mathbf{s}_i - \mathbf{s}_j) \\
&= \sum_{i=1}^{n} \lambda_i \gamma(\mathbf{s}_0 - \mathbf{s}_i) + \frac{m}{2} = \boldsymbol{\lambda}^{\mathrm{T}}\boldsymbol{\gamma},
\end{aligned}$$

where the third equality follows from the above linear system.

The kriging predictor is an exact interpolator in the sense that the kriging equations give $\widehat{Z}(\mathbf{s}_i) = Z(\mathbf{s}_i)$, for all observed variables, for $i = 1, \ldots, n$. This can be seen by noting that the square prediction error is identically 0 when all weight is put on the observed location which coincides with the desired prediction location.

2.2.1 An example: phosphorus prediction

We illustrate the kriging methodology by finding the kriging predictor of Phosphorus at an unobserved location, $\mathbf{s}_0$, based on two observed values, $\mathbf{s}_1$ and $\mathbf{s}_2$. Consider the two phosphorus measurements in log(ppm) given in Figure 2.1.

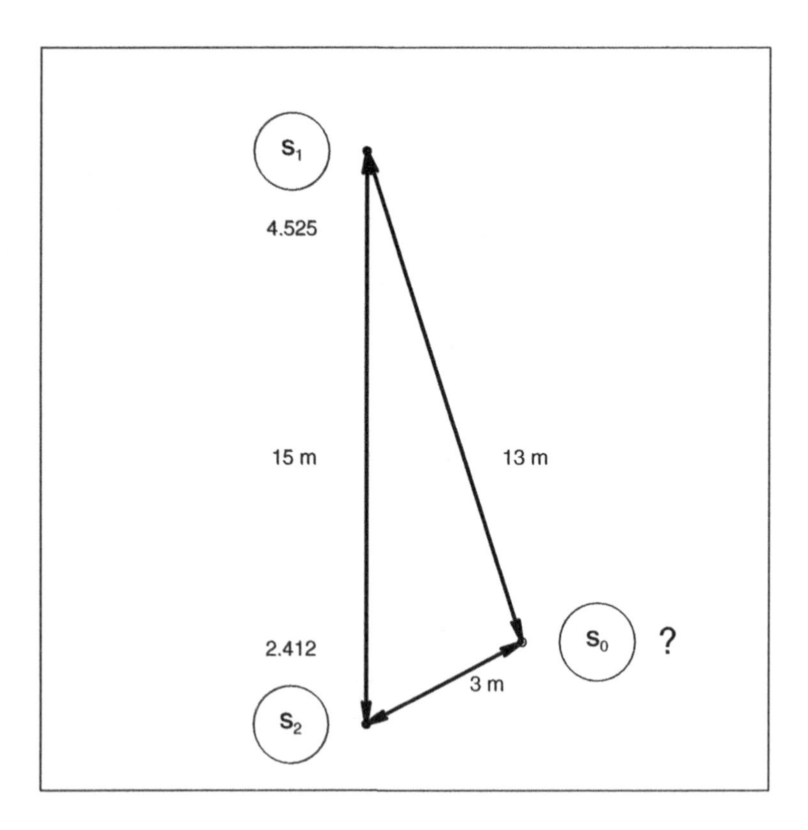

Figure 2.1 Kriging example: two observed measurements and a third to be predicted.

Assume the variogram function for any two locations, $\mathbf{s}_i$ and $\mathbf{s}_j$, is known to be $\gamma(d) = C_1[1 - \exp(-C_2 d)]$, where $d(\mathbf{s}_i, \mathbf{s}_j) = d$ and $C_1 = 0.5$ and $C_2 = 0.2$.

Then we have

$$\boldsymbol{\lambda} = \boldsymbol{\Gamma}^{-1}\boldsymbol{\gamma} = \begin{bmatrix} 0.0000 & 0.4751 & 1 \\ 0.4751 & 0.0000 & 1 \\ 1.0000 & 1.0000 & 0 \end{bmatrix}^{-1} \begin{bmatrix} 0.4629 \\ 0.2256 \\ 1.0000 \end{bmatrix} = \begin{bmatrix} 0.2503 \\ 0.7497 \\ 0.1067 \end{bmatrix}.$$

Thus, the kriging weights are $\lambda_1 = 0.2503$ and $\lambda_2 = 0.7497$. The predicted value of log(Phosphorus) at $\mathbf{s}_0$, $Z(\mathbf{s}_0)$, is

$$\widehat{Z}(\mathbf{s}_0) = 0.2503(4.525) + 0.7497(2.412) = 2.94\,\text{ppm}.$$

The kriging variance is

$$\sigma_{z0}^2 = (0.4629)(0.2503) + (0.2256)(0.7497) + (1)(0.1067) = 0.392\,\text{ppm}^2.$$

We know that no other weights (that sum to one) on the two points can have a lower variance than this. It is, however, informative to see how other reasonable predictors perform in this example in terms of predictions and their prediction variances.

Other predictors:

i. *The sample mean*

The weights are $\lambda_1 = \lambda_2 = 0.5$. The predictor is

$$\overline{Z}(\mathbf{s}_0) = \frac{1}{2}(4.525 + 2.412) = 3.47\,\text{ppm},$$

with a prediction variance of (from Equation 2.3):

$$E\{[Z(\mathbf{s}_0) - \widehat{Z}(\mathbf{s}_0)]^2\} = 2\sum_{i=1}^{2}\lambda_i\gamma(\mathbf{s}_0 - \mathbf{s}_i) - \sum_{i=1}^{2}\sum_{j=1}^{2}\lambda_i\lambda_j\gamma(\mathbf{s}_i - \mathbf{s}_j).$$

Taking $\lambda_1 = \lambda_2 = \frac{1}{2}$ in this formula gives

$$E\{[Z(\mathbf{s}_0) - \overline{Z}(\mathbf{s}_0)]^2\} = 2\left[\frac{\gamma(13) + \gamma(3)}{2}\right] - \frac{1}{4}[2\gamma(0) + 2\gamma(15)] = 0.451\,\text{ppm}^2.$$

Although this predictor is an unbiased predictor of $Z(\mathbf{s}_0)$, the prediction variance of 0.451 ppm^2 is approximately 15 percent larger than $\sigma^2_{z_0}$, the prediction variance of the optimal kriging predictor.

This is a rather modest improvement of the kriging predictor over the sample mean. This, however, is due to the form of the covariance function, and to the locations of the observed data relative to the prediction location. Various numerical experiments show that, for more smooth covariance functions (see, e.g., the Matérn family in Section 3.3), the mean becomes much less competitive with the kriging predictor. This can also be seen in the next section, Section 2.2.2, where we consider the power family of variograms.

ii. *Inverse distance interpolation*

A popular method in drawing maps is to set the weights as

$$\lambda_i = \frac{d^{-r}(\mathbf{s}_i, \mathbf{s}_0)}{\sum_{i=1}^{n} d^{-r}(\mathbf{s}_i, \mathbf{s}_0)},$$

where r is usually a positive integer power, taken to be $r = 1$ or $r = 2$. This inverse distance method is often called a linear interpolator. See, for example, Burrough and McDonnell (1998) for more on this method.

For $r = 1$, we have $\lambda_1 = 0.1875$ and $\widehat{Z}(\mathbf{s}_0) = 2.81$ ppm, with a prediction variance of 0.395 ppm^2. For $r = 2$ we have $\lambda_1 = 0.0506$ and $\widehat{Z}(\mathbf{s}_0) = 2.52$ ppm, with a prediction variance of 0.430 ppm^2.

The prediction and the prediction variance (when $r = 1$) are quite close to those of the kriging predictor. The results are not quite as good when $r = 2$, which is a more popular value of r when inverse distance predictions are made. This is one drawback of this method: the lack of a clear choice of r. Also, by definition, inverse distance predictors are less efficient than the kriging predictor.

Other local interpolators, like kernel smoothers and spline fits, similarly depend on the amount of smoothing, that is, the manner in which the interpolant is calculated. One virtue of the inverse distance method (and kernel smoothing or spline methods) is that knowledge of the covariance or variogram function is not necessary in order to compute the predictor. It seems somewhat optimistic, however, to compute predictions with no measure of the errors associated with these predictions. If an estimate of the accuracy of these predictions is desired, then knowledge of the variogram or covariance function is still necessary. Given that an estimate of the error is required, it is appropriate to use the optimal predictor based on the variogram, $\gamma(\cdot)$, whenever it is computationally feasible to do so.

iii. *Biased linear estimators*

The predictors in (i) and (ii) are both unbiased. Due to the focus on minimizing squared prediction error, it is natural to consider predictors that are biased to possibly reduce MSE. Towards this end, consider minimizing

$$MSE\left[\sum_{i=1}^{n} \lambda_i Z(\mathbf{s}_i) - Z(\mathbf{s}_0)\right]$$

$$= Var\left[\sum_{i=1}^{n} \lambda_i Z(\mathbf{s}_i) - Z(\mathbf{s}_0)\right] + E^2\left[\sum_{i=1}^{n} \lambda_i Z(\mathbf{s}_i) - Z(\mathbf{s}_0)\right],$$

without the restriction that $\sum_{i=1}^{n} \lambda_i = 1$. Letting $C_{ij} = Cov[Z(\mathbf{s}_i), Z(\mathbf{s}_j)]$, and using the results of Section 1.2.1, we have that:

$$Var\left[\sum_{i=1}^{n} \lambda_i Z(\mathbf{s}_i) - Z(\mathbf{s}_0)\right]$$

$$= \sum_{i=1}^{n} \sum_{j=1}^{n} \lambda_i \lambda_j C_{ij} + C_{00} - 2 \sum_{i=1}^{n} \lambda_i C_{i0}$$

$$= \boldsymbol{\lambda}^{\mathrm{T}} \boldsymbol{\Sigma} \boldsymbol{\lambda} + \sigma^2 - 2\boldsymbol{\lambda}^{\mathrm{T}} \boldsymbol{\sigma}_0,$$

where $\boldsymbol{\lambda}$ is now a vector of length n, $\sigma^2 = Var[Z(\mathbf{s}_0)]$, $\boldsymbol{\Sigma}$ is the $n \times n$ matrix C_{ij}, $i = 1, \ldots, n$, and $\boldsymbol{\sigma}_0$ is the vector of length n with row j equal to C_{j0}. Now,

$$E^2\left[\sum_{i=1}^{n} \lambda_i Z(\mathbf{s}_i) - Z(\mathbf{s}_0)\right] = \mu^2(\mathbf{1}^{\mathrm{T}}\boldsymbol{\lambda} - 1)^2,$$

and differentiating the MSE with respect to $\boldsymbol{\lambda}$, we find that

$$\boldsymbol{\lambda} = [\mu^2 \mathbf{J} + \boldsymbol{\Sigma}]^{-1}[\mu^2 \mathbf{1} + \boldsymbol{\sigma}_0],$$

where $\mathbf{J}$ denotes an $n \times n$ matrix of all 1s. We see that the MSE optimal prediction weights for linear predictors depend on μ. For illustration, if we assume that $\mu = 0$, then $\boldsymbol{\lambda} = \boldsymbol{\Sigma}^{-1}\boldsymbol{\sigma}_0$. In the example, the weights are then $\lambda_1 = 0.0471$ and $\lambda_2 = 0.5465$, giving a prediction of $\widehat{Z}(\mathbf{s}_0) = 1.53$ ppm. The MSE in this case is 0.348 ppm^2. Note that the variance using the biased estimator, 0.348 ppm^2, is smaller than the kriging variance of 0.392 ppm^2, and thus there is a possibility of an improvement on the kriging predictor in terms of MSE. We see, however, that the efficacy of this procedure depends on an assumed value of μ. Given that a reasonable estimator of μ is 3.47, however, this suggests that this predictor (with $\mu = 0$) has more than five-times the prediction MSE of the kriging predictor. If we take $\mu = 4$, for example, then the prediction weights become $\lambda_1 = 0.247$ and $\lambda_2 = 0.746$, which gives a predictor and MSE quite close to those of the kriging predictor. The dependence of the MSE on the generally unknown value of μ is somewhat unsatisfying. Note that the best unbiased estimator of μ is implicitly defined, through the equations in Section 2.2, as

$$\widehat{\mu} = (\mathbf{1}^{\mathrm{T}}\boldsymbol{\Sigma}^{-1}\mathbf{1})^{-1}\mathbf{1}^{\mathrm{T}}\boldsymbol{\Sigma}^{-1}\mathbf{Z}.$$

This is the generalized least squares estimator of the mean.

2.2.2 An example in the power family of variogram functions

To see how prediction depends critically on the choice of the variogram (or covariance) function, consider $n = 2$ observations in $d = 1$ dimension, at locations $\mathbf{s}_1$ and $\mathbf{s}_2$ being used to predict at $\mathbf{s}_0$ with $s_1 < s_2 < s_0$. Set $a = d(s_2, s_0)$, $b = d(s_1, s_2)$ and thus $a + b = d(s_1, s_0)$. For any set of $n = 2$ locations, the kriging equations in Section 2.2 give

$$\lambda_1 = \frac{1}{2}\left[1 + \frac{\gamma_{20} - \gamma_{10}}{\gamma_{21}}\right],$$

where γ_{ij} denotes $Var(Z_i - Z_j)$.

Assume that the variogram is given by the family of power models in $d = 1$ dimension for which $\gamma(h) = h^\alpha$ for $\alpha \in (0, 2)$. A specific example is the case $\alpha = 1$, which corresponds (in the discrete index situation) to an AR(1) model with $\rho = 1$. Using this variogram function in the above kriging equations we have:

$$\lambda_1 = \frac{1}{2}\frac{[a^\alpha + b^\alpha - (a+b)^\alpha]}{b^\alpha}.$$

From this formula, the weight on the observation at s_1, and its dependence on the value of α, can be seen. To see limiting behavior, note that as s_1 approaches s_2 ($b \to 0$), we have that, for $\alpha \neq 1$, the weight on Z_1 is approximately

$$\frac{1}{2}\left[1 - \frac{\alpha a^{\alpha-1}}{b^{\alpha-1}}\right].$$

Thus, for $\alpha < 1$, $\lambda_1 \to \frac{1}{2}$, while for $\alpha > 1$, $\lambda_1 \to -\infty$. This is an extreme example of a negative weight! At the value $\alpha = 1$ we have that $\lambda_1 = 0$ whenever $s_1 < s_2 < s_0$, that is, for all b. This represents an extreme example of 'screening.' All locations separated from the prediction location by another observation receive kriging weights identically equal to 0. As s_1 becomes very far away from both points ($b \to \infty$), we have that $\lambda_1 \to 0$ for *any* value of $\alpha \in (0, 2)$.

All these conclusions depend not only on the value of α, but also on the correctness of the power family of variogram functions. To see fundamentally different predictors, note that, under SOS (the power variogram corresponds to a non-SOS model), whenever $C(h) \to 0$ as $h \to \infty$, we have $\lambda_1 \to \frac{1}{2}[1 - C(a)/C(0)]$, as $s_1 \to -\infty$ ($b \to \infty$). Note this does not correspond to λ_1 in the power family for any value of α. It seems initially counterintuitive: as s_1 becomes further away from the prediction location it retains a nonzero weight. Note, however, that if the observations are independent, $\lambda_1 = \frac{1}{2}$, no matter how far away s_1 is located from s_0. In general, all observations have the same expectation, $E(Z)$, and thus Z_1 is always of some use to predict Z_0, if correlation is not too strong. As s_1 approaches s_2 ($b \to 0$), we have for $C(\cdot)$ differentiable that $\lambda_1 \to \frac{1}{2}[1 - C'(a)/C'(0)]$. Note that this also does not correspond to the power family for any α. For the specific example $C(h) = \exp(-h)$, we have that $\lambda_1 = \frac{1}{2}[1 - \exp(-a)]$ for any value of b. This is also a type of screening: as s_1 approaches s_2, λ_1 does not increase as long as $s_1 < s_2 < s_0$.

For general results on screening, see Stein (2002). A summary of his definition of screening is as follows. Consider locations on a grid. Consider a predictor based on observations in a neighborhood around the prediction location. Consider a second predictor based on all observations. If the first predictor becomes as good as the second as grid observations become more dense, then screening is defined to be present. Further, Stein describes which specific covariance functions have this property in terms of their spectral densities. We discuss spectral methodology in Section 3.3.1.

In conclusion, the kriging predictor is optimal amongst all linear predictors. However, prediction depends critically on the exact locations of the points and on the correlation between them. This means that good estimation of the variogram or covariance function is necessary to obtain efficient predictions. Variogram and covariance estimation are the main topics of Chapter 3.

2.3 Prediction intervals

Prediction intervals for the unobserved value $Z(\mathbf{s}_0)$ can be difficult to find, in general. If, however, observations are approximately normally distributed, then prediction intervals are straightforward. A standard result from multivariate analysis states that if $Z_1, \ldots, Z_n$ are jointly multivariate normally distributed, $\mathbf{Z} \sim N(\mu, \mathbf{\Sigma})$, then $\sum_{i=1} \lambda_i Z_i$ has a normal distribution for *any* weights λ_i. This means that, under multivariate normality,

$$\widehat{Z}(\mathbf{s}_0) \pm 1.96\sqrt{\sigma^2_{z0}}$$

is an accurate 95% prediction interval for $Z(\mathbf{s}_0)$. Further, if the spatial observations are multivariate normally distributed, then the best mean square predictor is linear [see, e.g., Mardia *et al.* (1979)]; that is,

$$E[Z(\mathbf{s}_0)|Z(\mathbf{s}_1), \ldots, Z(\mathbf{s}_n)] = \sum_{i=1}^{n} \lambda_i Z(\mathbf{s}_i).$$

In other words, the kriging predictor is also correct; that is, not just the best linear predictor, but the best MSE predictor (linear or nonlinear). For the above example in Figure 2.1, under joint normality we have

$$\left(2.94 \text{ ppm} \pm 1.96\sqrt{0.392 \text{ ppm}^2}\right) = (1.71 \text{ ppm}, 4.17 \text{ ppm})$$

is an accurate and correct 95% prediction interval for log(Phosphorus) at $\mathbf{s}_0$. We note that, in practice, the kriging variance, σ^2_{z0}, needs to be estimated. In this case, the Gaussian quantile, 1.96, should be replaced with the upper 2.5th percentile of the t-distribution, t_{n-p}, where n is the number of spatial locations and p denotes the number of parameters used to estimate $\sigma_{z_0}^2$.

The importance of Gaussian distributions in prediction suggests that a reasonable approach to prediction inference is to transform non-Gaussian observations to approximate normality, construct the prediction interval, and then transform the interval to the original scale. It should be noted, however, that the back-transformed predictions are no longer unbiased for the target predictand. This is because, for a nonlinear transformation, $g(\cdot)$, $E[g(Z)] \neq g[E(Z)]$. We next discuss this further in the special case of lognormal observations.

2.3.1 Predictions and prediction intervals for lognormal observations

We consider the lognormal distribution due to its popularity in modeling nonnegative observations. Also, we can make more specific analysis in this situation than for other non-Gaussian distributions. The development here follows that in de Oliveira (2006).

Denote the multivariate lognormal distribution by

$$\mathbf{Z} \sim \mathbf{\Lambda_n}(\mu, \mathbf{\Sigma}),$$

where we consider observations with possibly nonconstant means. Formally, this definition means that $\mathbf{Z} = \exp(\mathbf{Y})$, where $\mathbf{Y} \sim N(\mu, \mathbf{\Sigma})$. Denote the componentwise means and covariances of $\mathbf{Y}$ by $E[Y(\mathbf{s})] = \mu(\mathbf{s})$, and $Cov[Y(\mathbf{s}), Y(\mathbf{u})] = \Sigma_{\mathbf{s},\mathbf{u}}$. The mean and the covariance of $\mathbf{Z}$ in terms of those of $\mathbf{Y}$ are given by:

$$E[Z(\mathbf{s})] = \exp\left[\mu(\mathbf{s}) + \frac{\sigma^2}{2}\right]$$

and

$$Cov\big[Z(\mathbf{s}_i), Z(\mathbf{s}_j)\big] = \mu(\mathbf{s}_i)\mu(\mathbf{s}_j)\big[\exp(\Sigma_{\mathbf{s}_i,\mathbf{s}_j}) - 1\big].$$

The goal is to minimize:

$$E\{\big[\widehat{Z}(\mathbf{s}_0) - Z(\mathbf{s}_0)\big]^2\}.$$

If the predictand, $Z(\mathbf{s}_0)$, is defined jointly with the observations, $\mathbf{Z}$, then:

$$[Z(\mathbf{s}_0), \mathbf{Z}^{\mathrm{T}}]^{\mathrm{T}} \sim \Lambda_{n+1}\left[\begin{pmatrix}\mu_{Y_0} \\ \mu_{\mathbf{Y}}\end{pmatrix}, \begin{pmatrix}\sigma^2_{Y_0} & C^{\mathrm{T}}_{0Y} \\ C_{0Y} & \Sigma_{\mathbf{Y}}\end{pmatrix}\right],$$

where C_{0Y} denotes the $n \times 1$ vector $Cov[Y(\mathbf{s}_0), \mathbf{Y}]$. Further, the optimal 'predictor,' $E\left[Z(\mathbf{s}_0)|\mathbf{Z}\right]$, is given by

$$\widehat{Z}^*(\mathbf{s}_0) = \exp\left[\widehat{Y}^*(\mathbf{s}_0) + \frac{\widehat{\sigma}^*_{0Y}}{2}\right],$$

where

$$\widehat{Y}^*(\mathbf{s}_0) = E\left[Y(\mathbf{s}_0)|\mathbf{Y}\right] = \mu_0 + C^{\mathrm{T}}_{0Y}\Sigma^{-1}_Y\left(\mathbf{Y} - \mu_{\mathbf{Y}}\right), \tag{2.4}$$

$$\widehat{\sigma}^*_{0Y} = Var\left[Y(\mathbf{s}_0)|\mathbf{Y}\right] = \sigma^2_Y - C^{\mathrm{T}}_{0Y}\Sigma^{-1}_Y C_{0Y}. \tag{2.5}$$

Equations 2.4 and 2.5 are the simple kriging predictor and kriging variance of $Y(\mathbf{s}_0)$, assuming that μ_Y is known. In general, as assumed in ordinary kriging in Section 2.2, the mean response μ_Y is unknown. This presents the following options for point prediction:

i. Lognormal kriging:

   a. First, find the best linear unbiased predictor of $Y(\mathbf{s}_0)$.
   b. Then, exponentiate this predictor using a (multiplicative) bias correction to make $E[\widehat{Z}(\mathbf{s}_0)] = E[Z(\mathbf{s}_0)]$.

   Specifically:

$$\widehat{Y}^{0K}(\mathbf{s}_0) = \lambda^{\mathrm{T}}_{0Y}\mathbf{Y},$$

   where λ_{0Y} is the vector of ordinary kriging weights to predict $Y(\mathbf{s}_0)$ based on $\mathbf{Y}$.

   Then the lognormal kriging predictor is given by:

$$\widehat{Z}^{LK}(\mathbf{s}_0) = \exp\left[\widehat{Y}^{0K}(\mathbf{s}_0) + \frac{1}{2}\left(\sigma^2_Y - \lambda^{\mathrm{T}}_{0Y}\Sigma\lambda_{0Y}\right)\right].$$

   This can also be written, as usual, in terms of the ordinary kriging variance, $\sigma^2_{Y_0}$, and Lagrange multiplier, m_{OY}, as:

$$\exp\left[\widehat{Y}^{0K}(\mathbf{s}_0) + \sigma^2_{Y_0} - m_{OY}\right].$$

Note the property that the lognormal kriging predictor, $\widehat{Z}^{LK}(\mathbf{s}_0)$, has: it minimizes

$$E\left(\left\{\ln[\widehat{Z}(\mathbf{s}_0)] - \ln[Z(\mathbf{s}_0)]\right\}^2\right)$$

over

$$\widehat{Z}(\mathbf{s}_0) = \exp\left[a^{\mathrm{T}}Y + K_0\right], \quad K_0 \in R, \quad \mathbf{a} \in R^n,$$

such that:

$$E\left\{\ln[\widehat{Z}(\mathbf{s}_0)]\right\} = E\{\ln[Z(\mathbf{s}_0)]\} \quad \text{and} \quad E[\widehat{Z}(\mathbf{s}_0)] = E\,[Z(\mathbf{s}_0)]\,.$$

Note that this is not the original goal of prediction.

ii. Another approach is to try to attain the L_2 optimal predictor (for predicting $Z(\mathbf{s}_0)$ and not $\ln[Z(\mathbf{s}_0)]$) over

$$\widehat{Z}(\mathbf{s}_0) = \exp\left[a^{\mathrm{T}}Y + K_0\right], \quad a^{\mathrm{T}}1 = 1.$$

Algebraic manipulations show that the optimal predictor is:

$$\tilde{Z}(\mathbf{s}_0) = \exp\left[\widehat{Y}^{0K}(\mathbf{s}_0) + \frac{1}{2}\left(\sigma_Y^2 - \lambda_{0Y}^{\mathrm{T}}\Sigma_Y\lambda_{0Y} - 2m_{OY}\right)\right],$$

or

$$\exp\left[\widehat{Y}^{0K}(\mathbf{s}_0) + \sigma_{Y_0}^2 - 2m_{OY}\right].$$

Comparing the lognormal kriging predictor with this, we have simply replaced -1 by -2 in the term multiplying m_{OY}.

To see the practical effects of choice of predictor, consider the data in Section 2.2.1. Recall that we have the prediction interval:

$$2.94 \pm 1.96\sqrt{0.392} = [1.71 \text{ log(ppm)}, 4.17 \text{ log(ppm)}]\,,$$

for true phosphorus in log(ppm). We then have the natural prediction interval for phosphorus in terms of ppm as

$$[\exp(1.71), \exp(4.17)] = [5.53 \text{ ppm}, 64.72 \text{ ppm}]\,.$$

For point prediction, the biased prediction of phosphorus is $\exp(2.94) = 18.92$ ppm, while the lognormal kriging predictor is $\exp\left(2.94 + \frac{0.392}{2} - 0.1067\right) = 13.98$ ppm. Finally, the L_2 optimal point predictor is $\exp\left(2.94 + \frac{0.392}{2} - 2 \times 0.1067\right) = 12.56$ ppm. As discussed, the prediction variance of the predictors depends on the unknown mean, μ. In general, the two methods, unbiased lognormal or L_2 optimal, give similar prediction MSEs. The biased predictor, however, should be avoided.

For other transformations to a Gaussian distribution, corrections can be made using Taylor series methodology for general smooth functions $g(\cdot)$. This is known as 'trans-Gaussian kriging.'

We have discussed how to account for specific non-Gaussian distributions. If we believe that our observations are approximately Gaussian, we may wish to assume Gaussianity and account for the inappropriateness of this assumption by appropriately widening our prediction interval. In this case, we need to address the question: how badly (how inaccurately) do prediction intervals perform in the case where observations come from a nonnormal distribution? A very useful result is given in Pukelsheim (1994), which states that *for all* continuous unimodal distributions (e.g., normal, double exponential, chisquared, t, lognormal), it holds that

$$P\left[|\widehat{Z}(\mathbf{s}_0) - Z(\mathbf{s}_0)| > t\sigma^2_{zo}\right] \leq \frac{4}{9t^2},$$

for all $t > 1.633$. For example, if $t = 1.96$, then better than 88% coverage is obtained across this wide class of distributions. On the other hand, if 95% coverage is required across all continuous unimodal distributions, we can take $t = 2.98$, and the desired coverage is obtained. Note that the above inequality is much more useful than the celebrated Chebyshev inequality (which holds for *all* distributions with finite variance), which gives a much less informative bound of $1/t^2$.

2.4 Universal kriging

Until now we have assumed that the mean function is constant, that is, that $E[Z(\mathbf{s})] = \mu$. In many cases a more appropriate model is

$$Z(\mathbf{s}) = \mu(\mathbf{s}) + \delta(\mathbf{s}).$$

One of the main examples of a nonconstant mean function is where the mean of the spatial field depends on location. For example, we might assume that the mean function is a smooth function of spatial coordinates. Specifically, let $\mathbf{s} = (x, y)$ in the usual coordinate system. Then a reasonable model is that, for some positive integer r, we have

$$\mu(\mathbf{s}) = \sum_{0 \leq k+l \leq r} \beta_{k,l} x^k y^l.$$

For example, a quadratic surface takes $r = 2$, in which case

$$\mu(\mathbf{s}) = \beta_{0,0} + \beta_{1,0}x + \beta_{0,1}y + \beta_{1,1}xy + \beta_{2,0}x^2 + \beta_{0,2}y^2.$$

For observations on a regular grid, another, less smooth mean function that depends on spatial coordinates is given by

$$\mu(\mathbf{s}) = a + c(x) + r(y),$$

where $c(x)$ denotes a mean for all observations with x-coordinate equal to x, and $r(y)$ denotes a mean for all observations with y-coordinate equal to y.

More generally, we have

$$\mu(\mathbf{s}) = \sum_{k=1}^{p} f_{k-1,\mathbf{s}} \beta_{k-1},$$

where $f_{j,\mathbf{s}}$ are known functions and β_j are unknown parameters, for $j = 0, \ldots, p-1$. We always assume that $f_{0,\mathbf{s}} = 1$ for all $\mathbf{s}_i$. The $f_{j,\mathbf{s}}$ can include spatial coordinates as well as actual covariates, that is, explanatory variables associated with the observation at location $\mathbf{s}$. For example, altitude is often an important explanatory variable in spatial temperature or rainfall analysis.

An example: phosphorus data Consider fitting a mean surface to the $15 \times 5 = 75$ observations separated by 20 m in the x direction and 15 m in the y direction on the lattice in Figure 1.1. A smooth quadratic surface requires 6 parameters, while fitting separate row and column means in the mean function requires $1 + (15 - 1) + (5 - 1) = 19$ parameters. We discuss estimation in universal kriging in Chapter 3 along with variogram estimation, and in Chapter 4.

2.4.1 Optimal prediction in universal kriging

To simplify the nonconstant mean model, again let $\mathbf{Z} = [Z(\mathbf{s}_1), \ldots, Z(\mathbf{s}_n)]$, and let $\mathbf{X}$ be a $n \times p$ matrix with (i, k) element equal to $f_{k-1,\mathbf{s}_i}$. Then

$$\mathbf{Z} = \mathbf{X}\boldsymbol{\beta} + \boldsymbol{\delta}.$$

Note that this is equivalent to the observations, $Z(\mathbf{s}_i)$, $i = 1, \ldots, n$, following a general linear model. In the usual GLM, the errors are assumed to be independent, while here the errors, $\delta(\mathbf{s}_i)$, are assumed to be IS.

The predictand under this GLM is $Z(\mathbf{s}_0) = \mathbf{x}_0^{\mathrm{T}}\boldsymbol{\beta} + \delta(\mathbf{s}_0)$, where $\mathbf{x}_0$ denotes the p covariates at the location where prediction is desired. The linear predictor is still assumed to be of the form $\widehat{Z}(\mathbf{s}_0) = \sum_{i=1}^{n} \lambda_i Z(\mathbf{s}_i)$. Unbiasedness now requires that

$$E[\widehat{Z}(\mathbf{s}_0)] = \sum_{i=1}^{n} \lambda_i \mu(\mathbf{s}_i) = \sum_{i=1}^{n} \lambda_i \mathbf{x}_i^{\mathrm{T}}\boldsymbol{\beta} = \mu(\mathbf{s}_0) = \mathbf{x}_0^{\mathrm{T}}\boldsymbol{\beta},$$

for all $\boldsymbol{\beta}$. In other words, we require $\sum_{i=1}^{n} \lambda_i \mathbf{x}_i^{\mathrm{T}} = \mathbf{x}_0^{\mathrm{T}}$. This is a system of p equations. Note that in the case of ordinary kriging ($p = 1$), $\mathbf{x}_i = 1$, for all i, and so this constraint reduces to $\sum_{i=1}^{n} \lambda_i = 1$, as it must. Now, however, the assumption that $\sum_{i=1}^{n} \lambda_i \mathbf{x}_i^{\mathrm{T}} = \mathbf{x}_0^{\mathrm{T}}$ induces p constraints. Let $m_1, \ldots, m_p$ denote p Lagrange multipliers necessary to enforce these p constraints. Proceeding entirely analogously to the case of ordinary kriging, we find that

$$\boldsymbol{\lambda} = \boldsymbol{\Gamma}^{-1}\boldsymbol{\gamma},$$

where $\boldsymbol{\lambda} = (\lambda_1, \ldots, \lambda_n, m_1, \ldots, m_p)^{\mathrm{T}}$, $\boldsymbol{\gamma} = [\gamma(\mathbf{s}_0 - \mathbf{s}_1), \ldots, \gamma(\mathbf{s}_0 - \mathbf{s}_n), 1, f_{1,\mathbf{s}_0}, \ldots, f_{p-1,\mathbf{s}_0}]$, and $\boldsymbol{\Gamma}$ is a $(n + p) \times (n + p)$ matrix.

Note that optimal estimation of covariance requires knowledge of regression parameters. Conversely, knowledge of $\boldsymbol{\beta}$ is required to estimate covariance parameters properly. This is an inherent difficulty in universal kriging. In Section 3.6 we carry out estimation of mean and covariance parameters for the phosphorus data. In Chapter 4, details are given on estimation for Gaussian and binary observations.

2.5 The intuition behind kriging

Recall that the kriging weights, $\boldsymbol{\lambda} = \boldsymbol{\Gamma}^{-1}\boldsymbol{\gamma}$. We have seen in Section 2.2.1 that we can find $\boldsymbol{\lambda}$ in terms of the covariance function. Specifically, it can be shown (in a manner completely analogous to the derivation of the kriging weights through the variogram function) that

$$\boldsymbol{\lambda} = \boldsymbol{\Sigma}^{-1}\boldsymbol{\nu},$$

where $\boldsymbol{\Sigma}$ is an $(n + 1) \times (n + 1)$ matrix with elements $\sigma_{ij} = Cov[Z(\mathbf{s}_i), Z(\mathbf{s}_j)]$ for $i = 1, \ldots, n$, and $\nu_i = Cov[Z(\mathbf{s}_0), Z(\mathbf{s}_i)]$ for $i = 1, \ldots, n$. All remaining elements are in agreement with those in $\boldsymbol{\Gamma}$ and $\boldsymbol{\gamma}$ as defined in Section 2.2. The value of ν_i gives the *statistical distance* between variables $Z(\mathbf{s}_0)$ and $Z(\mathbf{s}_i)$. If ν_i is large, this implies that a relatively large weight should be put on location $\mathbf{s}_i$. Using only this statistical distance suggests predicting with weights $\omega_i = \nu_i / \sum_{i=1}^{n} \nu_i$, say. If $\boldsymbol{\Sigma}_{ij}$, the covariance between $Z(\mathbf{s}_i)$ and $Z(\mathbf{s}_j)$, is large, this suggests there is some redundancy in putting large weights on both $Z(\mathbf{s}_i)$ and $Z(\mathbf{s}_j)$ in prediction. Applying the inverse operator to the matrix $\boldsymbol{\Sigma}$ in the formula for $\boldsymbol{\lambda}$ appropriately down-weights these observations.

2.5.1 An example: the kriging weights in the phosphorus data

Assume that a mean function has been fit to the phosphorus data, $\mathbf{X}\widehat{\boldsymbol{\beta}}$, as detailed in Section 3.6, and then the residuals, $e(\mathbf{s}_i) := Z(\mathbf{s}_i) - \mathbf{x}_i^{\mathrm{T}}\widehat{\boldsymbol{\beta}}$, are to be treated as intrinsically stationary. We use the residuals to predict $e(\mathbf{s}_0)$, where $\mathbf{s}_0$ is the prediction location. Our final prediction of log(Phosphorus) is then $\mathbf{x}_0^{\mathrm{T}}\widehat{\boldsymbol{\beta}} + \widehat{e}(\mathbf{s}_0)$. Our interest focuses on the kriging weights in using $e(\mathbf{s}_i)$, $i = 1, \ldots, n$, to predict $e(\mathbf{s}_0)$.

Consider predicting log(Phosphorus) at the location $(40, 42.5)$ as depicted in Figure 2.2 using the isotropic variogram function developed in Chapter 3, Section 3.6. Note that this location is equally distant from the observed locations $(30, 35)$, $(50, 35)$, $(30, 50)$, and $(50, 50)$. Under isotropy, this implies that the weights using the statistical distance, ω_i, will be the same for all four locations. From Figure 2.2 we see, however, that the four weights are, successively, $-0.001, 0.052, 0.018, 0.067$. Although only a subset of the phosphorus locations are depicted in Figure 2.2, all $n = 103$ observations were used to determine these kriging weights. Note that the most isolated point, $(50, 35)$, receives the largest weight of the four, as suspected. Further, the location $(30, 35)$ is surrounded by other observations and becomes so redundant that its kriging weight is actually negative. Overall, in the entire data set, 8 out of 103 kriging weights are negative. Interestingly, the observation at location $(80, 290)$ (not shown in Figure 2.2) actually receives a larger weight than that at location $(30, 35)$, even though it is more than *twenty-times* further away from the desired location. One apparently surprising outcome is that the kriging weight at $(50, 35)$ is approximately three-times as large as that at $(30, 50)$ although they are the same distance (and thus the same statistical distance under isotropy) from the prediction location and appear to have locations at the

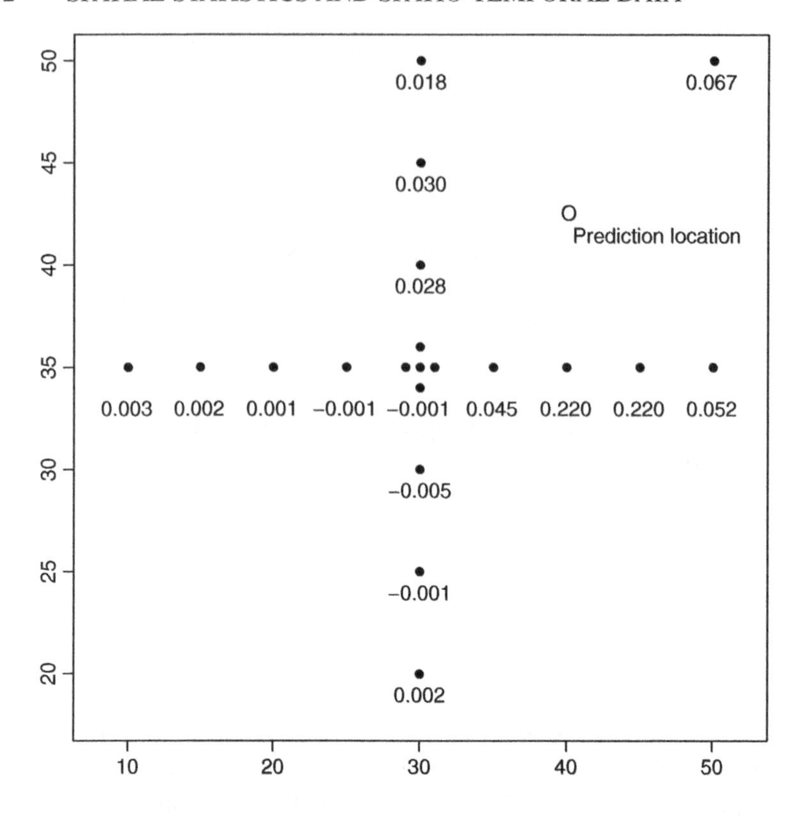

Figure 2.2 Kriging weights for the phosphorus data. All n $= 103$ *locations are used to predict the residual in universal kriging at the location* $(40.0, 42.5)$.

same distances surrounding them. Note, however, that the closest point to $(30, 50)$ on the regularly spaced grid is at location $(30, 65)$ (not shown) and is closer to $(30, 50)$ than is $(70, 35)$ to $(50, 35)$. The location $(70, 35)$ is the closest point to $(50, 35)$. This makes the data at location $(30, 50)$ more redundant than the data at location $(50, 35)$.

In summary, the weighting vector $\boldsymbol{\nu}$ ($\boldsymbol{\gamma}$ for the variogram) accounts for the correlation between each observation and the observation to be predicted while $\boldsymbol{\Sigma}$ ($\boldsymbol{\Gamma}$ for the variogram) accounts for the correlation between all observed variables. The kriging weights are a combination of these two correlations. Specifically, we have that a given kriging weight, λ_i, is large if: (i) $Z(\mathbf{s}_i)$ is heavily correlated with the value to be predicted; and/or, (ii) $Z(\mathbf{s}_i)$ is not too heavily correlated with other observed values.

Further, notice that to predict the value at a single location, we need not only the statistical relationship between all observed variables and the variable to be predicted, but also we need the relationship between all observed variables and all other observed variables.

It has been previously assumed that the variogram or covariance function is known. Using this knowledge we have seen that we can obtain optimal (minimum variance amongst linear unbiased predictors) predictions via kriging. In universal kriging it has been assumed that the trend parameters $\boldsymbol{\beta}$ and all covariance parameters in the general linear model (GLM) are known. In practice, however, all linear model parameters and correlation parameters are typically unknown, and need to be estimated from data observations. This is addressed in Chapter 3.

3

Variogram and covariance models and estimation

In the previous chapter we carried out optimal prediction when the variogram or covariance function is known. Although we assumed either IS or SOS, there is really no reason to do so under known correlations. For example, if we carefully look at the kriging predictor in Section 2.2, all that is necessary is knowledge of $Var[Z(\mathbf{s}_i) - Z(\mathbf{s}_j)]$ or $Cov[Z(\mathbf{s}_i), Z(\mathbf{s}_j)]$ for all pairs of locations, $\mathbf{s}_i$ and $\mathbf{s}_j$, necessary to compute $\boldsymbol{\lambda}$. In general, the variogram and covariance functions are unknown and need to be estimated from the data. In this case, an assumption of IS or SOS enables replication, and thus greatly aids in estimation.

3.1 Empirical estimation of the variogram or covariance function

Assume that IS (or SOS) holds and we need to estimate the variogram function $2\gamma(\mathbf{h})$ [or $C(\mathbf{h})$, the covariance function for SOS processes]. From Figure 1.2, we see that every pair of points separated by spatial lag $\mathbf{h}$ gives an observed estimate (replicate) of $2\gamma(\mathbf{h})$, namely $\{[Z(\mathbf{s}_i) - Z(\mathbf{s}_j)]^2\}$ with $\mathbf{s}_i - \mathbf{s}_j = \mathbf{h}$. This suggests an intuitive estimator

Spatial Statistics and Spatio-Temporal Data: Covariance Functions and Directional Properties Michael Sherman

of the semivariogram obtained via the average of all such pairs. The *moment estimator* or *sample semivariogram* is given by:

$$\widehat{\gamma}(\mathbf{h}) = \frac{1}{2|N(\mathbf{h})|} \sum_{N(\mathbf{h})} \{[Z(\mathbf{s}_i) - Z(\mathbf{s}_j)]^2\},$$

where $N(\mathbf{h}) = \{(\mathbf{s}_i, \mathbf{s}_j) : \mathbf{s}_i - \mathbf{s}_j = \mathbf{h})\}$, the set of all pairs of locations separated by vector $\mathbf{h}$. This assumes that for lag vectors $\mathbf{h}$, of interest, there are sufficient pairs of points separated by vector $\mathbf{h}$. For example, in the phosphorus data set there are several replicates at several spatial lags $\mathbf{h}$. To estimate the covariance function we use:

$$\widehat{C}(\mathbf{h}) = \frac{1}{|N(\mathbf{h})|} \sum_{N(\mathbf{h})} [Z(\mathbf{s}_i) - \overline{Z}_n][Z(\mathbf{s}_j) - \overline{Z}_n].$$

Notice that an estimator of μ, $\overline{Z}_n$, is used in the definition of $\widehat{C}(\mathbf{h})$. This makes it so that $E[\widehat{C}(\mathbf{h})] \neq C(\mathbf{h})$. It is, however, clear that $E[\widehat{\gamma}(\mathbf{h})] = \gamma(\mathbf{h})$. This need to estimate μ was our third reason (discussed in Chapter 2) to prefer the variogram function over the covariance function.

3.1.1 Robust estimation

The moment estimator is an average of squared differences, and thus can be greatly influenced by a small number of aberrant values. For this reason, it is sometimes preferable to consider a robust estimator, to lessen the importance of any large, squared differences. For example, Cressie and Hawkins (1980) proposed the following robust estimator:

$$\overline{\gamma}(\mathbf{h}) = \frac{1}{2|N(\mathbf{h})|} \frac{\sum_{N(\mathbf{h})} \{[Z(\mathbf{s}_i) - Z(\mathbf{s}_j)]^{1/2}\}^4}{0.457 + 0.494/N(\mathbf{h})}.$$

This is robust in the sense that it is resistant to contaminated normal distributions and to outliers possibly generated by heavy-tailed distributions. This can be seen by its use of square-root differences, in place of square differences, in the unbiased moment estimator given in Section 3.1. On the other hand, the constants in the denominator are approximately the correct constants for the contaminated normal model. They may not be appropriate for other heavy-tailed distributions. For estimators with a higher breakdown point, see, for example, Genton (2001).

3.1.2 Kernel smoothing

For data that are unequally spaced, there are often no pairs of points separated by any particular spatial lag, $\mathbf{h}$. To obtain an estimate in this case, some amount of smoothing is necessary. For this purpose, define a nonnegative symmetric bivariate density function, $w(\mathbf{u})$, centered at 0, with $\int w(\mathbf{u})d\mathbf{u} = 1$. Then the kernel estimator of $\gamma(\mathbf{h})$ is:

$$\widehat{\gamma}_\delta(\mathbf{h}) = \sum_{(i,j)} \frac{\{[Z(\mathbf{s}_i) - Z(\mathbf{s}_j)]^2\} w_\delta(\mathbf{h} - \mathbf{h}_{ij})}{\sum_{(i,j)} w_\delta(\mathbf{h} - \mathbf{h}_{ij})},$$

where $\mathbf{h}_{ij} = \mathbf{s}_i - \mathbf{s}_j$ is the observed spatial lag between i and j, and $w_\delta(\mathbf{u}) = \frac{1}{\delta} w\left(\frac{\mathbf{u}}{\delta}\right)$. The parameter δ is the bandwidth that determines the amount of averaging that goes into the estimate at each $\mathbf{h}$. When $w(\mathbf{u})$ is the uniform density on a circle of radius one, then we are averaging all squared differences between all pairs of observations separated by lags within distance δ of $\mathbf{h}$. Using, for example, a bivariate normal kernel, puts more weight on pairs of points separated by lag vectors that are closer to the desired spatial lag, $\mathbf{h}$.

Example Consider the estimated variogram of the residuals for the phosphorus data. The points in Figure 3.3 (page 65) display the values of the estimated semivariogram at 20 lags, $\|\mathbf{h}\|$, using a uniform kernel. This was obtained using the R-function variogram in geoR. Notice how the values increase until about 10–20 m, and then seem to become roughly constant. This suggests that the range of spatial dependence is in the neighborhood of 10–20 m. The number of squared average differences ranges from 8 ($\|\mathbf{h}\| \simeq 1$) to 129 ($\|\mathbf{h}\| \simeq 38$). The two smooth curves are parametric fits, and are discussed in Sections 3.5 and 3.6.

3.2 On the necessity of parametric variogram and covariance models

The moment estimator of the variogram is intuitive and unbiased, as discussed in Section 3.1. Still, there are two principal inadequacies of this estimator. The first stems from the fact that variances are positive. Specifically, this requires that for any constants a_i, $i = 1, \ldots, k$, and any set of spatial locations, $\mathbf{s}_i$, $i = 1, \ldots, k$, it holds that

$$Var\left[\sum_{i=1}^{k} a_i Z(\mathbf{s}_i)\right] = \sum_{i=1}^{k}\sum_{j=1}^{k} a_i a_j Cov[Z(\mathbf{s}_i), Z(\mathbf{s}_j)] \geq 0.$$

Analogously, we require that

$$\sum_{i=1}^{k}\sum_{j=1}^{k} a_i a_j \gamma(\mathbf{s}_i - \mathbf{s}_j) \leq 0,$$

for all a_i such that $\sum_{i=1}^{k} a_i = 0$. The moment estimator given in Section 3.1 does not satisfy this nonnegative definite property (for the covariance function) or the nonpositive definite property (for the variogram function).

The second reason that the moment estimator is not adequate is that it gives estimates of correlation only at *observed* spatial lags. However, to carry out kriging at an unobserved location, $\mathbf{s}_0$, one typically needs estimates of correlation at spatial lags that may not have been observed. Thus, in practice, an estimator of the spatial correlation at *any* spatial lag is required.

A common solution to these two difficulties is to fit valid parametric models for which the variogram function is nonpositive definite, which then allows for the computation of the variogram at any spatial lag. It should be noted, however, that there has been research into nonparametric estimation of the covariance function and variogram function that respects the nonnegative definite and nonpositive definite properties. For the covariance function, large sample results are given in, for example, Hall and Patil (1994) and Hall *et al.* (1994), while Huang *et al.* (2008) give computational details on achieving nonparametric nonpositive definite estimators of the variogram.

3.3 Covariance and variogram models

For SOS models we know that $C(\mathbf{h}) = C(\mathbf{0}) - \gamma(\mathbf{h})$, so given the covariance function, it is easy to express the corresponding variogram function. For SOS covariance functions, we will express parametric forms in terms of the covariance function. For intrinsically stationary, but not SOS processes we, necessarily, give the variogram function.

For now, assume that all covariances (and hence variograms) are *isotropic*. This means that the covariance between variables at two locations depends only on the distance between them and not on the direction or orientation between them. Formally, this means that

$C(\mathbf{h}) = C^*(\|\mathbf{h}\|)$ where $\|\mathbf{v}\|$ denotes the length of any vector $\mathbf{v}$ and $C^*(\cdot)$ denotes a function of a scalar variable. The importance of this assumption and the effects it has on predictions are discussed in Chapter 5.

We now discuss some well-known parametric covariance functions (and thus their associated variogram functions).

Exponential model

$$C(\mathbf{h}, \theta) = \sigma^2 \exp(-\nu\|\mathbf{h}\|) \quad \text{for } \sigma^2 > 0, \nu > 0,$$

where $\theta = (\sigma^2, \eta)$. The parameter σ^2 is the scale parameter giving the overall variability of the process, while the parameter ν determines the rate of decay of correlation as a function of distance. It is clear that the covariance approaches zero as the spatial lag length, $\|\mathbf{h}\|$, becomes large. The associated semivariogram function is $\gamma(\mathbf{h}, \theta) = \sigma^2[1 - \exp(-\nu\|\mathbf{h}\|)]$.

Gaussian model

$$C(\mathbf{h}, \theta) = \sigma^2 \exp(-\nu\|\mathbf{h}\|^2) \quad \text{for } \sigma^2 > 0, \nu > 0.$$

Again, σ^2 is the scale parameter giving the variability of the process, while the parameter ν determines the rate of decay as a function of distance. Although the Gaussian distribution is central to all of statistics, the same cannot be said for the Gaussian covariance model. This covariance function is infinitely differentiable, and this smoothness is not common in applications.

Spherical model

$$C(\mathbf{h}, \theta) = \sigma^2 \left(1 - \frac{3\|\mathbf{h}\|}{2m} + \frac{\|\mathbf{h}\|^3}{2m^3}\right) \quad \text{for } \sigma^2 > 0, \|\mathbf{h}\| \leq m.$$

While σ^2 has the same interpretation, the range of the correlation is now accounted for by m. Notice that while the exponential and Gaussian models have nonzero correlation at arbitrarily large spatial lags, the spherical model has 0 correlation between variables separated by more than the range parameter m. While this finite range correlation feature is appealing, it can sometimes lead to difficulties in estimation when maximum likelihood is used to fit this covariance model. In Section 3.5, we discuss fitting algorithms for covariance and variogram models.

Tent model

$$C(\mathbf{h}, \theta) = \sigma^2 \left(1 - \frac{\|\mathbf{h}\|}{\theta}\right) \quad \text{for } 0 \leq \|\mathbf{h}\| \leq \theta.$$

Although this function takes a very simple form, and might seem appealing for this reason, simple examples show that this covariance model has a serious drawback for spatial data. It can, however, be appropriate for temporal data, $d = 1$.

Example: is the tent model appropriate for spatial observations? Consider observations $Z(i, j)$ at equally spaced locations, on a $k \times k$ grid with nearest neighbors separated by distance $\theta 2^{-1/2}$, with the tent model covariance function holding. Consider $a_{ij} = 1$, for i and j equal, and $a_{ij} = -1$, otherwise. Consider the variance of the combination of variables given by

$$\widehat{Z} := \sum_{i=1}^{k} \sum_{j=1}^{k} a_{ij} Z(i, j).$$

Direct computation shows that:

$$Var(\widehat{Z}) = \sigma^2(k^2 - [4k(k-1)(1 - 2^{-1/2})]).$$

From this we see that, for k sufficiently large, this is negative, and so the tent model covariance is *not* nonnegative definite. This illustrates that we cannot define an apparently reasonable covariance function and hope that it is valid. We need to choose from a class of valid models.

The exponential model is a useful model in many practical applications. The following model class contains the exponential covariance as a special case and is a very flexible class of covariance models. Its functional form is more involved than the previous models, but its main flexibility is that this class allows for covariance functions of qualitatively different behavior.

Matérn model (K-Bessel model)

$$C(\mathbf{h}, \theta) = \frac{\sigma^2}{\Gamma(\eta + 1/2)} \left(\frac{\|\mathbf{h}\|}{2\nu}\right)^{\eta} K_{\eta}(\nu\|\mathbf{h}\|) \quad \text{for } \eta > 0, \nu > 0,$$

where K_η is a modified Bessel function of the second kind, of order η. The parameter σ^2 is the scale parameter, ν corresponds to the rate of decay of the covariance function, and η is typically known as the smoothness parameter.

To see the effect of the smoothness parameter, η, consider two special cases: namely, $\eta = 0.5$, and the limiting case of $\eta \to \infty$. In the former case, direct evaluation shows that

$$K_{0.5}(x) = \sqrt{\frac{\pi}{2x}} \exp(-x),$$

and thus that $C(\mathbf{h}, \theta) = \sigma^2 \exp(-\nu\|\mathbf{h}\|)$. This is exactly the exponential covariance function. As $\eta \to \infty$, analysis shows that $C(\mathbf{h}, \theta) \to \sigma^2 \exp(-\nu\|\mathbf{h}\|^2)$, and thus the Gaussian model is a limiting case of the Matérn model. Direct calculations also show that when $\eta = 1.5$, we have $C(\mathbf{h}, \theta) = (\sigma^2/2) \exp(-\nu\|\mathbf{h}\|)(1 + \nu\|\mathbf{h}\|)$. Using the recursive relationship for Bessel functions of the second kind, it holds that, for integer k:

$$K_{k+1}(x) = K_{k-1}(x) + \frac{2k}{x} K_k(x).$$

Thus, given the covariance function at two integer levels of smoothness, η, we can recursively compute the Matérn covariance function for any positive integer. This allows for the comparison of various models in terms of goodness of fit. Displayed in Figure 3.1 are two members of the Matérn family, along with the spherical and Gaussian covariance functions. Notice the quite different behavior of the functions when $\eta = 0.5$ and $\eta = 1.5$, particularly at the origin. To appreciate in what specific sense the smoothness parameter determines the smoothness, some discussion of spectral methods follows.

3.3.1 Spectral methods and the Matérn covariance model

It is not usually possible to characterize the smoothness of a spatial process via any specific realization from that process. We need to consider average behavior of the process. Towards this end, a spatial process is said to be Mean Square Continuous (MSC) at $\mathbf{s}$ if:

$$\lim_{h \to 0} E\left[Z(\mathbf{s} + \mathbf{h}) - Z(\mathbf{s})\right]^2 = 0.$$

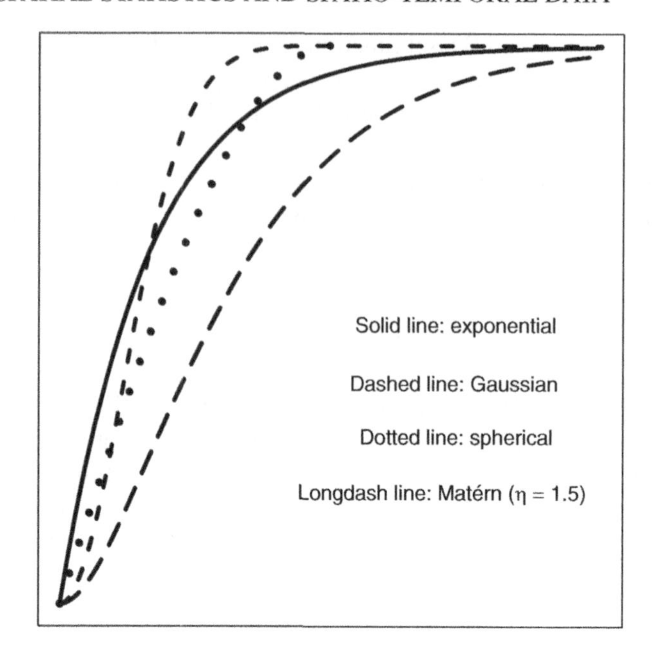

Figure 3.1 Four covariance functions. All four covariances are standardized to have the same sill. The exponential covariance is Matérn with $\eta = 0.5$.

Note that, under IS, if a process is MSC at the origin, $\mathbf{s} = \mathbf{0}$, then it is MSC at all $\mathbf{s}$. For mean square continuous processes, a further subset are Mean Square Differentiable (MSD). To define this concept, first consider the standardized differences:

$$\widetilde{Z_h}(\mathbf{s}) = \left[\frac{Z(\mathbf{s} + \mathbf{h}) - Z(\mathbf{s})}{\mathbf{h}}\right].$$

To consider the typical local behavior of the process, consider this standardized difference process for small $\mathbf{h}$. If

$$\lim_{\mathbf{h} \to 0} E\left[\widetilde{Z_h}(\mathbf{s}) - Z'(\mathbf{s})\right]^2 = 0,$$

for some $Z'(\mathbf{s})$, then $Z(\mathbf{s})$ is MSD with derivative $Z'(\mathbf{s})$. Higher order derivatives are defined by considering differences of lower order derivatives. To see clearly the issue of differentiability in the Matérn family, we need to consider, briefly, spectral methods.

If $Z(\mathbf{s})$ is MSC then $Z(\mathbf{s})$ has a spectral representation:

$$Z(\mathbf{s}) = \int_{R^2} \exp(i\boldsymbol{\omega}^T \mathbf{s}) M(d\boldsymbol{\omega}),$$

where M is a random measure.

Now assume isotropy of the covariance function. This allows for one dimensional representations of desired quantities. In the case where a spectral density exists, the relationship between the covariance function and the spectral density is given by:

$$C(\|\mathbf{h}\|) = \int_{-\infty}^{\infty} \exp(i\omega \|\mathbf{h}\|) f(\omega) dw.$$

There is also an inversion formula, expressing the spectral density as an integral equation in the covariance function:

$$f(\omega) = \frac{1}{(2\pi)} \int_R \exp(-i\omega \|\mathbf{h}\|) C(\|\mathbf{h}\|) d \|\mathbf{h}\|.$$

For the Matérn class, the spectral density has a relatively simple form, namely:

$$f(\omega) = \sigma^2(\nu^2 + \omega^2)^{-\eta - \frac{1}{2}}.$$

For nonnegative integers m in the Matérn family, it turns out that if Z is m-times differentiable, then:

$$\left.\begin{array}{c} \int_{-\infty}^{\infty} \omega^{2m} f(\omega) d\omega < \infty \\ \text{and} \\ \int_{-\infty}^{\infty} \omega^{2(m+1)} f(\omega) d\omega = \infty \end{array}\right\}$$

This gives the sense in which the parameter η determines the smoothness of the covariance function. We see that the smoothness (at the origin) depends on the behavior of the spectral density for large ω (high frequencies). In particular, we have for the Matérn family that $m < \eta < m + 1$. To see the implications, again consider the two special cases considered previously.

For the exponential covariance function, it holds that

$$f(\omega) = \sigma^2(\nu^2 + \omega^2)^{-1}.$$

From this it can be seen that the exponential covariance function is MSC, but not (even once) MSD.

For the Gaussian covariance we have

$$f(\omega) = Ce^{-\omega^2/4\nu}.$$

From this it can be seen that the Gaussian covariance function is MSD for *all* m. In other words, all derivatives of the covariance function exist.

We see that the value of η gives the number of mean square derivatives of the process Z. We have seen that a spatial process with Matérn covariance is m-times mean square differentiable if $\eta > m$ [Stein (1999)]. The exponential model has no mean square derivatives, while the Gaussian has infinitely many. When $\eta = 1.5$ holds, there is one mean square derivative, but not two.

The practical importance of this is that the Matérn class of covariance models allows the *data* to determine the smoothness. The other models discussed choose the smoothness a priori. So, if there are sufficient data to enable good estimation, the Matérn model is superior to those previously discussed.

We have seen that the Matérn model allows for qualitatively different behavior from the other covariance models, which predetermine the smoothness of the covariance function. Another class that allows for various levels of smoothness is the stable family of covariances given by:

$$C(\mathbf{h}, \theta) = \sigma^2 \exp\left[(-\nu\|\mathbf{h}\|)^\gamma\right] \quad \text{for } \sigma > 0, \gamma \in (0, 2], \nu > 0.$$

For $\gamma < 2$, the behavior of this covariance for small $\|\mathbf{h}\|$ is comparable to that of the Matérn family with $\gamma \simeq 2\eta$. In particular, the exponential covariance function corresponds to $\gamma = 1$, while for the Matérn covariance, $\eta = \frac{1}{2}$. For $\gamma = 2$, we have the Gaussian covariance function, which as we have seen corresponds to $\eta = \infty$ in the Matérn family. This shows that, for values of the Matérn smoothness parameter $\eta \in (1, \infty)$, there is no comparable covariance in the stable family. Both models have three parameters that need to be estimated, but the Matérn allows for a much wider range of covariance behavior. On the other hand, the stable family is practically more easily estimated, due to the simplicity of its functional form.

Other covariance functions include the circular covariance and the hole covariance. These and other functions are discussed, for example, in Chiles and Delfiner (1999). Variograms for IS processes will be discussed in Section 3.4.1.

3.4 Convolution methods and extensions

Often the use of a particular covariance model is deemed inflexible by some users. Process convolutions give an alternative method to construct very flexible classes of covariance functions. Specifically, consider a zero-mean white noise process, $X(\mathbf{s})$. In other words, $E[X(\mathbf{s})] = 0$ and

$$\begin{cases} Cov[X(\mathbf{s}+\mathbf{h}), X(\mathbf{s})] = 0, & \mathbf{h} \neq \mathbf{0}, \\ Cov[X(\mathbf{s}+\mathbf{h}), X(\mathbf{s})] = \sigma^2, & \mathbf{h} = \mathbf{0}. \end{cases}$$

Let $\phi_{\mathbf{s}}(\mathbf{u})$ denote a kernel function (a smoothing kernel) which is square integrable, $\int_D \phi^2(\mathbf{u})d\mathbf{u} < \infty$. Observations, $Z(\mathbf{s})$, are assumed to be generated from

$$Z(\mathbf{s}) = \int_D \phi_{\mathbf{s}}(\mathbf{u})X(\mathbf{u})d\mathbf{u}.$$

For now assume that the kernel is stationary. This means that $\phi_{\mathbf{s}}(\mathbf{u}) = \phi(\mathbf{s}-\mathbf{u})$. Then, the covariance function is given by:

$$C(\mathbf{h}) = \int_D \phi(\mathbf{u}-\mathbf{h})\phi(\mathbf{u})d\mathbf{u}.$$

If the kernel $\phi(\cdot)$ is isotropic, $\phi(\mathbf{u}) = \phi(\|\mathbf{u}\|)$, then there is a one-to-one correspondence between the kernel and the resulting covariance function. Given the flexibility in choice of kernels, we obtain a large class of stationary covariance functions. This approach has been advocated by, for example, Barry and Ver Hoef (1996) and Higdon (1998). For a novel application in the multivariate space–time setting (Chapter 9), see Calder (2007).

A discrete version replacing the integral equation defining $Z(\mathbf{s})$ has been proposed by Higdon (1998). Applications of this approach are given in, for example, Lee *et al.* (2005). Use of a nonstationary kernel function, $\phi(\mathbf{u})$ allows for nonstationary covariance functions. This is briefly discussed in Section 3.7.

Most spatial observations occur on a sphere, namely the earth. The models that have been considered until this point, however, do not explicitly account for this fact. Jun and Stein (2007) formulate space–time covariance models for observations on a sphere, with applications to global wind and air pollution monitoring. Their model allows for space–time asymmetries and is used to analyze ozone levels while respecting the shape of the earth.

3.4.1 Variogram models where no covariance function exists

Thus far we have discussed models for SOS processes. In this section, we discuss models which are IS, but not SOS. For covariance models, when $C(\|\mathbf{h}\|) \to 0$ as $\|\mathbf{h}\| \to \infty$, we have that $\gamma(\|\mathbf{h}\|) \to C(\|\mathbf{0}\|) = \sigma^2$. This value is known as the *sill* of the semivariogram function. The value $\|\mathbf{h}\|$, at which the sill is reached, is often known as the *range* of the covariance function. Note that the exponential model has sill equal to σ^2, but this sill is not reached at any finite range. For this reason, the lag at which some large percentage (e.g. 95%) of the sill is reached is often known as the practical range. For IS models there is no sill. We now consider IS models with no sill.

The logarithmic variogram is given by:

$$\gamma(\|\mathbf{h}\|) = \log(\|\mathbf{h}\|).$$

This is also known as the de Wigs model. As discussed, note that there is no limiting value of $\gamma(\|\mathbf{h}\|)$ as $\|\mathbf{h}\| \to \infty$. For an interesting discussion of the use of this model in ore exploration, see Chiles and Delfiner (1999).

Power model

$$\gamma(\|\mathbf{h}\|) = \|\mathbf{h}\|^{\alpha} \quad \text{for } \alpha \in (0, 2).$$

We have encountered this model in Section 2.2.2 in the case of $d = 1$ dimension. Recall that we obtain quite different behavior as α varies, and for this reason this is a flexible variogram model for IS processes. The reason for the range of α is given by the following necessary condition for any variogram model.

Any variogram model must satisfy the Intrinsic hypothesis:

$$\gamma(\|\mathbf{h}\|)/\|\mathbf{h}\|^2 \to 0, \quad \text{as } \|\mathbf{h}\| \to \infty.$$

It is immediate that the logarithmic and power models satisfy this property. Any covariance model such that $C(\|\mathbf{h}\|) \to 0$, as $\|\mathbf{h}\| \to \infty$, clearly satisfies this Intrinsic hypothesis. The covariance models above are all such that $\gamma(\|\mathbf{h}\|) \to C(\|\mathbf{0}\|)$, so the Intrinsic hypothesis is clearly satisfied.

3.4.2 Jumps at the origin and the nugget effect

Examining the functions in the previous section, we see that, for all SOS processes, we have $C(\|\mathbf{h}\|) \to \sigma^2$ as $\|\mathbf{h}\| \to 0$, while $\gamma(\|\mathbf{h}\|) \to 0$ as

$\|\mathbf{h}\| \to 0$. There are many cases where this smoothness of the covariance function at the origin is not appropriate.

There are several reasons why we may not have a smooth covariance function (or variogram function) at the origin. To appreciate how this can happen, consider the following fact concerning linear combinations of covariance functions:

If $C_1(\mathbf{h})$ and $C_2(\mathbf{h})$ are covariance functions, then for any positive constants a and b:

$$aC_1(\mathbf{h}) + bC_2(\mathbf{h}) \text{ is a valid covariance function.}$$

This can most easily be seen to hold from the nonnegative property of valid covariance functions from Section 3.2.

To see a practical use of this, consider the model

$$Z(\mathbf{s}) = \mu(\mathbf{s}) + \delta(\mathbf{s}) + e(\mathbf{s}),$$

where $\mu(\mathbf{s})$ denotes the mean function, and $\delta(\mathbf{s})$ has covariance function $C_1(\mathbf{h})$. The new term, $e(\mathbf{s})$, represents a noise term with $Var[e(\mathbf{s})] = \sigma_e^2$, and which is independent of the noise term at any other location, $Cov[e(\mathbf{s}), e(\mathbf{t})] = 0$, whenever $\mathbf{s} \neq \mathbf{t}$.

Thus, $C_2(\mathbf{h}) := Cov[e(\mathbf{s}), e(\mathbf{s}+\mathbf{h})] = \sigma_e^2$, if $\mathbf{h} = \mathbf{0}$, and 0 if $\mathbf{h} \neq \mathbf{0}$. Then the covariance function for $Z(\mathbf{s})$, $C(\mathbf{h}) = C_1(\mathbf{h}) + I\{\mathbf{h} = \mathbf{0}\}\sigma_e^2$, with semivariogram $\gamma_1(\mathbf{h}) + \sigma_e^2$ for $\mathbf{h} \neq \mathbf{0}$.

For example, the exponential semivariogram function, with a *nugget effect* of σ_e^2, is given by

$$\gamma(\mathbf{h}, \theta) = \sigma_e^2 + \sigma^2[1 - \exp(-\nu\|\mathbf{h}\|)], \tag{3.1}$$

for $\mathbf{h} \neq \mathbf{0}$, in which case the sill is $\sigma_e^2 + \sigma^2$. It is often advisable to allow for a nugget effect, due to small-scale noise. In particular, analysis of the phosphorus data supports a nonzero nugget effect. This will be seen in Section 3.6.

3.5 Parameter estimation for variogram and covariance models

Given that we desire a parametric covariance function, we need to choose a parametric family, for instance, the Matérn family. This is typically done based on the appropriateness of this family to model the phenomenon of interest. Given this parametric family, we require algorithms to estimate the appropriate model parameters from the data. We now discuss methods of data-based estimation of the parameters, for any of the models discussed in Section 3.3.

Let $\boldsymbol{\Lambda} = \{\mathbf{h}_1, \ldots, \mathbf{h}_m\}$ denote a set of m spatial lags where the variogram is to be estimated. Let $\widehat{\boldsymbol{\gamma}} = \{\widehat{\gamma}(\mathbf{h}_1), \ldots, \widehat{\gamma}(\mathbf{h}_m)\}$ denote the vector, of length m, of semivariogram estimates, at the m spatial lags, via any of the nonparametric estimators in Section 3.1. For example, $\widehat{\boldsymbol{\gamma}}$ can be any of the moment estimator, robust estimator or kernel smoothed estimator. Let $\widehat{\gamma}_i$ denote the ith component of this vector, $i = 1, \ldots, m$.

We seek a parametric function which is closest in some sense to this nonparametric moment estimator. Often, close means minimizing a (possibly standardized) square distance between the two. Specifically, assume that the true semivariogram function lies within a particular parametric family, and denote this as $[\gamma(\cdot, \boldsymbol{\theta}) : \boldsymbol{\theta} \in \boldsymbol{\Theta}]$, where the dimension of the parameter space $\boldsymbol{\Theta}$ is that of the number of parameters in the variogram model. Define, for each $\boldsymbol{\theta}$, $\boldsymbol{\gamma}(\boldsymbol{\theta})$ to be the m-vector of parametric semivariogram values at spatial lags in $\boldsymbol{\Lambda}$.

Define $Q(\boldsymbol{\theta}) = \widehat{\boldsymbol{\gamma}} - \boldsymbol{\gamma}(\boldsymbol{\theta})$ to be the errors in estimation we seek to minimize, and define the $m \times m$ variance–covariance matrix $\mathbf{V}(\boldsymbol{\theta}) := Var[Q(\boldsymbol{\theta})]$. The OLS proposal then seeks to minimize:

$$R(\boldsymbol{\theta}) := Q(\boldsymbol{\theta})^{\mathrm{T}} Q(\boldsymbol{\theta}) = \sum_{i=1}^{m} [\widehat{\gamma}_i - \gamma_i(\boldsymbol{\theta})]^2. \tag{3.2}$$

Although this minimization seems natural, and is computationally straightforward, it is not the most efficient estimator. In particular, note that it takes no account of

i. the number of squared differences averaged at each lag, $N(\mathbf{h}_i)$,

ii. the variability associated with each squared difference, or

iii. the covariance between elements in $\widehat{\boldsymbol{\gamma}}$.

The generalized least squares (GLS) estimator accounts for all of (i), (ii), and (iii), and seeks to minimize:

$$R(\boldsymbol{\theta}) := Q(\boldsymbol{\theta})^{\mathrm{T}} V(\boldsymbol{\theta})^{-1} Q(\boldsymbol{\theta}). \tag{3.3}$$

The value of $\boldsymbol{\theta}$ that minimizes $R(\boldsymbol{\theta})$ is the GLS estimator. Although this is the best of the L_2 minimum discrepancy estimators, there are some practical difficulties with implementation. The first is that the matrix

$\mathbf{V}(\boldsymbol{\theta})$ is difficult to find in practice. Some large sample approximations can be found for Gaussian processes [see, e.g., Priestley (1981), for the covariance of moment estimators], but it is not a simple task, in general. Second, minimizing this criterion is difficult in practice, as $R(\boldsymbol{\theta})$ is a nonlinear criterion function. Due to these difficulties with GLS, an appealing weighted least squares (WLS) proposal seeks to minimize:

$$R(\boldsymbol{\theta}) := Q(\boldsymbol{\theta})^{\mathrm{T}}\mathbf{W}Q(\boldsymbol{\theta}) = \sum_{i=1}^{m} w_i^2 Q_i^2(\boldsymbol{\theta}), \tag{3.4}$$

where $\mathbf{W}$ denotes a diagonal matrix with weights w_i^2 on the diagonals. For the weights, note that for Gaussian observations, $Z(\mathbf{0}) - Z(\mathbf{h})$ is distributed as $N[0, 2\gamma(\mathbf{h})]$, by the definition of the variogram function. This implies that $[Z(\mathbf{0}) - Z(\mathbf{h})]^2/2$ is distributed as $\gamma(\mathbf{h})\chi_1^2$, where χ_1^2 denotes the chisquared distribution with one degree of freedom. The moment estimator is simply the average of $N(\mathbf{h}_i)$ such elements, each with variance equal to $2\gamma^2(\mathbf{h}_i)$. If, for each i, we assume that the $N(\mathbf{h}_i)$ squared differences are independent, then $Var[\widehat{\gamma}(\mathbf{h}_i)] = 2\gamma^2(\mathbf{h}_i)/N(\mathbf{h}_i)$. This suggests setting $w_i^2 = N(\mathbf{h}_i)/2\gamma^2(\mathbf{h}_i)$. In other words, we seek to minimize

$$\sum_{i=1}^{m} N(\mathbf{h}_i)\left[\frac{\widehat{\gamma}_i}{\gamma_i(\boldsymbol{\theta})} - 1\right]^2. \tag{3.5}$$

This accounts for deficiencies (i) and (ii) suffered by ordinary least squares. In this sense, WLS is a compromise between the easy-to-compute OLS estimator and the efficient, yet difficult to compute, GLS estimator. In several geostatistical software packages, WLS is the default fitting criterion.

There are other attempts to use methods close to GLS. One approach attempts estimation of $\mathbf{V}(\boldsymbol{\theta})$. For example, Lee and Lahiri (2002) obtain a nonparametric estimator, $\widehat{V}_n$, of $\mathbf{V}(\boldsymbol{\theta})$, using block resampling (see Chapter 10). They demonstrate the large sample consistency of their variance estimation and then propose minimizing

$$Q(\boldsymbol{\theta})^{\mathrm{T}}\widehat{V}_n^{-1}Q(\boldsymbol{\theta}).$$

The resulting subsampling least squares (SLS) estimator, $\widehat{\theta}_{SLS}$, is then shown to be consistent for $\boldsymbol{\theta}$, asymptotically normally distributed, and asymptotically efficient amongst all estimators based on minimizing $R(\boldsymbol{\theta})$ in Equation 3.3.

Weighted least squares is based on the assumption that the underlying process is Gaussian. If this is approximately correct, then a strong competitor to WLS or GLS is using maximum likelihood (ML). This approach is discussed in Chapter 4 and applied to rainfall estimation in Chapter 9. Although this will not be a competitor in terms of a great visual fit to the empirical variogram, it often gives superior predictions (in terms of lower kriging variances), if the data come from a Gaussian, or approximately Gaussian, spatial process. Zimmerman and Zimmerman (1991) numerically compare OLS, WLS, REML (restricted maximum likelihood), and ML for linear and exponential variograms. They find that, for Gaussian observations, ML is only slightly better than OLS and WLS. REML, which often has the best bias properties, was found to have comparable bias to ML. This is due to the low dimension of the mean parameter vector considered. The choice between least squares and likelihood methods is not a clear one; both are useful in the proper context.

Another approach in the likelihood direction seeks to obtain efficient estimators without variance estimation. Nordman and Caragea (2008) consider using empirical likelihood. Empirical likelihood was originally proposed in Owen (1990) to obtain inferences for means and functions of means computed from independent observations. Let $\mathbf{w} = (w_1, \ldots, w_n)$ denote a weighting vector, with $\sum_{i=1}^{n} w_i = 1$, and let $F_{\mathbf{w}}$ denote the corresponding distribution function with mass w_i on observation X_i, $i = 1, \ldots, n$. The usual empirical distribution has $w_i = 1/n$ for $i = 1, \ldots, n$. Consider parameters of interest $\theta = t(F)$. The empirical likelihood, EL, is given by:

$$L_{\text{emp}}(\theta) = \max_{[\mathbf{w}:t(F_{\mathbf{w}})=\theta]} \prod_{i=1}^{n} w_i.$$

In the case where $\theta = \mu$, the population mean, the constraint $t(F_{\mathbf{w}}) = \theta$ becomes $\sum_{i=1}^{n} w_i x_i = \mu$. For each μ we find the maximum of $\prod_{i=1}^{n} w_i$ subject to this constraint; the maximizer over all values of μ is the empirical likelihood estimator of μ. In this case $\mu = \overline{x}_n$. We can obtain more information from the EL. Amazingly, in a large class of problems (smooth functions of means, statistics from estimating functions), we have:

$$-2\log\left[\frac{L_{\text{emp}}(\theta)}{L_{\text{emp}}(\widehat{\theta})}\right] \xrightarrow{D} \chi_1^2, \tag{3.6}$$

as in the parametric setting. For reasonable sample sizes we can use this for hypothesis tests and confidence intervals. Further, as in the parametric case there is a Bartlett correction to the likelihood ratio test given by Diciccio *et al.* (1991). In the spatial setting, the likelihood does not factor into a product likelihood. However, Kitamura (1997) suggested forming a product likelihood based on *blocks* of observations in the time series setting. Within blocks the dependence structure is maintained. Nordman and Caragea (2008) generalize this to parameter estimation in the spatial setting, using spatial blocks of observations. Specifically, for any spatial location, $\mathbf{s}$, let

$$Y(\mathbf{s}) = \left\{[Z(\mathbf{s}+\mathbf{h}_1) - Z(\mathbf{s})]^2, \ldots, [Z(\mathbf{s}+\mathbf{h}_m) - Z(\mathbf{s})]^2\right\}.$$

Under a parametric model we have $E[Y(\mathbf{s})] = 2\gamma(\boldsymbol{\theta})$. Now let $\mathbf{Y}_i$, $i = 1, \ldots, I_n$, denote overlapping (square) block means of $Y(\mathbf{s})$, each of dimension b_n by b_n. The empirical likelihood is then defined as:

$$L_{\text{emp}}(\boldsymbol{\theta}) = \max_{\left[\mathbf{w}:\sum_{i=1}^{I_n} w_i \mathbf{Y}_i = 2\boldsymbol{\gamma}(\boldsymbol{\theta})\right]} \prod_{i=1}^{n} w_i,$$

where $\sum_{i=1}^{n} w_i = 1$. As in the independent observation setting, the maximizer of $L_{\text{emp}}(\boldsymbol{\theta})$ is the EL estimator of $\boldsymbol{\theta}$. The authors show that a standardized version of Equation 3.6 holds, enabling inferences to be drawn on the variogram parameters $\boldsymbol{\theta}$. The authors show their inferences to be competitive with those of the SLS estimators, and superior to OLS or WLS over a wide range of block sizes. Note the similarity of the EL in the independent and spatially dependent settings. We see that any software that implements EL in the independent setting can be used for spatial observations by replacing the data observations with the block means, and the parameter $\boldsymbol{\theta}$ with the variogram values $2\boldsymbol{\gamma}(\boldsymbol{\theta})$.

In addition to the work of Nordman and Caragea (2008) on the distribution of parameter estimates arrived at through empirical likelihood estimation, there has been further distribution theory development for OLS, WLS, and GLS estimators. These results are arrived at through both increasing domain and infill asymptotic regimes, for example in Lee and Lahiri (2002). Zhang and Zimmerman (2005) compare ML estimation under increasing domain and infill asymptotics in the Matérn family of covariances. They find that finite sample behavior agrees more with the infill asymptotic inferences than with the increasing domain asymptotics.

3.5.1 Estimation with a nonconstant mean function

The assumption of a constant mean function is often appropriate when the true mean function is smooth. This is because information used in kriging is often very local; that is, relatively short spatial lags are used in the prediction. There are many cases, however, when the mean function is nonconstant in a significant and systematic way, and it becomes desirable to account for the varying mean function.

Specifically, as in Section 2.4, assume that the mean of the spatial process depends on location. Namely, that the model for observations is:

$$Z(\mathbf{s}) = \mu(\mathbf{s}) + \delta(\mathbf{s}).$$

To see specifically how a nonconstant mean function impacts estimation of covariance functions, note that

$$E[2\widehat{\gamma}(\mathbf{h})] = E[Z(\mathbf{s}+\mathbf{h}) - Z(\mathbf{s})]^2 = [\mu(\mathbf{s}+\mathbf{h}) - \mu(\mathbf{s})]^2 + 2\gamma(\mathbf{h}).$$

It is seen that for nonconstant mean functions, the empirical moment variogram estimator estimates the variogram and the squared difference in mean functions. In other words, it is biased. If the mean function were known, we could estimate the variogram from $\delta(\mathbf{s}) := Z(\mathbf{s}) - \mu(\mathbf{s})$. In practice, however, the mean function is typically unknown and needs to be estimated. Common examples of mean functions are discussed in Section 2.4. We now discuss some approaches to accounting for a nonconstant mean function.

i. *Use the linear model structure*

   In the general linear model we can

   a. Estimate $\boldsymbol{\beta}$ using OLS, obtaining $\widehat{\boldsymbol{\beta}}$.
   b. Estimate the variogram, $2\gamma(\mathbf{h})$, from the residuals,

   $$\hat{\delta}(\mathbf{s}) = Z(\mathbf{s}) - \sum_{k=1}^{p} x_{k,\mathbf{s}}\widehat{\beta}_k.$$

   c. If necessary, we can re-estimate $\boldsymbol{\beta}$ in (a), using GLS with covariance matrix from the estimated variogram.

   This can be iterated to a solution. The final predictions are then

   $$\widehat{Z}(\mathbf{s}_0) = \sum_{k=1}^{p} x_{k,\mathbf{0}}\widehat{\beta}_k + \hat{\delta}(\mathbf{s}_0),$$

where $\hat{\delta}(\mathbf{s}_0)$ is the ordinary kriging predictor of $\delta(\mathbf{s}_0)$, based on the residuals in (b).

ii. *Ignore the nonconstant mean*

This may seem inappropriate, but the kriging predictor typically puts most weight on observations that are close to the desired location prediction. For relatively smooth mean functions, we have that $\mu(\mathbf{s}+\mathbf{h}) - \mu(\mathbf{s})$ is small, for small $\mathbf{h}$. In this situation, it may be appropriate to ignore the nonconstant mean. If the nonconstant mean is such that it is constant over subdomains of the data domain, then the variogram can be estimated using the moment estimator on each subdomain, and combined. Local or general relative variograms use an assumed relation between the mean and the variance to account for the nonconstant mean. See, for example, Isaaks and Srivastava (1989), for more details on these approaches.

iii. *Assume Gaussianity*

If the data are jointly Gaussian, it is often preferable to find the maximum likelihood estimators (MLEs) by maximizing the data likelihood in the trend and covariance parameters. We give details on this approach in Chapter 4.

iv. The empirical likelihood method in Section 3.5 can be modified for use with a nonconstant mean function, as discussed in Nordman and Caragea (2008).

3.6 Prediction for the phosphorus data

We now illustrate fitting a nonconstant mean function and IS variogram model for the phosphorus data set. This will enable a comparison of predictions using approaches (i) and (ii) in the previous section for nonconstant mean functions.

First: mean function fitting

Consider a smooth function of the form

$$\mu(\mathbf{s}) = \sum_{0 \le k+l \le r} \beta_{kl} x^k y^l,$$

with $r = 2$, yielding

$$\widehat{\mu}(\mathbf{s}) = 1.722 - 0.00194x + 0.0693y + 0.0000164x^2 - 0.000672y^2.$$

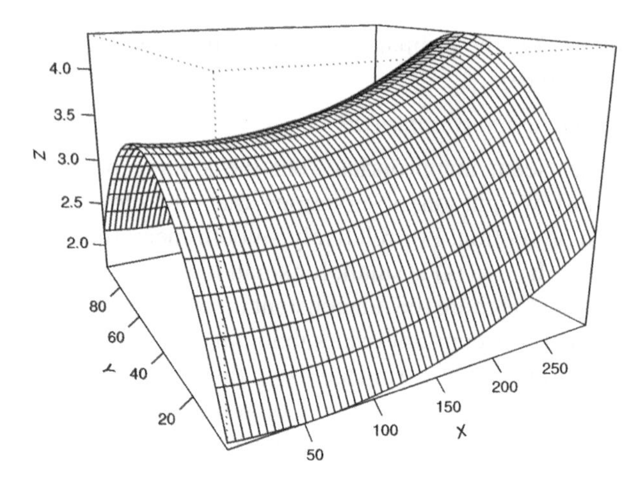

Figure 3.2 Mean function fit to the phosphorus data.

The coefficient for the xy term is quite nonsignificant. This estimated mean function is shown in Figure 3.2. The overall proportion of variation in log(Phosphorus) levels explained by this model is approximately 20%, and the p-value for the overall model is 0.003.

A possible explanation for the observed accumulations is the position of aerators in the ponds. An aerator is a mechanical device that increases the level of dissolved oxygen in the water. In this pond, four aerators are located along each side of the x axis of the pond. The paddle wheels are evenly spaced starting at 10 m from each edge. The accumulation in the deep ($x = 300$ m) and shallow ($x = 0$ m) ends of the pond may be a result of the reduced water current where no aerator is present. The relatively high accumulation in the center (along the y axis) can be explained in the same way. This effect has also been noted by Smith (1996) and Boyd (1992).

This model is fit using OLS using only the 75 equidistant observations (excluding the observations from the crosses in Figure 1.1) in fitting the mean surface.

Second: covariance function estimation

Next, consider parameter modeling of the covariance function of all residuals from the OLS fit in the previous section. The residuals are also discussed in Section 2.5. First, consider fitting a member of the exponential family. From the moment estimates displayed in Figure 3.3,

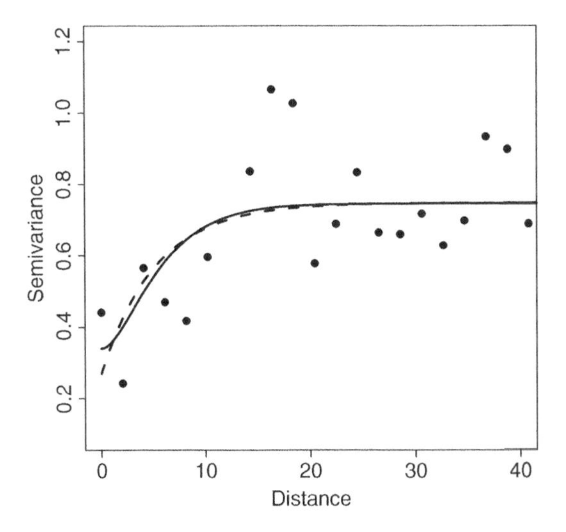

Figure 3.3 Variogram moment estimates and WLS fits of the exponential (dotted) and Matérn model (smooth) covariances to phosphorus data.

it seems appropriate to allow for a nugget effect. Thus, we seek to fit the variogram given in Equation 3.1. Here, the parameter vector is $\theta^{\mathrm{T}} = (\sigma_e^2, \sigma^2, \nu)$. Minimizing the OLS and WLS criteria given in Equations 3.2 and 3.5 gives the parameter estimates in Table 3.1.

The OLS and WLS estimates are in reasonable agreement for the exponential model. In terms of the value of the sill, $\sigma_e^2 + \sigma^2$, WLS gives $(0.27 + 0.48) = 0.75$. The OLS estimator, for reasons given previously, is generally not as reliable as WLS (under approximate normality). The fitted semivariogram using the WLS fit is

Table 3.1 Covariance parameter estimates for the phosphorus variogram.

Parameter	OLS_e	WLS_e	OLS_m	WLS_m
σ_e^2	0.30	0.27	0.33	0.34
σ^2	0.46	0.48	0.43	0.41
ν	0.14	0.19	0.27	0.33

OLS_e and WLS_e denote the ordinary least squares and weighted least squares estimates in the exponential model, while OLS_m and WLS_m are the estimates for the Matérn model with $\eta = 1.5$.

$$\widehat{\gamma}(\mathbf{h}, \widehat{\theta}) = 0.27 + 0.48[1 - \exp(-0.19\|\mathbf{h}\|)].$$

Note the effective range, where correlation is practically 0, is given by the distance such that $\widehat{\gamma}(\mathbf{h}, \widehat{\theta}) = 0.95(0.75) = 0.71$. The range is estimated to be $\|\mathbf{h}\| = 14$ m for the WLS estimates, and $\|\mathbf{h}\| = 18$ m for the OLS estimates. Note that either range indicates that without the densely located observations in the cross, in Figure 1.1, we would have no information until 15 m on the variogram, and thus would incorrectly report a complete lack of correlation. This would falsely suggest that increased sampling would linearly decrease variation in estimating the total amount of phosphorus. The locally dense design allows for a more accurate appraisal of correlation at this finer scale. Both the OLS and WLS estimates attribute a large proportion of total variation to the nugget effect. In particular, the OLS estimate attributes $0.30/(0.30 + 0.46) = 0.40$ of the total variation to measurement error, while the WLS estimate attributes $0.27/(0.27 + 0.48) = 0.36$ of the variation to the nugget effect. Either estimate assigns a significant proportion of the total variation in log(Phosphorus) to measurement error. After this study was finished, a followup study actually took replicate measurements at a few locations, and analyzed the results separately. The estimated measurement error turned out to be approximately 30% of $\sigma_e^2 + \sigma^2$, which is roughly compatible with either the WLS or OLS estimates.

The Matérn family (with smoothness parameter $\eta = 1.5$) gives apparently quite different estimates from the exponential family. Recall, however, that the interpretation of the two sets of parameter estimates is different for the two parametric families. Figure 3.3 plots the estimated semivariogram for the exponential and Matérn families where each is estimated via WLS. The two curves are quite similar over the majority of the range of the variogram. In particular, the sills are approximately the same for the two models. The main difference is for $\|\mathbf{h}\|$ close to 0. It is seen that the estimated Matérn covariance function approaches 0 much more smoothly than does the exponential function, as we have seen earlier in Section 3.3. The WLS fitting criterion mildly indicates that the Matérn fit is better than the exponential fit, due to a decrease in sum of squared error (SSE) of about 2%. We take this WLS fit to the Matérn family with smoothness parameter $\eta = 1.5$ to be the estimated covariance in what follows.

Third: prediction

As discussed in Section 3.5.1, there are a few different choices in how to handle a possibly nonconstant mean function. In the previous development we have laid out approach (i) in Section 3.5.1: estimating the mean function and estimating the variogram from the residuals from the fitted mean. When using this approach to make predictions, there is variability in both the mean function estimation and in the kriging prediction through the residuals. To illustrate this approach, consider prediction at a given location, $\mathbf{s}_0^{\mathrm{T}} = (x, y)$. The predictor is

$$\widehat{Z}(\mathbf{s}_0) = \mathbf{x}_0^{\mathrm{T}}\widehat{\boldsymbol{\beta}} + \hat{\delta}(\mathbf{s}_0),$$

where $\mathbf{x}_0 = (1, x, y, x^2, y^2)$, $\widehat{\boldsymbol{\beta}}$ is given above, and $\hat{\delta}(\mathbf{s}_0)$ is the kriged predictor of $\delta(\mathbf{s}_0)$ based on the residuals from the OLS fit. Noting that OLS parameter estimates are independent of the residuals, we take the prediction variance of this predictor to be

$$\mathbf{x}_0^{\mathrm{T}}\widehat{\Sigma}_{\beta}\mathbf{x}_0 + \widehat{\sigma}_{\delta}^2,$$

the estimated variance of the linear predictor, $\mathbf{x}_0^{\mathrm{T}}\widehat{\boldsymbol{\beta}}$, plus the kriging variance of $\hat{\delta}(\mathbf{s}_0)$ in predicting $\delta(\mathbf{s}_0)$.

In approach (ii) in Section 3.5.1, we simply calculate the ordinary kriging predictor and kriging variance as in Chapter 2. In assuming that the linear predictor is of the correct form, we have that the linear predictor is unbiased for its predictand. This suggests that the predictor in approach (ii) has a bias that can be estimated by the difference in prediction using approach (i) and the prediction using approach (ii). In Table 3.2, for approaches (i) and (ii), we give the predictions of log (Phosphorus), the prediction variance, and the estimated MSE in prediction for five locations.

For these locations, the estimated variance of the linear predictor, $\mathbf{x}_0^{\mathrm{T}}\widehat{\Sigma}_{\beta}\mathbf{x}_0$, ranges between 0.04 and 0.18, and so this variation is a relatively small percentage of the total variation. For the location $(40, 42.5)$ the predictions using (i) or (ii) are quite similar, and thus estimating the mean function did not help, and approach (ii) is superior. For locations $(275, 45)$ and $(10, 10)$, however, it is seen that the predictor that ignores the trend induces a bias that greatly increases the overall prediction MSE. Of course, the bias calculation assumes that the mean function is correct, which makes MSE1 possibly optimistically small. Overall, the mean function is of moderate use in

Table 3.2 Prediction and kriging variances for five phosphorus locations.

x	y	Pred1	Var1 = MSE1	Pred2	Var2	MSE2
40	42.5	3.43	0.76	3.31	0.65	0.66
31	36	3.37	0.47	3.24	0.59	0.60
275	45	4.67	0.73	3.38	0.92	2.60
30	21	3.01	0.58	3.16	0.63	0.66
10	10	2.38	0.93	3.28	0.84	1.66

For each location, 'Pred1' and 'Var1' give the predictions and kriging variances under a mean function. 'Pred2' and 'Var2' give the predictions and kriging variances under an assumed constant mean function. 'MSE2' includes the estimated square bias under this approach.

this example, and there is a suggestion that approach (i) is worthwhile, although approach (ii) seems reasonable here as well. In Chapter 4, using approach (iii) (maximum likelihood), we see an example where the mean function explains a larger percentage of overall variability.

A further question arising in this phosphorus study is: what is the increased utility of additional sampling? One way to address this question is to find the estimated variance of the mean level of phosphorus using the estimated variogram assuming that the mean model is accurately fit. In the observed data there are a total of $15 \times 5 = 75$ observations on an equally spaced lattice in each dimension. To compare the sampling actually performed to alternative sampling rates, consider equally spaced designs of the form:

$$D_n := \{\mathbf{s} : \mathbf{s} = (x, y) \quad \text{where} \quad x = (300i)/(3n + 1),$$
$$y = (100j)/(n + 1)\},$$

for $i = 1, \ldots, 3n,\ j = 1, \ldots, n$. Our observed data are approximately from this design with $n = 5$. It is found that increasing the sampling by a factor of 4, that is, $n = 10$, makes the estimated variance of the mean decrease to 62% of the value when $n = 5$. Note that this decrease is far less than the 25% we would expect under independent observations. This suggests that little is gained by significant increased sampling beyond that performed.

3.7 Nonstationary covariance models

Ordinary kriging in Section 2.1 and universal kriging in Section 2.4 differ in their assumptions on the underlying mean function. In both cases, however, we have assumed that the errors are intrinsically stationary (or second-order stationary). These models successfully model a wide range of observed data. Despite this, there are many cases in ecology and geosciences where a stationary covariance function is untenable. What are the effects of using a nonstationary covariance function? First, recall that the ordinary kriging predictor is the optimal predictor for *any* assumed-to-be-known variogram or covariance function that is valid. As long as $\gamma(\mathbf{s}_i - \mathbf{s}_j)$ is known for $i, j = 0, 1, \ldots, n$, then the optimal weights are the kriging weights from Section 2.2. This is also the case for universal kriging. When the mean function is unknown, it can be optimally estimated using weighted least squares. The difficulty is that the variogram function is generally unknown, and the moment estimators in Section 3 and the variogram model estimates based on them (OLS,WLS, GLS) depend on replication. General nonstationary covariance functions are not naturally estimable. If, however, there is local stationarity, at a certain spatial scale, then estimation is still possible.

One way to allow for local estimation is to fit a global covariance function, but to make the kriging neighborhood local. Let $C(\mathbf{h}, \boldsymbol{\theta})$ denote a parametric covariance function, and let $\boldsymbol{\Sigma}(\widehat{\boldsymbol{\theta}})$ denote the $(n+1) \times (n+1)$ estimated covariance matrix, and $\boldsymbol{\nu}(\widehat{\boldsymbol{\theta}})$ the vector of covariances between observed variables and the unobserved prediction location as in Section 2.5. To predict an unobserved variable at location $\mathbf{s}_j$, consider the reduced kriging neighborhood $Z[\mathbf{s}_i : i \in N(j)]$. The kriging predictor is then based on the $|N_j| \times |N_j|$ matrix $\boldsymbol{\Sigma}_{N_j}(\widehat{\boldsymbol{\theta}})$ and $|N_j| \times 1$ vector, $\boldsymbol{\nu}_{N_j}(\widehat{\boldsymbol{\theta}})$. The main motivation of this predictor is to reduce the computations associated with the full kriging predictor. If there are sufficient data in N_j to allow for estimation of $\boldsymbol{\theta}$, then the covariance parameter can be estimated separately on each N_j neighborhood, yielding $\boldsymbol{\Sigma}_{N_j}(\widehat{\theta}_j)$. This fitting of separate covariance models on subsets of data can account for large amounts of nonstationarity due to a varying covariance function. Further, this approach makes the assumption of a constant mean in ordinary kriging more appropriate, as now only a local mean stationarity assumption is being made. This approach to prediction was proposed and developed in Haas (1990). Another approach to nonstationary covariance functions is through kernel convolution as discussed in Section 3.4, where $\phi_s(\mathbf{u})$ now depends on the location $\mathbf{s}$. This is pursued in Barry and Ver Hoef (1996) and Higdon (1998).

4

Spatial models and statistical inference

In Chapter 1, we briefly discussed the conditional spatial autoregressive model. In this chapter, we discuss this model for both continuous and discrete spatial variables. In 1910 two researchers, Mercer and Hall, conducted an experiment and observed the yields of wheat in each of 500 plots on an (apparently) equally spaced 20×25 lattice of plots [Mercer and Hall (1911)]. In Figure 4.1, we see the relative locations of each plot and the spatial distribution of yields. Notice that this data set is 'complete,' in the sense that no additional data observations will, or could ever be, available within this domain. For this reason, it is not of interest to fit a variogram or covariance function as in Chapter 3, for use in optimal prediction (as discussed in Chapter 2). It is of interest, however, to describe how the yield at one location depends on nearby yields, and whether this dependence is direction dependent. In order to address these types of questions we consider spatial models. The idea is to have models analogous to autoregressive time series models as discussed in Section 1.2.1.

An apparently natural way to do this is via

$$Z(\mathbf{s}) - \mu(\mathbf{s}) = \sum_{\mathbf{t}} \beta_{\mathbf{st}}[Z(\mathbf{t}) - \mu(\mathbf{t})] + \epsilon_{\mathbf{s}},$$

Spatial Statistics and Spatio-Temporal Data: Covariance Functions and Directional Properties Michael Sherman

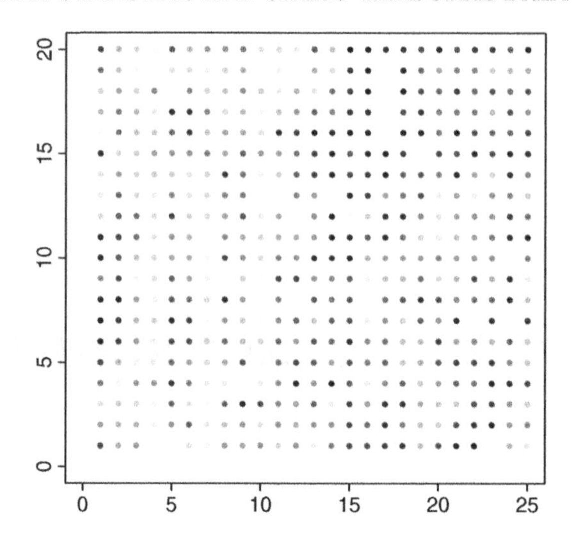

Figure 4.1 The wheat yields on 500 plots on a 20 × 25 *lattice of plots. Dark points indicate the larger yields and white points indicate the smaller.*

where β_{st} are spatial regression parameters and ϵ_s are spatially indexed errors. In matrix notation this can be written as

$$(\mathbf{I} - \mathbf{B})(\mathbf{Z} - \boldsymbol{\mu}) = \boldsymbol{\epsilon},$$

where $\mathbf{Z} = [Z(\mathbf{s}_1), \ldots, Z(\mathbf{s}_n)]^{\mathrm{T}}$, and $\mathbf{B} = [\beta_{st}]$, $\beta_{ss} = 0$ for all $\mathbf{s}$. It is assumed that $E(\boldsymbol{\epsilon}) = \mathbf{0}$, and let $\boldsymbol{\Lambda} = Var(\boldsymbol{\epsilon})$ denote the $n \times n$ variance–covariance matrix of the errors.

This is known as a *simultaneous* spatial regression model, as proposed by Whittle (1963). Writing $\mathbf{Z} = \boldsymbol{\mu} + (\mathbf{I} - \mathbf{B})^{-1}\boldsymbol{\epsilon}$, we have

i. $E(\mathbf{Z}) = \boldsymbol{\mu}$;

ii. $$\begin{aligned} Var(\mathbf{Z}) &:= E[(\mathbf{Z} - \boldsymbol{\mu})(\mathbf{Z} - \boldsymbol{\mu})^{\mathrm{T}}] \\ &= (\mathbf{I} - \mathbf{B})^{-1} E(\boldsymbol{\epsilon}\boldsymbol{\epsilon}^{\mathrm{T}})(\mathbf{I} - \mathbf{B}^{\mathrm{T}})^{-1} \\ &= (\mathbf{I} - \mathbf{B})^{-1} \boldsymbol{\Lambda} (\mathbf{I} - \mathbf{B}^{\mathrm{T}})^{-1}; \end{aligned}$$

iii. If the errors $\boldsymbol{\epsilon}$ have a multivariate normal distribution with mean $\mathbf{0}$, and variance–covariance matrix $\boldsymbol{\Lambda}$, then $\mathbf{Z}$ has a multivariate normal distribution with mean and variance as given in (i) and (ii); and

iv. The covariance between the observations and the errors is

$$E(\boldsymbol{\epsilon}\mathbf{Z}^{\mathrm{T}}) = \boldsymbol{\Lambda}(\mathbf{I} - \mathbf{B}^{\mathrm{T}})^{-1}.$$

From (iv), it is seen that the observations and the errors are *not* independent, even when $\boldsymbol{\Lambda}$ is diagonal, that is, when the errors are independent. This is unlike the situation in AR(p) time series models, where ϵ_i is independent of X_{i-j}, for $j = 1, \ldots, p$. In this sense, the simultaneous spatial regression model is not analogous to temporal autoregression. In particular, least squares estimation of model parameters is not necessarily consistent for the true model parameters.

Another way to model the observations is through the *conditional* spatial regression model. As discussed in Section 1.2.1, we model all observations conditionally given their neighbors. The conditional mean and variance are given by:

1. $E[Z(\mathbf{s})|Z(\mathbf{t}) : \mathbf{t} \neq \mathbf{s}] = \mu(\mathbf{s}) + \sum_{\mathbf{t}} \gamma_{\mathbf{st}}[Z(\mathbf{t}) - \mu(\mathbf{t})]$, and
2. $Var[Z(\mathbf{s})|Z(\mathbf{t}) : \mathbf{t} \neq \mathbf{s}] = \tau_{\mathbf{s}}^2$.

The example in Chapter 1 is the case where $\mu(\mathbf{s}) = \mu$ and $\gamma_{\mathbf{st}} = \gamma$ when $\mathbf{s}$ and $\mathbf{t}$ are nearest neighbors.

In the case where the conditional distribution $[Z(\mathbf{s})|Z_{\mathbf{t}} : \mathbf{t} \neq \mathbf{s}]$ is a normal distribution, with mean and variance given in (i) and (ii), then it can be shown using the results of Besag (1974) that this set of conditional distributions corresponds to a joint, *unconditional* distribution of $\mathbf{Z}$, which is multivariate normal with mean μ and variance–covariance matrix $(\mathbf{I} - \mathbf{G})^{-1}\boldsymbol{\tau}$. The matrix $\mathbf{G} = [\gamma_{\mathbf{st}}]$, and $\boldsymbol{\tau}$ is an $n \times n$ diagonal matrix with diagonal elements $\tau_{\mathbf{s}_1}, \ldots, \tau_{\mathbf{s}_n}$.

If we let $\boldsymbol{\nu} := (\mathbf{I} - \mathbf{G})(\mathbf{Z} - \boldsymbol{\mu})$ denote the *pseudo*-errors, then it is seen that $E(\boldsymbol{\nu}\mathbf{Z}^{\mathrm{T}}) = (\mathbf{I} - \mathbf{G})E[(\mathbf{Z} - \boldsymbol{\mu})(\mathbf{Z} - \boldsymbol{\mu})^{\mathrm{T}}] = \boldsymbol{\tau}$. Thus, if $\boldsymbol{\tau}$ is diagonal, then the pseudo-errors are independent of the observations. This is in stark contrast to the situation in simultaneous regression, and in this fundamental regard, conditional spatial regression is more analogous to autoregression than is simultaneous spatial regression.

Figure 4.2 shows the histogram and a q-q plot for the wheat yield data. The histogram seems quite compatible with an assumption of (marginal) normality, as does the q-q plot (except for a few observations in the lower and upper quantiles). The sample mean is 3.95, and the sample median is 3.94, suggesting a symmetry in the observations. The sample skewness of 0.036 further attests to this, as does the (centered) sample kurtosis of −0.254. The visual summaries and the numerical

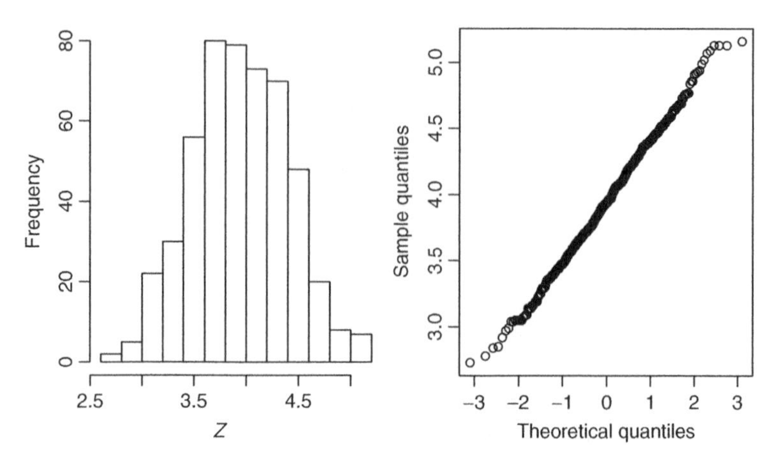

Figure 4.2 Histogram and q-q plot of the Mercer and Hall wheat yields.

measures all point towards a rough compatibility with a marginal normal distribution. Although this does not guarantee joint normality, an attempt to model this data set via a conditional normal model seems appropriate. Next is a discussion of how such an estimation can be carried out.

4.1 Estimation in the Gaussian case

Consider $\mathbf{Z}$ multivariate normally distributed, with mean vector $\boldsymbol{\mu}$ and variance–covariance matrix $\boldsymbol{\Sigma}$. Then, the likelihood is given by

$$L(\mathbf{Z};\boldsymbol{\mu},\boldsymbol{\Sigma}) = (2\pi)^{-n/2}|\boldsymbol{\Sigma}|^{-1/2}\exp\left[\left(\frac{-1}{2}\right)(\mathbf{Z}-\boldsymbol{\mu})^{\mathrm{T}}\boldsymbol{\Sigma}^{-1}(\mathbf{Z}-\boldsymbol{\mu})\right]. \tag{4.1}$$

In the general linear model we have $\boldsymbol{\mu} = \mathbf{X}\boldsymbol{\beta}$, while in the stationary Gaussian model we have $\boldsymbol{\mu} = \mu\mathbf{1}$, that is, constant. The matrix $\boldsymbol{\Sigma} = (\mathbf{I}-\mathbf{G})^{-1}\boldsymbol{\tau}$, in the conditional spatial regression model. Often this matrix is parameterized, for example, by setting $\gamma_{\mathbf{st}} = \alpha/d_{\mathbf{st}}^{k}$, for integer k, where $d_{\mathbf{st}}$ denotes the distance between locations $\mathbf{s}$ and $\mathbf{t}$. The parameter α determines the strength of spatial correlation. In matrix notation, $\mathbf{G} = \alpha\mathbf{D}$ where $\mathbf{D} = \left[d_{\mathbf{st}}^{-k}\right]$, for $\mathbf{s} \neq \mathbf{t}$. For the purpose of illustration, we take $\mathbf{G}$ to be parameterized in this way, and take $\boldsymbol{\tau} = v^2\mathbf{I}$. This implies that $\gamma_{\mathbf{st}}\tau_{\mathbf{t}}^2 = \gamma_{\mathbf{ts}}\tau_{\mathbf{s}}^2$, which is required in the conditional model. Denote the loglikelihood $\ln[L(\mathbf{Z};\boldsymbol{\mu},\boldsymbol{\Sigma})]$ as $l(\boldsymbol{\beta}, v, \alpha)$. The goal is to maximize this loglikelihood in the parameters.

i. For any fixed value of α, we have that

$$\widehat{\boldsymbol{\beta}} = [\mathbf{X}^{\mathrm{T}}(\mathbf{I} - \alpha\mathbf{D})\mathbf{X}]^{-1}\mathbf{X}^{\mathrm{T}}(\mathbf{I} - \alpha\mathbf{D})\mathbf{Z}.$$

This is simply the GLS estimator of $\boldsymbol{\beta}$, and is a function of α.

ii. Given this value of $\widehat{\boldsymbol{\beta}}$, we have

$$\widehat{v}^2 = (\mathbf{Z} - \mathbf{X}\widehat{\boldsymbol{\beta}})^{\mathrm{T}}(\mathbf{I} - \alpha\mathbf{D})(\mathbf{Z} - \mathbf{X}\widehat{\boldsymbol{\beta}}).$$

which is also a function of α.

iii. Now, substitute $\widehat{\boldsymbol{\beta}}$ and $\widehat{v}^2$ into the expression for $l(\boldsymbol{\beta}, v, \alpha)$, which can now be maximized as a function of α only, obtaining $\widehat{\alpha}$.

iv. Finally, $\widehat{\boldsymbol{\beta}}$ and $\widehat{v}^2$ are recomputed from $\mathbf{X}$, $\mathbf{Z}$ and $\widehat{\alpha}$.

For large sample size, n, the determinant of $\boldsymbol{\Sigma}$ often becomes difficult to compute. An approximate solution, more simply, minimizes $(\mathbf{Z} - \boldsymbol{\mu})^{\mathrm{T}}$ $\boldsymbol{\Sigma}^{-1}(\mathbf{Z} - \boldsymbol{\mu})$. This is the GLS solution, discussed in Section 3.5, and amounts to ignoring the term $|\boldsymbol{\Sigma}|^{-1/2}$ in the Gaussian likelihood.

In Section 3.5 we considered OLS, WLS, and GLS methods of estimation for covariance functions and variogram functions. We see that for multivariate normal observations we have $\boldsymbol{\Sigma} = \boldsymbol{\Sigma}(\boldsymbol{\theta})$, and we can maximize over $\boldsymbol{\theta}$ as above. One benefit to this method of fitting is that no binning or smoothing, as in Section 3.1.2, is necessary to perform model fitting. Of course, the appropriateness of this method depends on the appropriateness of the assumption of Gaussian observations. We now illustrate maximum likelihood fitting for the wheat yield data set.

4.1.1 A data example: model fitting for the wheat yield data

As discussed above, all indications are that the wheat yield distribution is compatible with Gaussianity. Thus, we are reasonably comfortable with assuming that the observations come from a joint Gaussian distribution and now proceed to fit conditional models based on the above Gaussian likelihood.

Specifically, under consideration are six different models for the wheat yield data:

1. Constant mean, $\mu = \mu$, with first-order nearest neighbor correlation. It is found that $\widehat{\gamma} = 0.094$, with loglikelihood $l(\widehat{\mu}, \widehat{\gamma}) = -563.7$.

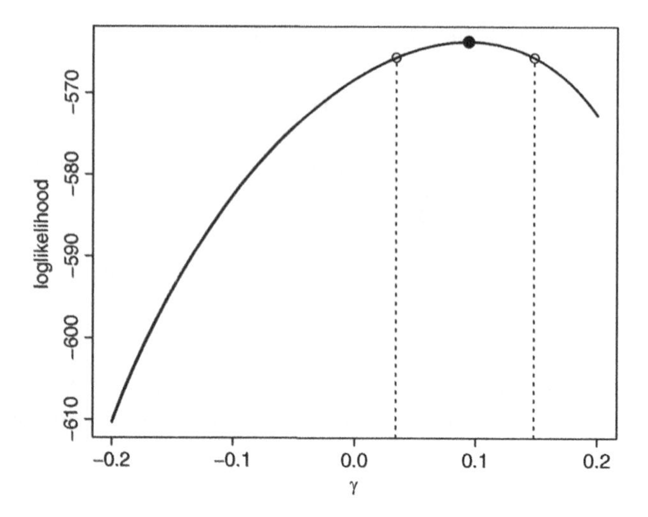

Figure 4.3 Loglikelihood for the correlation parameter γ in the constant mean model. The MLE of γ, $\hat{\gamma}_{\text{ML}} = 0.094$, is at the maximum of the loglikelihood. The two dotted vertical lines give the endpoints for a 95 percent confidence interval.

The loglikelihood under independence is $l(\widehat{\mu}, 0) = -568.3$. To assess significancewe use

$$\chi^2 := 2 \times [l(\widehat{\mu}, \widehat{\gamma}) - l(\widehat{\mu}, 0)] = 9.2.$$

Comparing this to a chisquared distribution, with 1 degree of freedom, under the null hypothesis we have p-value$=0.002$. Figure 4.3 shows the loglikelihood for γ. A large sample 95 percent confidence interval based on the chisquare approximation for the loglikelihood statistic is then $(0.034, 0.148)$. This interval has a slight left skewness with interval asymmetry equal to $(0.148 - 0.094)/(0.094 - 0.034) = 0.90$.

2. Constant mean, $\mu = \mu$. Now consider separate effects in the north–south and east–west directions. Towards this, define $Z(\mathbf{s}) = Z(i, j)$, and the conditional mean is now assumed to be:

$$E[Z(i, j)|Z(k, l)] = \mu + \gamma_1[Z(i, j+1) + Z(i, j-1)] \\ + \gamma_2[Z(i+1, j) + Z(i-1, j)].$$

Fitting this model, we find $\widehat{\gamma}_1 = 0.246$ and $\widehat{\gamma}_2 = -0.052$, with loglikelihood $l(\widehat{\mu}, \widehat{\gamma}_1, \widehat{\gamma}_2) = -549.8$, and $\widehat{\tau}^2 = 0.169$. Note that the correlation is much stronger in the east–west direction than in the north–south direction, and there is some suggestion that the latter relationship is negative.

Comparing Model (1) with Model (2), we find $\chi^2 = 2 \times [-549.8 - (-563.7)] = 27.8$ on 1 d.f. This is very significant. We strongly prefer the model allowing for different correlation in different directions, and use this correlation model for all further modeling.

3. Linear trend: $\mu = \mu(x, y) = \mu + \beta_1 x + \beta_2 y$. In this case, the loglikelihood is

$$l\left(\widehat{\mu}, \widehat{\beta}_1, \widehat{\beta}_2, \widehat{\gamma}_1, \widehat{\gamma}_2\right) = -535.2,$$

   and so $\chi^2 = 2 \times [-535.2 - (-549.8)] = 29.2$ on 2 d.f. It is seen that allowing a linear trend significantly improves the overall fit.

4. Linear and quadratic trends: this adds two additional parameters to the mean fit, but increases the loglikelihood to only -534.9. This is clearly not a significant improvement to the overall fit.

5. Considering a model that allows for separate means for each row and column, it is found that the variability in column means is approximately five-times as large as that in the row means. This suggests fitting a model with separate column means. This model has a total of 25 mean function parameters. The loglikelihood is $l\left(\widehat{\mu}_{C_j}, j = 1, \ldots, 25, \widehat{\gamma}_1, \widehat{\gamma}_2\right) = -462.2$. Comparing this to the model in (3), we have $\chi^2 = 2 \times [-462.0 - (-535.2)] = 146.4$, on $(27 - 5) = 22$ d.f. This is a very significant difference, and we accept this model as the current best. The variance is estimated at: $\widehat{v}^2 = \widehat{\tau}^2 = 0.129$.

6. Despite the fact that the column means are much more variable than are the row means, it is worthwhile to consider the model with both row and column effects. Under this model we find a loglikelihood of -441.1. Comparing this to the previous model, we have $\chi^2 = 41.8$, on 19 d.f., with an associated p-value $= 0.002$. We conclude, somewhat reluctantly, that this is the best overall model. The previous model has many fewer parameters, yet has roughly comparable explainability of the overall variation.

The final model for the observations is then

$$E[Z(i, j)|Z(k, l)] = \mu_{R_i} + \mu_{C_j} + \gamma_1[Z(i, j + 1) + Z(i, j - 1)] + \gamma_2\big[Z(i + 1, j) + Z(i - 1, j)\big],$$

with fitted values of $\gamma_1 = 0.213$, $\gamma_2 = 0.002$, and an estimated conditional variance of

$$\widehat{\upsilon}^2 = \widehat{\tau}^2 = 0.122.$$

The sample variance of the wheat yields is 0.210, and thus the model containing column and row mean effects, and nearest neighbor correlation effects, explains approximately 40 percent of the total variation in wheat yields. From Models (2) and (6) (and a model fit with mean effects only), we find that approximately 20–30% is attributed to the mean effects, and approximately 10–20% to the nearest neighbor correlation.

4.2 Estimation for binary spatial observations

In this section, we give some details on estimation for a non-Gaussian conditional model. Specifically, we consider binary outcomes that are spatially located. As before, let $\{Z(\mathbf{s}) : \mathbf{s} \in D\}$ denote the observed data, where D denotes the set of indices where data are observed, and $|D|$ denotes the total number of observations. It will be seen shortly that we need to employ less standard methods of estimation in this setting, and more generally, in other non-Gaussian settings.

Cancer rates of the liver and gallbladder (including bile ducts) for white males, during the decade from 1950 to 1959, for each county in the eastern USA, are shown in Figure 4.4. Here D denotes the set of county locations in the data set, of which there are 2003 total locations, so that $|D| = 2003$. For each location $\mathbf{s}$, as in Section 1.2.1, the neighborhood $N(\mathbf{s})$ denotes the four nearest neighbors of $\mathbf{s}$. Code $Z(\mathbf{s}) = +1$ if the cancer mortality rate is 'high' at site i, and code $Z(\mathbf{s}) = -1$ if the rate is 'low.' This was done using the quantiles of the observed rates from maps in Riggan *et al.* (1987). The given cutoffs classify approximately 27 percent of all counties as having a high rate. There is some apparent geographic clustering, particularly in the south east of the USA. The question is, whether this is a significant feature of the data.

Figure 4.4 County cancer rates for cancer of the liver and gallbladder in the eastern USA. Solid circle = 'high' rate, empty circle = 'low' rate.

Assume the data were generated by an autologistic model [Besag (1974)]. The model has conditional probabilities given by

$$P[Z(\mathbf{s}) = z(\mathbf{s})|Z(\mathbf{t}) : \mathbf{s} \neq \mathbf{t}] =$$
$$\frac{\exp\{z(\mathbf{s})[\alpha + \beta \sum_{\mathbf{t}\in N(\mathbf{s})} Z(\mathbf{t})]\}}{\exp[\alpha + \beta \sum_{\mathbf{t}\in N(\mathbf{s})} Z(\mathbf{t})] + \exp[-\alpha - \beta \sum_{\mathbf{t}\in N(\mathbf{s})} Z(\mathbf{t})]}, \tag{4.2}$$

so that the conditional distribution of $Z(\mathbf{s})$ depends on all other observations only through the sum of its four nearest neighbors. This is the same neighborhood structure as in the previous Gaussian

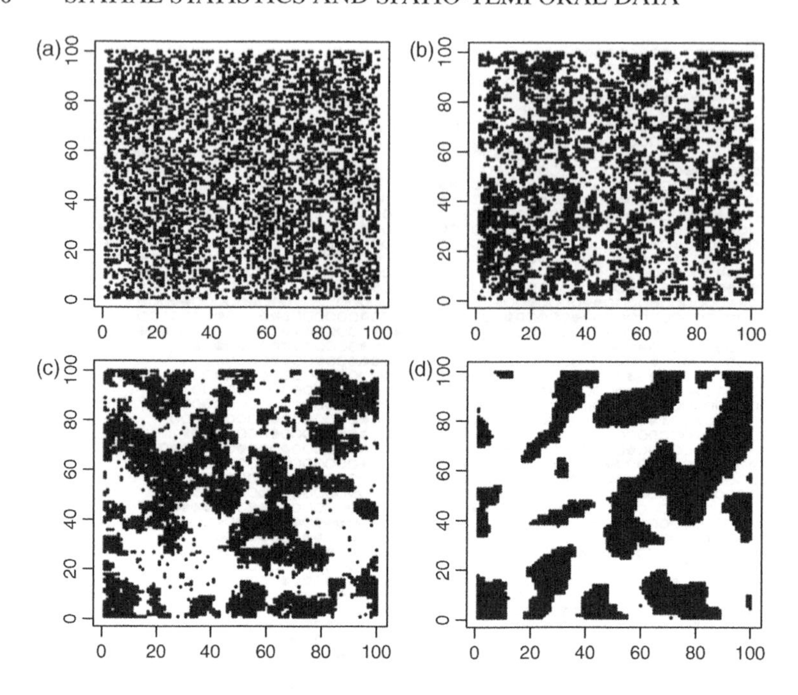

Figure 4.5 Four binary spatial fields with varying strengths of correlation. Correlation parameter $\beta = 0.1$ *(a),* 0.3 *(b),* 0.5 *(c), and* 1.0 *(d).*

model. Note, that for every location $\mathbf{s}$, it holds that $P[Z(\mathbf{s}) = 1| Z(\mathbf{t})] + P[Z(\mathbf{s}) = -1|Z(\mathbf{t})] = 1$ for any values of $Z(\mathbf{t})$, as it must. As in the Gaussian case, one can define the neighborhood as appropriate to the application. In this case, the parameter α determines the overall proportion of high mortality rates, while the parameter β determines the strength of clustering in the data. Clearly $\beta = 0$ corresponds to no clustering or no dependence of a county's rate with neighboring county rates.

To see the effect of the value of β, four data sets each generated from this model with $\alpha = 0$ for four values of β are displayed in Figure 4.5. This model is known as the Ising model [Ising (1925)] in statistical mechanics, and image analysis. In statistical mechanics, this models whether a 'spin up' or 'spin down' state is present, while in image analysis this models whether some entity is present or absent (for example, black or white on a white surface).

Note that, as β increases, we see increased clustering of $Z(\mathbf{s}) = +1$ (and of $Z(\mathbf{s}) = -1$). In particular, note that if a given location has all four neighbors with a positive response, then the chance of a positive response at this location is given by $\exp(4\beta)/[\exp(4\beta)+\exp(-4\beta)]$, which is 0.6900, 0.9168, 0.9821, and 0.9997 for $\beta = 0.1, 0.3, 0.5$, and 1.0, respectively. In other words, in 10 000 such situations in a model with $\beta = 1.0$, a response of $Z(\mathbf{s}) = -1$ is expected to be seen approximately three times. This increasing dearth of sign 'flipping' is clearly observed in Figure 4.5.

For the cancer data, given the model, the goal becomes to estimate the parameters α and β, and to draw inferences on these parameters. Due to its asymptotic optimality in a variety of settings, and to its tractability in the spatial Gaussian case, it is natural to attempt to fit the model and assess accuracy using maximum likelihood. It can be shown [see, e.g., Besag (1974)] that the joint likelihood of the parameters given the observations is

$$L(\alpha,\beta) := \frac{\exp\{\alpha\sum_{\mathbf{s}\in D} z(\mathbf{s}) + (\beta/2)\sum_{\mathbf{s}\in D} z(\mathbf{s})[\sum_{\mathbf{t}\in N(\mathbf{s})} z(\mathbf{t})]\}}{\sum_{z(\mathbf{s}_1),\ldots,z(\mathbf{s}_n)} \exp\{\alpha\sum_{\mathbf{u}\in D} z(\mathbf{u}) + (\beta/2)\sum_{\mathbf{u}\in D} z(\mathbf{u})[\sum_{\mathbf{v}\in N(\mathbf{u})} z(\mathbf{v})]\}}, \tag{4.3}$$

where $\sum_{z(\mathbf{s}_1),\ldots,z(\mathbf{s}_n)}$ denotes the sum over all $2^{|D|}$ possible realizations. For our data set, the denominator has 2^{2003} summands. This example shows why it is not, in contrast to the Gaussian case, practical to carry out maximum likelihood estimation for moderate to large spatial binary data sets by direct enumeration.

For this reason, alternative methods of estimation have been sought. In particular, Besag (1974) noted that $[Z(\mathbf{s})|Z(\mathbf{t}) : \mathbf{t} \neq \mathbf{s}]$ and $[Z(\mathbf{u})|Z(\mathbf{t}) : \mathbf{t} \neq \mathbf{u}]$ are conditionally independent if $\mathbf{s} \notin N(\mathbf{u})$ and $\mathbf{u} \notin N(\mathbf{s})$. In the current nearest neighborhood model for the cancer data, this conditional independence holds whenever $d(\mathbf{s},\mathbf{u}) > 1$. Thus, the data grid can be 'coded' into two groups of observations, such that within each group individual components are conditionally independent. Using this coding, the usual likelihood theory for data in each group applies. These estimates can be obtained in a standard fashion, and this gives two sets of estimates which can be combined, by averaging, for example, to form one estimator.

Instead of combining the coding estimators, it appears reasonable to more simply pool all conditional probability components together to form a single likelihood. The resulting 'pseudo-likelihood' [Besag (1975)] is

$$PL(\alpha,\beta) = \prod_{\mathbf{s}\in D} \frac{\exp\{z(\mathbf{s})[\alpha + \beta\sum_{\mathbf{t}\in N(\mathbf{s})} z(\mathbf{t})]\}}{\exp[\alpha + \beta\sum_{\mathbf{t}\in N(\mathbf{s})} z(\mathbf{t})] + \exp[-\alpha - \beta\sum_{\mathbf{t}\in N(\mathbf{s})} z(\mathbf{t})]}.$$

The corresponding maximum pseudo-likelihood estimators (MPLEs) are the parameter values that maximize $PL(\alpha, \beta)$. Note that this is simply the product over all sites of the conditional probabilities given in Equation 4.2. Although this is not a likelihood in the traditional sense, the large sample consistency and asymptotic normality of MPLEs have been demonstrated by, for example, Geman and Graffigne (1987), Comets (1992), and Guyon and Künsch (1992), and the method has been applied in a variety of settings. For practical implementation, note that Equation 4.2 can be written as

$$\frac{1}{2}\ln\left\{\frac{P[Z(\mathbf{s})=1|Z(\mathbf{t})]}{1-P[Z(\mathbf{s})=1|Z(\mathbf{t})]}\right\}=\alpha+\beta\sum_{\mathbf{t}\in N(\mathbf{s})}Z(\mathbf{t}),$$

and thus the parameter estimates can be found using a standard logistic regression for independent observations with responses $Z(\mathbf{s})$ and covariates $\sum_{\mathbf{t}\in N(\mathbf{s})} Z(\mathbf{t})$. The transformation of the observations from $[-1, 1]$ to $[0, 1]$ is given by $Z^*(\mathbf{s}) = \frac{1}{2}[Z(\mathbf{s}) + 1]$, and then the parameter estimates are related by $\widehat{\beta} = \widehat{\beta}^*/4$ and $\widehat{\alpha} = \widehat{\alpha}^*/2 + \widehat{\beta}^*$.

For the cancer of the liver and gallbladder data set, the MPLEs are $\widehat{\alpha} = -0.322$ and $\widehat{\beta} = 0.106$. The parameter values suggest an overall majority of low-rate counties ($\widehat{\alpha} < 0$), and that there is some indication of clumping or positive association between neighboring counties ($\widehat{\beta} > 0$).

Although it is straightforward to find parameter estimates via pseudo-likelihood, standard errors for the PL estimators usually have no closed form solution, and so it is a nontrivial problem to estimate them. Nonparametric resampling gives that $s.e.(\widehat{\alpha}) = 0.044$ and $s.e.(\widehat{\beta}) = 0.026$. See Chapter 10 for further details on how these are calculated, and large sample justification of their use.

Another method to obtain standard errors is through parametric resampling (details are given in Chapter 10) from the conditional distribution using the observed parameter estimates and Gibbs sampling. Using this approach across a wide range of tuning parameters suggests $s.e.(\widehat{\alpha}) \simeq 0.036$, and that the $s.e.(\widehat{\beta}) \simeq 0.019$. Using the (more conservative) nonparametric standard errors we see that the indications of an overall majority of low-rate counties ($t = -0.322/0.044 = 7.3$), and positive association between neighboring counties ($t = 0.106/0.026 = 4.1$), are statistically sound. Using the parametric standard errors gives qualitatively similar, and slightly stronger, conclusions.

An alternative method of estimation is the generalized pseudo-likelihood (GPL) of Huang and Ogata (2002). They note that the full likelihood is maximizable whenever $2^{|D|}$ is not prohibitively large; for instance, up to $|D| = 13$, say. Then, we can form the product of full likelihoods over subsets of the full data set. Pseudo-likelihood is then a special case, where each subset is of size one. Huang and Ogata show modest increased efficiency for their method over that obtained by pseudo-likelihood. In the cancer data set $|D| = 5$ gives $\widehat{\alpha} = -0.332$ and $\widehat{\beta} = 0.101$, while $|D| = 13$ gives $\widehat{\alpha} = -0.341$ and $\widehat{\beta} = 0.096$.

Another alternative to PL estimation is via the Monte Carlo maximum likelihood (MCML) of Geyer and Thompson (1992). Their methodology tries to approximate the full likelihood in Equation 4.3, by approximating the denominator in the full likelihood via Monte Carlo sampling. Some details are given in Chapter 10, Section 10.5.2. In the cancer data set, MCML finds that $\widehat{\alpha} = -0.303$ with $s.e.(\widehat{\alpha}) \simeq 0.034$, and $\widehat{\beta} = 0.117$ with $s.e.(\widehat{\beta}) \simeq 0.018$. Note that the parameter estimates and standard error estimates are very close to those arrived at via PL, using parametric resampling from the estimated conditional distributions. The MCML method has better efficiency then either MPL or GPL, as it approaches the efficiency of the MLE. There is a slight suggestion of this in this example, with both standard errors slightly smaller than those using PL. To translate this theoretical efficiency into practice in general, however, the user needs good initial parameter estimates and the recognition that extensive computation is often necessary. Further, computations grow as the strength of correlation (large β) increases. For example, when $\beta = 0.33$, the MCML estimates are more variable than the PL estimates, even after 200 000 Gibbs resampling steps. When $\beta = 0.5$, an even larger number of resampling steps is necessary in the MCML. See Section 10.5.2 for some details comparing PL, GPL, and MCML. For the use of MCML in geostatistics see, for example, Christensen (2004).

4.2.1 Edge effects

For a temporal process of length n, there is only one observation, the first one, that does not have an immediate predecessor in time. The effect of excluding this observation in estimation is usually minimal. For a spatial process on an $n \times n$ grid, however, there are $4(n - 1)$ locations (those on the edges and corners) that do not have all 4 nearest neighbors. For the wheat data, the 25×20 grid has 86 edge sites. Ignoring these observations leads to a substantial loss of information. On the other hand, in either maximum likelihood or in pseudo-likelihood estimation, it is not clear how to make use of these edge sites.

In the cancer data set there are actually 2293 locations in Figure 4.4. Only 2003 of these locations have all 4 nearest neighbors and can then be included in the PL. This amounts to using 87.4% of the data sites directly in the pseudo-likelihood. A natural question is: can we use the partial information present for observations with 1–3 neighbors. One simple approach is to impute an estimate at each site where a neighbor is unobserved. A natural value is the overall mean of the spatial process. For the Gaussian model this approach sets each unobserved value equal to μ. This effectively removes this observation from $\sum_{\mathbf{t}} \gamma_{st}[Z(\mathbf{t}) - \mu(\mathbf{t})]$. In other cases the mean must be estimated. In the binary model, we set the unobserved responses equal to the empirical mean of the data, $\widehat{E} = \widehat{P}[Z = 1] - \widehat{P}[Z = -1]$. In the cancer data this gives $\widehat{E} = -0.46$. The PL estimates with these additional imputed values are $\tilde{\alpha} = -0.309$ and $\tilde{\beta} = 0.113$. These estimates are both approximately 7 percent larger than those that ignored these edge points. Note that accounting for edge effects has moved the PL estimates closer to the, likely more efficient, MCML estimates.

In general, the increased sample size makes these estimates less variable, but the imputed values are not actual observed values and thus these estimates are often more biased. To see the effects on bias and variance of correcting for edge sites using this approach, and a more formal EM-algorithm-based approach in the autologistic model, see Lim *et al.* (2007).

4.2.2 Goodness of model fit

We have assessed the model via parameter estimation and standard error estimation, and seen that the model is useful. We now address the question of the *adequacy* of our fitted model. Table 4.1 shows the counts of observations and the counts of the corresponding neighboring sum for the 2003 observations with all 4 nearest neighbors.

For the model with $\beta = 0$ we have $\widehat{\alpha} = -0.499$. Under this model of independence, we have that the expected number of positive observations for each neighbor sum is simply $\exp(\widehat{\alpha})/[\exp(\widehat{\alpha}) + \exp(-\widehat{\alpha})] = 0.269$ times the column sum. This gives the expected counts shown in Table 4.2.

The chisquared statistic is $\chi^2 = 68.4$. If we compare this to the chisquared distribution with four degrees of freedom, we thoroughly reject

Table 4.1 Observed cancer data counts. Cross classification of observations with the corresponding nearest neighbor sum.

NSum	−4	−2	0	2	4	Total
−1	524	554	296	77	12	1463
+1	113	200	153	60	14	540
Total	637	754	449	137	26	2003

Table 4.2 Expected cancer data counts: Model I. Expected counts for the cancer data under independence between all observations.

NSum	−4	−2	0	2	4	Total
−1	465.3	550.7	328.0	100.1	19.0	1463.1
+1	171.7	203.3	121.1	36.9	7.0	540
Total	637.0	754.0	449.1	137.0	26.0	—

Table 4.3 Expected cancer data counts: Model II. Expected counts for the cancer data under fitted autologistic model.

Nsum	−4	−2	0	2	4	Total
−1	520.0	561.1	294.4	76.0	11.7	1463.2
+1	117.0	192.9	154.6	61.0	14.3	540
Total	637.0	754.0	449.0	137.0	26.0	—

the null hypothesis. This is in complete agreement with our strongly rejecting $\beta = 0$.

For the model fitted by pseudo-likelihood, we have $\widehat{\alpha} = -0.322$ and $\widehat{\beta} = 0.106$. The expected number of positive observations for each neighbor sum, k, is now simply $\exp(\widehat{\alpha} + \widehat{\beta}k)/[\exp(\widehat{\alpha} + \widehat{\beta}k) + \exp(-\widehat{\alpha} - \widehat{\beta}k)]$ times the column sum. This gives the expected counts in Table 4.3.

Note that the expected and observed counts under this model are quite close overall. The chisquared statistic is $\chi^2 \simeq 0.6$. If we compare this to the chisquared distribution with three degrees of freedom, we have essentially no evidence that this model is not adequate.

The conclusions based on the chisquared statistics are not formally correct, because the observations are not mutually independent. If we analyze the data exactly as above but separately for the two coding schemes discussed in Section 4.2, then the observations are conditionally independent, and thus this analysis is formally justifiable. The two chisquared statistics are $\chi_4^2 = 35.99$ and $\chi_4^2 = 40.06$ for the model with $\beta = 0$, and $\chi_3^2 = 1.84$ and $\chi_3^2 = 1.46$ for the fitted model. The conclusions are still that the data completely reject $\beta = 0$, while the data are completely compatible with the autologistic model fitted by PL.

For count data there is an auto-Poisson model, where the conditional Poisson mean depends on its neighbors. Unfortunately, in order for this model to be valid, all correlation parameters need to be negative. One way to allow for positive correlations is to use a truncated Poisson distribution. This is pursued in Kaiser and Cressie (1997).

We have discussed psuedo-likelihood, GPL, and MCML for estimation in non-Gaussian models. Several other approaches have been proposed. For example, several researchers have proposed and implemented composite likelihood in the spatial setting. For example, Heagerty and Lele (1998) consider a composite likelihood approach to modeling binary spatial observations. This is also a computational advantage against full ML and can handle nonconstant mean functions in a natural way. Gibbs regression similarly extends the conditional likelihood model in Equation 4.2 to include covariates.

5

Isotropy

We have assumed in covariance model formulation, and in data analyses in Chapter 3, that all covariance (and hence variogram) functions are isotropic. Formally, this means that $C(\mathbf{h}) = C^*(\|\mathbf{h}\|)$ for SOS processes, while $\gamma(\mathbf{h}) = \gamma^*(\|\mathbf{h}\|)$ for any IS process, where $\|\mathbf{v}\|$ denotes the length of any vector $\mathbf{v}$, and $C^*(\cdot)$ and $\gamma^*(\cdot)$ denote functions of a scalar variable.

The assumption of isotropy is typically made out of convenience. This convenience is given in terms of both simple model formulation, and ease of estimation. For model formulation, any anisotropic (not isotropic) model will need additional model parameters to describe how the covariance depends on direction. The additional parameters make model estimation a more difficult task. A natural question is: to what extent does a lack of isotropy influence predictions? In other words, what if isotropy is not a reasonable assumption? How does it matter?

To see concretely how an assumption of isotropy impacts an analysis, consider the following covariance model

$$C(x, y) = \exp(-x^2 - \gamma y^2), \tag{5.1}$$

where $C(x, y)$ denotes the covariance between two observations separated by spatial lag (x, y). Data coming from this spatial process (with $\gamma \neq 1.0$) follow an anisotropic correlation, whose spatial dependence in the x direction is stronger than that in the y direction, when $\gamma > 1$.

Spatial Statistics and Spatio-Temporal Data: Covariance Functions and Directional Properties Michael Sherman

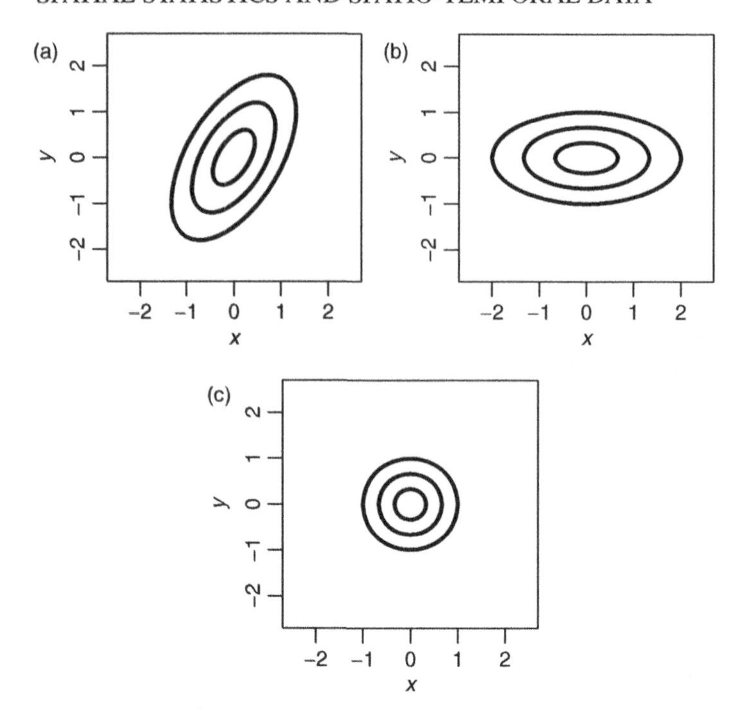

Figure 5.1 A depiction of geometric isotropy: (a) anisotropic covariance; (b) rotated; (c) rotated and rescaled to isotropy. Solid curves depict contours with equal covariance.

Figure 5.1 (b) shows the contours of this covariance function with $\gamma = 4$. Note that, for example, observations at locations $(0, 1)$ and $(2, 0)$ have the same correlation with an observation at the origin $(0, 0)$.

Suppose four observations have been made of a stationary random field following this covariance structure at locations $\left(0, \frac{2}{3}\right)$ $\left(0, -\frac{2}{3}\right)$, $\left(\frac{2}{3}, 0\right)$ and $\left(-\frac{2}{3}, 0\right)$. The situation is depicted in Figure 5.2, and the goal is prediction of the value at the location $(0, 0)$, which is equally distant from the observed four locations.

Under an assumed isotropic model, the optimal weights are equal to 0.25 for each of the four observations. This is true for *any* isotropic covariance model.

Ignoring the correct anisotropic structure has three potentially undesirable effects:

i. A data analyst who assumes isotropy a priori would not detect this dependence of correlation on direction, and thus would miss

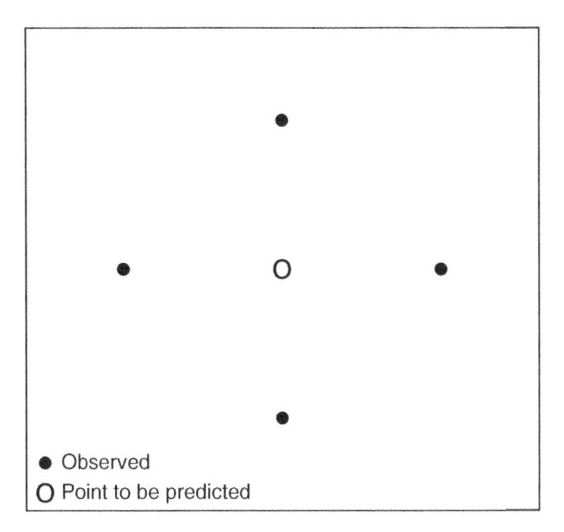

Figure 5.2 Four observed locations (filled circles) with prediction location (open circle).

the true underlying spatial structure. This could lead to improper understanding of the physical phenomenon under study. In many cases, the lack of isotropy leads to a search for reasons for directional effects. For example, analysis of the locations of a desert plant by Rosenberg (2004) strongly suggests that the spatial distribution of the *Ambrosia dumosa* plant depends on direction. Schenk and Mahall (2002) suggest that directional effects of shading on seed germination may be the cause of the anisotropy present.

ii. Kriging weights and variances are greatly affected by anisotropy. Specifically, using Equation 2.3, we can calculate the squared prediction error for *any* prediction weights $\boldsymbol{\lambda}$. Using the knowledge of Equation 5.1, we can calculate the optimal kriging weights for various values of the anisotropy parameter γ. These are given in the second and third columns of Table 5.1. Further, we can calculate the squared prediction error using this (assumed to be) correct anisotropic covariance using the optimal kriging weights as given in the fourth column. To see the effect of ignoring the true anisotropic structure, we also compute the prediction variance obtained from Equation 2.3 using the incorrect equal weights under an assumed isotropy. These variances are given in the fifth column.

Table 5.1 The effect of isotropy on kriging weights and prediction variance.

γ	$\lambda_1 = \lambda_3$	$\lambda_2 = \lambda_4$	Krvar	Isovar	Ratio
1.0	0.25	0.25	0.2155	0.2155	1.00
2.0	0.42	0.08	0.2877	0.3542	0.81
3.0	0.47	0.03	0.3000	0.4514	0.67
4.0	0.498	0.002	0.3021	0.5152	0.59
6.0	0.516	−0.016	0.3011	0.5827	0.52
10.0	0.526	−0.026	0.2993	0.6267	0.48

The parameter γ gives the degree of anisotropy, λ_is are the kriging weights. 'Krvar' denotes the prediction variance using the true anisotropic covariance, while 'Isovar' denotes the true prediction variance using an incorrectly assumed isotropic covariance.

There are several key effects due to anisotropy in this case. We observe the following:

a. The kriging weights become more unequal as the degree of anisotropy increases. Note, in particular, that a large degree of anisotropy ($\gamma = 6$ and $\gamma = 10$) actually leads to negative weights on observations at locations 2 and 4.
b. As the degree of isotropy increases, the kriging variances become relatively worse under an incorrectly assumed isotropy.
c. For large degrees of anisotropy (large γ) the correct prediction variances actually become smaller as the degree of anisotropy increases, while assuming isotropy causes prediction variance to increase as anisotropy increases.

iii. In a spatial regression model

$$\mathbf{y} = \mathbf{X}\boldsymbol{\beta} + \boldsymbol{\epsilon}, \quad \mathrm{Var}(\boldsymbol{\epsilon}) = \boldsymbol{\Sigma},$$

the OLS estimator, $\widehat{\boldsymbol{\beta}}_{\mathrm{OLS}} := (\mathbf{X}^{\mathrm{T}}\mathbf{X})^{-1}\mathbf{X}^{\mathrm{T}}\mathbf{y}$ has

$$Var(\widehat{\boldsymbol{\beta}}_{\mathrm{OLS}}) = (\mathbf{X}^{\mathrm{T}}\mathbf{X})^{-1}\mathbf{X}^{\mathrm{T}}\boldsymbol{\Sigma}\mathbf{X}(\mathbf{X}^{\mathrm{T}}\mathbf{X})^{-1},$$

while the GLS estimator, $\widehat{\boldsymbol{\beta}}_{\mathrm{GLS}} := (\mathbf{X}^{\mathrm{T}}\boldsymbol{\Sigma}^{-1}\mathbf{X})^{-1}(\mathbf{X}^{\mathrm{T}}\boldsymbol{\Sigma}^{-1}\mathbf{y})$, has

$$Var(\widehat{\boldsymbol{\beta}}_{\mathrm{GLS}}) = (\mathbf{X}^{\mathrm{T}}\boldsymbol{\Sigma}^{-1}\mathbf{X})^{-1}.$$

The OLS estimator does not require knowledge of $\boldsymbol{\Sigma}$, while the GLS estimator does. We see, however, that for both GLS and OLS, in order to draw inferences using $\widehat{\boldsymbol{\beta}}$ concerning the regression parameter vector $\boldsymbol{\beta}$, the matrix $\boldsymbol{\Sigma}$ needs to be estimated. The accuracy of inferences depends on the accuracy in estimation of the matrix $\boldsymbol{\Sigma}$. Thus, we see that the appropriateness of inferences drawn to regression parameters depends on correctly specifying the correlation structure. Specifically, if an anisotropic covariance function is estimated parametrically under an incorrectly assumed isotropic model, then the standard errors can be badly incorrect. This will have the effects detailed in Section 1.2.1.

For these three reasons, it is important to detect anisotropy when present and to, if possible, properly account for anisotropy in our analyses.

5.1 Geometric anisotropy

The previous section details the reasons why it is important to account for anisotropy, when it is present. This importance leads to the desire for models for anisotropic covariances, and methods to account for anisotropy in spatial analyses. A simple, and often useful, model for anisotropic covariances or variograms is that of geometric anisotropy.

Geometric anisotropy occurs when the covariance function can be converted to an isotropic covariance through rotation and rescaling of the coordinate axes. Specifically, let $\mathbf{h} = (h_x, h_y)$ be a spatial lag of interest. The vector $\mathbf{h}$ is rotated to a vector in a new coordinate system parallel to the main axes of concentric ellipses. To form a rotation of the coordinate axes, we transform to:

$$\mathbf{h}^* = \mathbf{R}\mathbf{h},$$

where

$$\mathbf{R} = \begin{bmatrix} \cos(\theta) & \sin(\theta) \\ -\sin(\theta) & \cos(\theta) \end{bmatrix},$$

and θ is the angle of rotation.

After rotation, the two main axes of the ellipsoid coincide with those of the original coordinate axes. In general, the two axes of the ellipsoid are not of equal length and thus the major axis needs to be shrunk. Let

$$\mathbf{T} = \begin{bmatrix} \frac{1}{b_1^{1/2}} & 0 \\ 0 & \frac{1}{b_2^{1/2}} \end{bmatrix},$$

where b_1 and b_2 denote the lengths of the two principal axes. Finally,

$$\tilde{\mathbf{h}} = \mathbf{TRh}.$$

The anisotropic variogram is now completely given by the isotropic variogram, in the rotated and rescaled coordinate system. Specifically, for $Z(\mathbf{s}_i)$ and $Z(\mathbf{s}_j)$ with $\mathbf{s}_i - \mathbf{s}_j = \mathbf{h}$, we transform lag $\mathbf{h}$ in the original space into $\tilde{\mathbf{h}} = \mathbf{TRh}$, and we have $\gamma(\mathbf{h}) = \gamma^*(\|\tilde{\mathbf{h}}\|)$. An isotropic covariance or variogram function can now be fitted using the transformed spatial lags.

Figure 5.1 (a) illustrates the contours of a covariance function that is not isotropic, but is geometrically anisotropic. In this example, $\theta = \pi/3$, and $b_1 = 4$, $b_2 = 1$. We see how the rotation matrix $\mathbf{R}$ aligns the main axes of the ellipsoid with the original coordinate system axes [in Figure 5.1 (b)], and how the matrix $\mathbf{T}$ rescales to a coordinate system where isotropy holds [in Figure 5.1 (c)]. Note that we can say that geometric anisotropy holds if for all lag vectors $\mathbf{h}$ it holds that $\gamma(\mathbf{h}) = \gamma^*(\sqrt{\mathbf{h}^T\mathbf{Bh}})$, for some positive definite matrix $\mathbf{B}$. In our development, $\mathbf{B} = \mathbf{R}^T\mathbf{T}^T\mathbf{TR}$.

To see the effects of the rotation and rescaling in this example, consider lags $\mathbf{h}_1 = (0, 1)^T$ and $\mathbf{h}_2 = (1, 0)^T$. After rotation and rescaling, these two lags estimate the isotropic variogram at $\tilde{\mathbf{h}}_1 = (0.433, 0.5)^T$ and $\tilde{\mathbf{h}}_2 = (0.25, -0.866)^T$ at (transformed) distances 0.66 and 0.91, respectively.

Doing this for all observed lag distances enables (isotropic) moment estimation of the variogram. Further, any of the (isotropic) models in Section 3.3 can now be fitted using the original data locations on the transformed distance scale.

Note that geometric anisotropy assumes that the variogram has the same sill in all directions, but that the range of correlation is not the same in different directions. This is a common departure from isotropy, but not the only type.

5.2 Other types of anisotropy

Geometric anisotropy is not the only way that a covariance function can depend on direction. Note that in geometric anisotropy it is assumed that

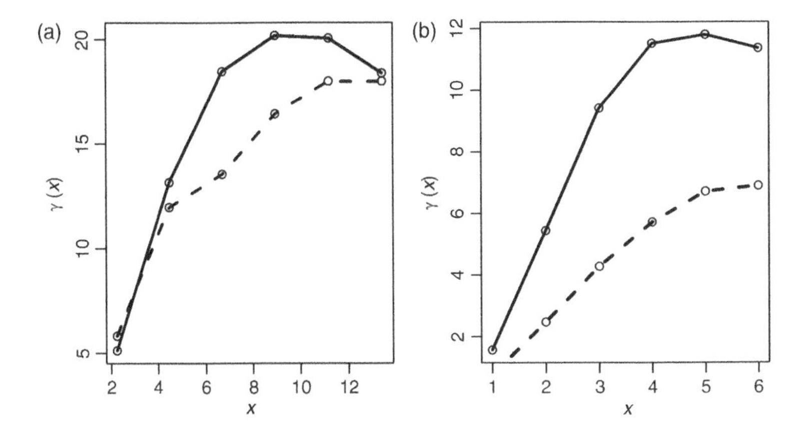

Figure 5.3 Two different types of anisotropic semivariograms. (a) Range anisotropy. (b) Sill anisotropy.

only the range is direction dependent. This means, in particular, that the sill and any possible nugget effect is the same in all directions. Zimmerman (1993) recommended that anisotropy due to a specific feature be named after that feature that leads to anisotropy. So, for example, sill anisotropy refers to a variogram that has the same range and nugget effect but with a different sill, that is, a different total level of variability. As another example, if geometric anisotropy does not hold, that is, the covariance contours are not ellipsoidal, but the ranges are not the same, we call this range anisotropy. Then, geometric anisotropy is a specific type of range anisotropy. Figure 5.3 displays semivariograms in two separate directions for data on wind speeds described in Section 5.6.5. In the first case (a) we see two semivariograms reaching approximately the same sill with quite different ranges, thus range anisotropy is exhibited. In the second figure (b), we see that the semivariograms have approximately the same shape, but that one sill is approximately twice the other, thus sill anisotropy is exhibited. In the former case, geometric anisotropy may hold while in the second case it cannot.

5.3 Covariance modeling under anisotropy

We have already seen that under geometric anisotropy, one isotropic model suffices to completely describe spatial dependence. In modeling nongeometric anisotropy behavior, we need to account for the specific correlation behavior in each direction separately.

For example, consider the four exponential semivariograms in the north/south, northeast/southwest, east/west, and southeast/northwest directions:

$$\gamma_{\mathrm{N}} = c_1 + \sigma_1^2[1 - \exp(-\nu_1\|\mathbf{h}\|)]$$
$$\gamma_{\mathrm{NE}} = c_2 + \sigma_2^2[1 - \exp(-\nu_2\|\mathbf{h}\|)]$$
$$\gamma_{\mathrm{E}} = c_3 + \sigma_3^2[1 - \exp(-\nu_3\|\mathbf{h}\|)]$$
$$\gamma_{\mathrm{SE}} = c_4 + \sigma_4^2[1 - \exp(-\nu_4\|\mathbf{h}\|)].$$

Under nugget anisotropy, we have $c_i \neq c_j$, for some $i \neq j$, with all other parameters equal. Likewise, $\sigma_i \neq \sigma_j$, for some i and j, corresponds to sill anisotropy [as in Figure 5.3 (b)], while $\nu_i \neq \nu_j$, for some i and j, corresponds to range anisotropy [as in Figure 5.3 (a)]. Of course, we can have more than one type of anisotropy in any given situation.

In many cases, covariance modeling can be made more simple by a judicious use of sparse mean function models. For example, different sills in different directions are often due to an underlying nonconstant, direction-dependent mean function. That being said, it is also the case that intensive mean fitting may be necessary to make a covariance model relatively simple. For this reason, it is often useful to try to detect and account for a direction-dependent covariance function.

5.4 Detection of anisotropy: the rose plot

In the previous section, we have discussed how to treat different types of anisotropies, when present. We now discuss a step that is actually performed prior to those in modeling. The data analyst first needs to ascertain whenever the covariance model generating the data can be safely assumed to be isotropic. If this seems to be the case, then one can be comfortable modeling the covariance assuming isotropy. Towards this end, we first consider a graphical method to assess isotropy: the rose plot.

A popular method to assess the assumption of isotropy is to visually examine plots of direction-specific sample variograms or covariance functions. While being used for a graphical assessment of isotropy, it will also be the basis for a more formal assessment to be discussed in the following sections of Chapter 5.

Let $\mathbf{h} = (h_x, h_y)$ denote an arbitrary spatial lag of interest. Then, the variogram can be estimated at observed spatial lags $\mathbf{h}_{ij} = (h_{x(i)}, h_{y(j)})$,

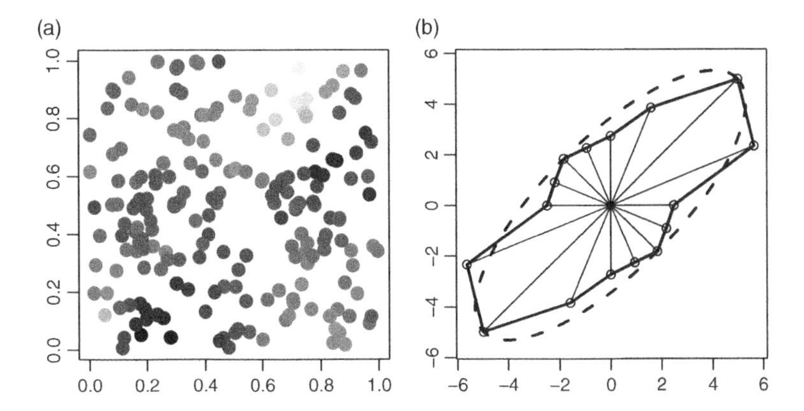

Figure 5.4 Rose plot illustration: the data are 200 observations from a Gaussian process (a), and the rose plot with eight directions (b). Darker points in the data plot correspond to larger response values. In the rose plot, the solid line segments are empirical direction-dependent estimates of covariance; the dotted curve is the estimated ellipsoid under an assumed geometric anisotropy.

for $i = 1, \ldots, n_x$, $j = 1, \ldots, n_y$. The idea is to compute the sample variogram in each of k directions. Then, choose a given value of the variogram, and draw a line in each of the k directions indicating the distance at which this given value is reached. A common value is the effective range of the variogram or a substantial proportion of the range. If this distance is the same for all directions, this suggests isotropy. In other words, if we connect the endpoints of the lines, we should see an approximate circle. If, on the other hand, we see an approximate noncircular ellipse, this suggests geometric anisotropy, which can be handled as discussed in Section 5.1. For small k, an alternative is to simply plot the direction-specific sample variograms on one plot, as illustrated in the previous section. Using the rose plot indicates whether isotropy is reasonable, and, if not, whether geometric anisotropy is reasonable. Further, the rose plot gives estimates of the angle of rotation ($\mathbf{R}$) and the shrinking factor ($\mathbf{T}$).

Figure 5.4 (a) shows the location and response values of $n = 200$ observations. Figure 5.4 (b) is the rose plot based on $k = 8$ directions. Specifically, the angle between adjacent lines is $\pi/8$. To estimate the best fitting potential ellipse, we find that the range is the largest at $\pi/4$, with length of major axis equal to 7.05. The minor axis, in the direction $3\pi/4$, is of length 2.59. This gives an anisotropy ratio of $7.05/2.59 = 2.72$.

In truth, these $n = 200$ observations came from a stationary Gaussian process with covariance contours as given in Figure 5.1, and discussed at the end of Section 5.1. In this case, the true ratio of the lengths of the major and minor axes is 2.00. The correct angle of rotation is $\pi/3$. The rose plot from the data overestimates the anisotropy ratio by approximately 36 percent, and the angle of rotation is in error by $\pi/4 - \pi/3 = \pi/12$. Note that the latter is the second smallest possible error, given that the variogram was estimated in directions $k\pi/8$ for $k = 0, \ldots, 7$.

We can see the benefits and difficulties in using the rose plot. Firstly, from the data plot, it is difficult to see any possible anisotropic indications. Further, there can be some comfort that the estimated rose plot is closer to the truth than are the circular contours that would be implied by an assumption of isotropy. On the other hand, we did not get as close as was possible. Further, in general, we do not know that the relatively good behavior in this example is reproducible.

In summary, this graphical diagnostic and others like it are very useful in suggesting whether isotropy is reasonable, and, if not, what types of anisotropy are reasonable. They can be, however, sometimes difficult to assess, and any conclusions drawn are open to interpretation. For this reason, it is worthwhile to consider more formal testing procedures. This we address in the following section.

5.5 Parametric methods to assess isotropy

Given that isotropy is suspect, a natural approach is to attempt to model anisotropy through inclusion of a low-dimensional parameter to capture the anisotropic effects in the covariance. For example, Baczkowski and Mardia (1990) assume that, for observations $Z(i, j)$, on a regular lattice, a simultaneous spatial regression model holds. Specifically, they employ the doubly geometric process of Martin (1979):

$$\begin{aligned} Z(i,j) &= \beta_1 Z(i-1,j) + \beta_2 Z(i,j-1) \\ &\quad - \beta_1\beta_2 Z(i-1,j-1) + e(i,j), \end{aligned}$$

where the $e(i, j)$s are independent $N(0, \sigma^2)$ variables. In this case, the covariance function is given by

$$Cov[Z(i,j), Z(i+u,j+v)] = V^2 \beta_1^{|u|} \beta_2^{|v|},$$

where $V^2 = \sigma^2 \left[\left(1-\beta_1^2\right)\left(1-\beta_2^2\right)\right]^{-1}$.

When $\beta_1 = \beta_2$, we have that the process is symmetric, in that $C(u, v) = C(-u, v)$. Although this symmetry does not imply isotropy, it is necessary for isotropy to hold, and so the rejection of $\beta_1 = \beta_2$ rejects isotropy under the given model. Baczkowksi and Mardia give large sample distribution theory for a test of $\beta_1 = \beta_2$, and discuss how to extend to a test of isotropy.

While useful, one needs to assess the appropriateness of the assumption of normality of the errors, and the assumption that the observations follow the given doubly geometric process.

Other testing procedures for investigating isotropy have been proposed by, for example, Cabana (1987), Jona-Lasinio (2001), and Molina and Feito (2002). Cabana (1987) tests isotropy for a class of 'affine processes,' for which he develops an F-test by combining statistics obtained on a partition of the original field. Jona-Lasinio (2001) tests for isotropy of multivariate spatial observations by applying principal component analysis to reduce the dimensionality, and obtains statistical significance via simulation. Molina and Feito (2002) consider testing for isotropy in images. Lu and Zimmerman (2005) assess the closely related hypothesis of directional symmetry in the spectral domain.

5.6 Nonparametric methods of assessing anisotropy

The moment estimator in Section 3.1 is nonparametric in that the variogram function is estimated unbiasedly, for any marginal distribution. In this spirit, it is useful to assess isotropy while imposing minimal assumptions on the underlying distribution and on the underlying strength of dependence. It turns out that equally spaced data locations and unequally spaced locations lead to tests of qualitatively different behavior. As in Guan *et al.* (2004), we initially consider regularly spaced locations and then consider unequally spaced locations. Lu and Zimmerman (2001) consider the former case. Their approach is based on the asymptotic distribution of the sample (semi)variogram, and performs well for equally spaced Gaussian processes.

5.6.1 Regularly spaced data case

Consider a strictly stationary random field $\{Z(\mathbf{s}) : \mathbf{s} \in \mathbb{Z}^2\}$, where $\mathbb{Z}^2$ denotes the two-dimensional space of integer lattice points. Let $D \subset \mathbb{Z}^2$

be a finite set of lattice points at which observations are taken. For any given spatial lag $\mathbf{h}$, we define the variogram function as in Chapter 2:

$$2\gamma(\mathbf{h}) \equiv \text{Var}\,[Z(\mathbf{s}) - Z(\mathbf{s}+\mathbf{h})].$$

As discussed in Chapter 2, the variogram function is considered in place of the covariance function due to the facts that it can be estimated more accurately for a variety of data structures, and that for stationary processes $2\gamma(\mathbf{h}) = 2[\text{C}(\mathbf{0}) - \text{C}(\mathbf{h})]$, and is thus more generally defined.

Recall that if two lags have the same length, that is, they have the same Euclidean distance from the origin, the corresponding values of $2\gamma(\mathbf{h})$ will be equal under isotropy. This suggests that a test for isotropy may be obtained by comparing variogram values at lags with the same length, but in different directions. The rose plot of Section 5.4 uses this idea informally. Now, a formal test is given, using data-based estimates in each direction. The moment estimator of the variogram, the sample variogram, defined in Section 3.1, will be used. Other robust estimators can certainly be used, if appropriate.

In order to test, the user must first decide at which spatial lags an assessment of the variogram is desired. Towards this end, let $\mathbf{\Lambda}$ be a set of lags for which we want to calculate and compare the sample variogram values. Define $\mathbf{G} \equiv \{2\gamma(\mathbf{h}) : \mathbf{h} \in \mathbf{\Lambda}\}$ to be the vector of variogram values for the spatial lags in $\mathbf{\Lambda}$. In order to formally justify methodology, consider a sequence of increasing index sets D_n, with $\{Z(\mathbf{s}) : \mathbf{s} \in D_n\}$. Now, for each n, let $2\widehat{\gamma}_n(\mathbf{h})$ and $\widehat{\mathbf{G}}_\mathbf{n} \equiv \{2\widehat{\gamma}_n(\mathbf{h}) : \mathbf{h} \in \mathbf{\Lambda}\}$ be the vector of moment estimators of $2\gamma(\mathbf{h})$ and $\mathbf{G}$ obtained over D_n, respectively.

Observed differences in $\widehat{\mathbf{G}}_\mathbf{n}$, for lag vectors of the same length, indicate potential anisotropy. To obtain formal inferences, however, the distribution of $\widehat{\mathbf{G}}_\mathbf{n}$ is required. Finite sample properties are difficult to obtain, in general, so, in what follows, consider the large sample properties of $\widehat{\mathbf{G}}_\mathbf{n}$ under an increasing domain setting.

To formally state the asymptotic properties of $\widehat{\mathbf{G}}_\mathbf{n}$, it is necessary to quantify the strength of dependence in the random field. Towards this end, consider a model-free mixing condition. Consideration of a mixing coefficient allows for any subsequent results to be valid over a broad class of covariance functions. The first mixing coefficient, for temporal observations, was proposed by Rosenblatt (1956). Following Rosenblatt, consider a particular type of strong mixing coefficient defined by

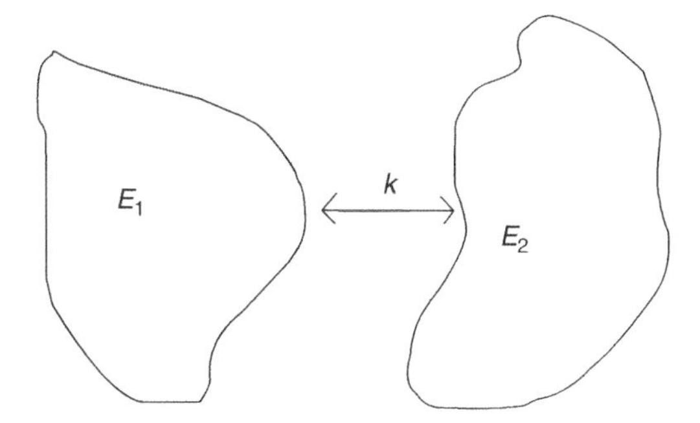

Figure 5.5 A depiction of the mixing coefficient, $\alpha_p(k)$. The two sets of indices, E_1 and E_2, of cardinality no larger than p, separated by distance k.

$$\alpha_p(k) \equiv \sup\{|P(A_1 \cap A_2) - P(A_1)P(A_2)| : A_i \in \mathcal{F}(E_i),$$
$$|E_i| \leq p, d(E_1, E_2) \geq k\},$$

for $i = 1, 2$, where $|E|$ is the cardinality of the index set E, $\mathcal{F}(E)$ is the σ-algebra generated by the random variables $\{Z(\mathbf{s}) : \mathbf{s} \in E\}$, and $d(E_1, E_2)$ is the minimal 'city block' distance between E_1 and E_2. The mixing coefficient quantifies and bounds the correlation between sets of observations in the sets E_1 and E_2, separated by spatial distance k. The situation is depicted in Figure 5.5.

Note that if the observations are independent, then $\alpha_p(k) = 0$ for all $k > 0$. It is desirable, however, to allow for correlation at arbitrarily large spatial lags, so that any results that follow will hold for several of the covariance functions in Section 3.3; for example, for the Matérn class of covariances. The following mixing condition requires $\alpha_p(k)$ to approach 0 for large k, at a rate depending on the cardinality, p. Specifically, assume the following mixing condition:

$$\sup_p \frac{\alpha_p(k)}{p} = \mathrm{O}(k^{-\epsilon}), \quad \text{for some } \epsilon > 2. \tag{5.2}$$

The importance of allowing the mixing coefficient to depend on the number of observations, p, was pointed out by Bradley (1993). Condition 5.2 says that, at a fixed distance k, as the cardinality increases, we allow dependence to increase at a rate controlled by p. As the distance increases, the dependence must decrease at a polynomial rate in k. Any m-dependent random field (i.e., observations separated by a distance larger than m are independent) satisfies this condition. Thus, the spherical model, in particular, satisfies the condition. Further, it can be seen from Doukhan (1994) that a stronger condition, $\sup_p \alpha_p(k) = O(k^{-\epsilon})$, holds for a stationary Gaussian random field, given that its correlation decays at a polynomial rate faster than $2 + \epsilon$, and has a spectral density which is bounded below. Other examples of spatial processes satisfying this mixing condition can also be found in Doukhan (1994) and Sherman (1996).

It is desirable to have a large class of domains, that is, shapes, where we observe data. Towards this end, define the boundary of a set D to be the set

$$\partial D \equiv \{\mathbf{s} \in D : \text{ there exists } \mathbf{s}' \notin D \text{ with } d(\mathbf{s}, \mathbf{s}') = 1\},$$

where $d\left[(s_x, s_y), (s'_x, s'_y)\right] \equiv \max\left(|s_x - s'_x|, |s_y - s'_y|\right)$. Let $|\partial D|$ denote the number of points in ∂D. We assume the conditions:

$$|D_n| = O(n^2) \quad \text{and} \quad |\partial D_n| = O(n). \tag{5.3}$$

Condition 5.3 basically says that the domain containing the data observations is truly spatial, and that the domain is not too irregularly shaped. These assumptions on the spatial domain are satisfied by many commonly encountered field sequences. For example, let $A \subset (0, 1] \times (0, 1]$ be the interior of a simple closed curve with nonempty interior. Now multiply the set A by n, to obtain the set $A_n \subset (0, n] \times (0, n]$; that is, A_n is the shape A inflated by a factor n. Define $D_n \equiv \{\mathbf{s} : \mathbf{s} \in A_n \cap \mathbb{Z}^2\}$, the locations on the grid contained in A_n. Then D_n satisfies Condition 5.3. This formulation allows for a wide variety of shapes on which the data can be observed, including rectangular, circular, elliptical regions, and star shapes, as well as more general shapes.

Finally, the following moment condition is required:

$$\sup_n \mathrm{E}\left\{\left|\sqrt{|D_n|} \times [2\widehat{\gamma}_n(\mathbf{h}) - 2\gamma(\mathbf{h})]\right|^{2+\epsilon}\right\} \leq C_\epsilon \quad \text{for some } \epsilon > 0, \; C_\epsilon < \infty. \tag{5.4}$$

Condition 5.4 is only slightly stronger than the existence of the large sample variance of the (standardized) variogram estimator, $2\widehat{\gamma}_n(\mathbf{h})$. For better interpretation, often this condition can be reduced to a marginal moment condition under sufficient weak dependence. For example, if the random field is m-dependent, it can be shown that the finiteness of $\mathrm{E}\left[|Z(\mathbf{s})|^{4+2\epsilon}\right]$, for some ϵ, is sufficient for Condition 5.4 to hold. Note that the fourth moment is required for the existence of $\mathbf{\Sigma_R}$, the large sample variance of the variogram estimators, to be defined in the following section. We can now state the large sample properties of the direction-dependent variogram estimators.

Theorem 5.6.1 *Asymptotic normality of sample variogram estimates Let $\{Z(\mathbf{s}) : \mathbf{s} \in \mathbb{Z}^2\}$ be a strictly stationary random field, which is observed at locations in D_n, satisfying Condition 5.3. Assume*

$$\sum_{\mathbf{s}\in\mathbb{Z}^2} \left|Cov\left\{[Z(\mathbf{0}) - Z(\mathbf{s}_1)]^2, [Z(\mathbf{s}) - Z(\mathbf{s} + \mathbf{s}_2)]^2\right\}\right| < \infty \quad \textit{for all } \mathbf{s}_1,\ \mathbf{s}_2. \tag{5.5}$$

Then $\mathbf{\Sigma_R} \equiv \lim_{n\to\infty} |D_n| \times Cov(\widehat{\mathbf{G}}_\mathbf{n}, \widehat{\mathbf{G}}_\mathbf{n})$ exists, the (i, j)th element of which is

$$\sum_{\mathbf{s}\in\mathbb{Z}^2} Cov\left\{[Z(\mathbf{0}) - Z(\mathbf{h}_i)]^2, [Z(\mathbf{s}) - Z(\mathbf{s} + \mathbf{h}_j)]^2\right\}.$$

If we further assume that $\mathbf{\Sigma_R}$ is positive definite, and that Equations 5.2 and 5.4 hold, then the limiting distribution of $\sqrt{|D_n|}(\widehat{\mathbf{G}}_\mathbf{n} - \mathbf{G})$ is multivariate normal, with mean $\mathbf{0}$ and covariance matrix $\mathbf{\Sigma_R}$.

The two Conditions, 5.2 and 5.4, are often referred to as the mixing and moment conditions for stationary random fields, respectively. They are natural generalizations of the mixing and moment conditions used to obtain large sample normality for the sample mean computed from a stationary time series or from a spatial field. Some of these results are discussed in Chapter 10.

For a Gaussian process, it can be shown that the absolute integrability of its covariance function, that is, $\int_{\mathbb{R}^2} |C(\mathbf{h})|d\mathbf{h} < \infty$, is sufficient for Equation 5.5 to hold.

Several of the covariance models discussed in Section 3.3 can be shown to satisfy the integrability condition, and thus satisfy Equation 5.5. For example, the exponential, Gaussian, and spherical models all satisfy Equation 5.5 for Gaussian random fields.

5.6.2 Irregularly spaced data case

Due to the commonness of irregularly spaced observations, we now consider this setting. Again, consider a strictly stationary random field $\{Z(\mathbf{s}) : \mathbf{s} \in \mathbb{R}^2\}$ and let $D \subset \mathbb{R}^2$ denote the domain of interest in which observations are taken. Now, the locations at which $Z(\cdot)$ are observed are viewed as *random* in number and location; specifically they are generated from a homogeneous two-dimensional Poisson process with intensity parameter ν. This process is described in Chapter 7. Karr (1986) makes a strong case for the plausibility of the Poisson assumption in many practical situations, for instance, data arising from meteorological studies and geological explorations.

In what follows, denote the random point process by N, and the random number of points of N contained in any (Borel) set B by $N(B)$. For now, assume the process driving the locations, N, to be independent of the responses at these locations, $Z(\cdot)$. This assumption can be evaluated: see, for example, Schlather *et al.* (2004). To construct a test statistic, an estimate of the variogram is needed. Here we consider one based on kernel smoothing.

In an adaption of the notation in Section 5.6.1, let $|D|$ now denote the volume (not the cardinality) of D, ∂D denote the boundary of D, and $|\partial D|$ denote the length (not the number of points) of ∂D. In the unequally spaced location setting, the spatial lag separating any two locations is typically unique. For this reason, a natural approach is to use local spatial averaging to estimate the covariance or variogram function.

Towards this goal, let δ be a positive constant, and let $w(\cdot)$ be a bounded, nonnegative, symmetric density function which takes positive values only on a finite support, C. Let $d\mathbf{x}$ denote an infinitesimally small disc centered at $\mathbf{x}$. Define $D + \mathbf{y} := \{\mathbf{s} + \mathbf{y} : \mathbf{s} \in D\}$ for arbitrary $\mathbf{y} \in \mathbb{R}^2$, and $N^{(2)}(d\mathbf{x}_1, d\mathbf{x}_2) := N(d\mathbf{x}_1)N(d\mathbf{x}_2)I(\mathbf{x}_1 \neq \mathbf{x}_2)$, where $I(\mathbf{x}_1 \neq \mathbf{x}_2) = 1$ if $\mathbf{x}_1 \neq \mathbf{x}_2$ and 0 otherwise. The kernel variogram estimator is given, as in Section 3.1.2, by:

$$2\widehat{\gamma}(\mathbf{h}) = \frac{1}{\nu^2} \int_{\mathbf{x}_1 \in D} \int_{\mathbf{x}_2 \in D} \delta^{-2} w\left(\frac{\mathbf{h} - \mathbf{x}_1 + \mathbf{x}_2}{\delta}\right) \times \frac{[Z(\mathbf{x}_1) - Z(\mathbf{x}_2)]^2}{|D \cap (D - \mathbf{x}_1 + \mathbf{x}_2)|} N^{(2)}(d\mathbf{x}_1, d\mathbf{x}_2),$$

where the term $|D \cap (D - \mathbf{x}_1 + \mathbf{x}_2)|$ serves as a boundary correction for the estimator. In practice, the location intensity parameter, ν, is usually replaced by $N(D)/|D|$, which is a consistent estimator of ν [Illian *et al.* (2008)]. We adopt the definitions of $\mathbf{\Lambda}$, $\mathbf{G}$, $2\widehat{\gamma}_n(\mathbf{h})$, and $\widehat{\mathbf{G}}_\mathbf{n}$ in Section 5.6.1, with the understanding that the variogram estimator is now defined by the kernel estimator above.

To account for dependence in the random field, we modify the mixing coefficient introduced in 5.6.1. Following Politis *et al.* (1998), consider a particular type of strong mixing coefficient defined by

$$\alpha_p(k) \equiv \sup\{|P(A_1 \cap A_2) - P(A_1)P(A_2)| : A_1 \in \mathcal{F}(E_1), A_2 \in \mathcal{F}(E_2)\},$$

where $E_2 = E_1 + \mathbf{s}$, $|E_1| = |E_2| \le p$, $d(E_1, E_2) \ge k$, and where the supremum is taken over all convex subsets $E_1 \subset \mathbb{R}^2$, and over all $\mathbf{s} \in \mathbb{R}^2$ such that $d(E_1, E_2) \ge k$. In the above, $\mathcal{F}(E)$ denotes the σ-algebra generated by the random variables $\{Z(\mathbf{s}) : \mathbf{s} \in E\}$. Again, assume that Mixing Condition 5.2 holds (for this alternative mixing coefficient). The interpretation of this mixing coefficient is as in the equally spaced case.

There is also a need to account for the shape of the random field and the choice of bandwidth. Consider a sequence of domains of interest, D_n, and a sequence of bandwidths, δ_n. Assume

$$|D_n| = \mathrm{O}(n^2), \; |\partial D_n| = \mathrm{O}(n), \quad \text{and} \quad \delta_n = \mathrm{O}(n^{-\beta}) \text{ for some } \beta \in (0, 1). \tag{5.6}$$

The first two assumptions are analogous to those in the equally spaced case. The assumption on the bandwidth allows for the bandwidth to be relatively smaller when a larger number of observations allows for more spatial averaging in the estimator. Let $\gamma^{(4)}(\mathbf{h}) := \mathrm{E}\{[Z(\mathbf{h}) - Z(\mathbf{0})]^4\}$. We require the following moment conditions:

$$2\gamma(\mathbf{h}), \gamma^{(4)}(\mathbf{h}) \text{ are bounded and continuous,} \quad \text{and} \tag{5.7}$$

$$\sup_n \mathrm{E}\left(|\sqrt{|D_n|} \times h_n \times \{2\widehat{\gamma}_n(\mathbf{h}) - \mathrm{E}[2\widehat{\gamma}_n(\mathbf{h})]\}|^{2+\omega}\right) \le C_\omega, \tag{5.8}$$

for some $\omega > 0, C_\omega < \infty$. The following theorem states that $2\widehat{\gamma}_n(\mathbf{h})$ is a consistent estimator for $2\gamma(\mathbf{h})$, and that $\widehat{\mathbf{G}}_\mathbf{n}$ has a large sample normal distribution under certain conditions.

Theorem 5.6.2 *Let $\{Z(\mathbf{s}) : \mathbf{s} \in \mathbb{R}^2\}$ be a strictly stationary random field observed on a general shaped region D_n satisfying Condition 5.6, where the points at which $Z(\cdot)$ is observed are generated by a homogeneous Poisson process. Assume Condition 5.7, and that*

$$\int_{\mathbf{s}\in\mathbb{R}^2} |Cov\{[Z(\mathbf{0}) - Z(\mathbf{s}_1)]^2, [Z(\mathbf{s}) - Z(\mathbf{s} + \mathbf{s}_2)]^2\}| d\mathbf{s} < \infty \quad \textit{for all } \mathbf{s}_1, \, \mathbf{s}_2.$$

Then, $\mathrm{E}[2\widehat{\gamma}_n(\mathbf{h})] \to 2\gamma(\mathbf{h})$ *and* $\mathbf{\Sigma}_{\mathbf{IR}} \equiv \lim_{n\to\infty} |D_n| \times h_n^2 \times Cov(\widehat{\mathbf{G}}_\mathbf{n}, \widehat{\mathbf{G}}_\mathbf{n})$ *exists, the (i, j)th element of which is*

$$\int_C w^2(\mathbf{s}) d\mathbf{s} \times \gamma^{(4)}(\mathbf{h}_i) \times I(\mathbf{h}_i = \pm\mathbf{h}_j)/\nu^2,$$

where $I(\mathbf{h}_i = \pm\mathbf{h}_j) = 1$ *if* $\mathbf{h}_i = \pm\mathbf{h}_j$ *and 0 otherwise. If it holds that* $\gamma^{(4)}(\mathbf{h}) > 0$ *for all* $\mathbf{h} \in \mathbf{\Lambda}$, *and that Conditions 5.2 and 5.8 hold, then*

$$\sqrt{|D_n|} \times \delta_n \times \left[\widehat{\mathbf{G}}_\mathbf{n} - \mathrm{E}(\widehat{\mathbf{G}}_\mathbf{n})\right]$$

is asymptotically normally distributed, with mean $\mathbf{0}$ *and covariance matrix* $\mathbf{\Sigma}_{\mathbf{IR}}$.

As in the case of equally spaced observations, the large sample normality will be the basis of testing. From the above covariance matrix, $\mathbf{\Sigma}_{\mathbf{IR}}$, it is seen that $2\widehat{\gamma}_n(\mathbf{h}_1)$ and $2\widehat{\gamma}_n(\mathbf{h}_2)$, for arbitrary $\mathbf{h}_1 \neq \pm\mathbf{h}_2$, are asymptotically uncorrelated. This is in complete contrast to the equally spaced situation, where the two variogram estimators are, in general, correlated. This can be explained as follows: in the irregularly spaced case, for a large number of locations, n, and a small bandwidth, δ_n, a relatively small number of data points compared to the total number of points in D_n will be used to calculate the sample variogram at any given spatial lag. Due to the randomness of the locations, the chance that the same or even nearby data points are used to calculate $2\widehat{\gamma}_n(\mathbf{h}_1)$ and $2\widehat{\gamma}_n(\mathbf{h}_2)$, where $\mathbf{h}_1 \neq \pm\mathbf{h}_2$, becomes small, as n becomes large. Thus, $2\widehat{\gamma}_n(\mathbf{h}_1)$ and $2\widehat{\gamma}_n(\mathbf{h}_2)$ tend to be based on observations separated by large distances, and thus approximately uncorrelated.

5.6.3 Choice of spatial lags for assessment of isotropy

To formally assess the hypothesis of isotropy, the spatial lag set, $\mathbf{\Lambda}$, needs to be specified. Recall, this was also required in the construction of the graphical rose plot. In general, the choice of $\mathbf{\Lambda}$ depends on a number of factors, including the configuration of the data set, the goal of the study, and the underlying physical/biological phenomena of interest. For regularly spaced observations, note that the two components of any lag $\mathbf{h}$, h_x and h_y, should both be integers so that the estimated variogram, $2\widehat{\gamma}(\cdot)$, can be calculated. Thus, only lags with integer components can be included in $\mathbf{\Lambda}$. For example, we can take $\mathbf{\Lambda} = \{(1, 0), (0, 1), (1, 1), (-1, 1)\}$ if we desire to compare the sample variograms in two directions, at each of two distances. For irregularly spaced observations, in principle, sample variogram values at any lags can be calculated, and thus more options for $\mathbf{\Lambda}$ are available. Beyond these general statements, however, there is no unique rule for choosing $\mathbf{\Lambda}$, for either regularly spaced or irregularly spaced observations. Nevertheless, it is worthwhile to make some general recommendations.

In considering the choice of lag sets to evaluate the variogram, smaller lags are often preferable to larger ones. This is due to two principle observations. Often $2\widehat{\gamma}(\cdot)$ for large $\mathbf{h}$ is based on fewer observations than estimates at smaller lags, and is consequently more variable. Secondly, as has been seen, observations at smaller lags are usually more correlated, and identification of the correlation at these lags is more important for effective spatial prediction. On the other hand, in order to detect certain types of departures from isotropy, a sufficient number of lags is necessary. For example, the anisotropy associated with matrix **B2** in Section 5.6.5 would not be detected if the variogram values are compared only at lags $(1, 1)$ and $(-1, 1)$, because $2\gamma(1, 1) = 2\gamma(-1, 1)$ for the given anisotropic covariance function. Comparing lags $(1, 0)$ vs. $(0, 1)$ *and* $(1, 1)$ vs. $(-1, 1)$ simultaneously, the user can better detect departures from isotropy induced by any of the matrices **B2**, **B3**, **B4**, or **B5** in Section 5.6.5, under sufficiently strong correlation. It is also the case, however, that a relatively large number of spatial observations is necessary to assess isotropy for a large number of lags. Numerical evidence demonstrates the validity of an increasing number of lags for increasingly large data grids. This can be seen, for example, in the experiments in Section 5.7.

In addition to the size of the data grid, the choice of the spatial testing lags should rely on knowledge of the underlying physical/biological process generating the observations. For example, wind is a natural possible source of anisotropy in an air pollution study. Thus a natural choice for $\mathbf{\Lambda}$ is to include lags in the major wind direction and those perpendicular to that direction.

5.6.4 Test statistics

Recall that the hypothesis of isotropy can be expressed as H_0 : $2\gamma(\mathbf{h}) = 2\gamma_0(\|\mathbf{h}\|)$ for some function $2\gamma_0(\cdot)$, where $\|\mathbf{h}\| = \sqrt{\mathbf{h}^T\mathbf{h}}$. It can be rewritten, in terms of variogram values at lags belonging to $\mathbf{\Lambda}$, as

$$H_0 : 2\gamma(\mathbf{h}_1) = 2\gamma(\mathbf{h}_2), \quad \mathbf{h}_1, \mathbf{h}_2 \in \mathbf{\Lambda}, \mathbf{h}_1 \neq \mathbf{h}_2, \quad \text{but } \|\mathbf{h}_1\| = \|\mathbf{h}_2\|.$$

Thus, under the hypothesis of isotropy, there exists a full row rank matrix $\mathbf{A}$ such that $\mathbf{AG} = \mathbf{0}$. This observation was made in a preprint by Lu and

Zimmerman (2001). For example, if $\mathbf{\Lambda} = \{(1,0), (0,1), (1,1), (-1,1)\}$, that is, $\mathbf{G} = \{2\gamma(1,0), 2\gamma(0,1), 2\gamma(1,1), 2\gamma(-1,1)\}^{\mathrm{T}}$, then we may set

$$\mathbf{A} = \begin{bmatrix} 1 & -1 & 0 & 0 \\ 0 & 0 & 1 & -1 \end{bmatrix}.$$

Here and henceforth, let d denote the row rank of the contrast matrix $\mathbf{A}$. For regularly spaced observations, it follows from Theorem 5.6.1, that if H_0 is true, then

$$|D_n| \times (\mathbf{A}\widehat{\mathbf{G}}_{\mathbf{n}})^{\mathrm{T}} (\mathbf{A}\mathbf{\Sigma}_{\mathbf{R}} A^{\mathrm{T}})^{-1} (\mathbf{A}\widehat{\mathbf{G}}_{\mathbf{n}}) \xrightarrow{D} \chi^2_d \quad \text{as } n \to \infty.$$

The covariance matrix $\mathbf{\Sigma}_{\mathbf{R}}$ is estimated using a subsampling estimator $\widehat{\mathbf{\Sigma}}_{\mathbf{R,n}}$, which is discussed in Chapter 10. Then, the test statistic is:

$$TS_{\mathrm{R},n} \equiv |D_n| \times (\mathbf{A}\widehat{\mathbf{G}}_{\mathbf{n}})^{\mathrm{T}} (\mathbf{A}\widehat{\mathbf{\Sigma}}_{\mathbf{R,n}} A^{\mathrm{T}})^{-1} (\mathbf{A}\widehat{\mathbf{G}}_{\mathbf{n}}).$$

Since $\widehat{\mathbf{\Sigma}}_{\mathbf{R,n}} \xrightarrow{L_2} \mathbf{\Sigma}_{\mathbf{R}}$, $TS_{\mathrm{R},n} \xrightarrow{D} \chi^2_d$ as $n \to \infty$, by the multivariate Slutsky's theorem [Ferguson (1996)].

The above result suggests that, for regularly spaced observations, an approximate size-α test for isotropy is to reject H_0 if $TS_{\mathrm{R},n}$ is bigger than $\chi^2_{d,\alpha}$; that is, the upper α percentage point of a χ^2 distribution with d degrees of freedom. The convergence, however, can be somewhat slow. Another possibility is to obtain p-values using subsampling as discussed in Politis and Romano (1994), see also Guan *et al.* (2004).

For irregularly spaced observations, $\mathrm{E}(\mathbf{A}\widehat{\mathbf{G}}) = 0$, under H_0. The above test statistic can be extended naturally as follows:

$$TS_{\mathrm{IR},n} \equiv |D_n| \delta_n^2 (\mathbf{A}\widehat{\mathbf{G}}_{\mathbf{n}})^{\mathrm{T}} (\mathbf{A}\widehat{\mathbf{\Sigma}}_{\mathbf{IR,n}} \mathbf{A}^{\mathrm{T}})^{-1} (\mathbf{A}\widehat{\mathbf{G}}_{\mathbf{n}}),$$

where $\widehat{\mathbf{\Sigma}}_{\mathbf{IR,n}}$ is a subsampling variance estimator, and δ_n is the bandwidth defined in Equation 5.6. Similarly,

$$TS_{\mathrm{IR},n} \xrightarrow{D} \chi^2_d, \quad \text{as } n \to \infty.$$

Thus the limiting χ^2 approach and the subsampling approach can both be applied in the irregularly spaced case, with the understanding that the test now is based on $TS_{\mathrm{IR},n}$. The approach of $TS_{\mathrm{IR},n}$ to the asymptotic distribution is relatively rapid for irregularly spaced observations, so subsampling to estimate p-values is less necessary than in the equally spaced setting.

5.6.5 Numerical results

To study the performance of the nonparametric isotropy test for regular spaced observations, consider the wind-speed data. The wind-speed data consist of the east–west component of the wind speed, in meters per second, over a region in the tropical western Pacific Ocean (145° E–175° E, 14° S–16° N). The data are given on a regular spatio-temporal grid of 17×17 sites with a grid spacing of about 210 km, and temporal spacing of six hours, for the period November 1992 through February 1993. This yields data at 289 locations and 480 time points. Cressie and Huang (1999) examined the second-order stationary assumption, and did not find evidence against it. They further fitted a stationary spatio-temporal variogram to the data with the spatial component being isotropic.

We next assess the validity of the isotropy assumption for these data. To visually investigate isotropy, plots of the empirical variograms at spatial lags from 1 to 11 in the E–W direction and N–S direction were inspected. This is a natural choice of directions to study due to the directional nature of the observations. The empirical variogram at each lag was calculated by averaging all the sample variograms over the 480 time periods. Some interesting features of the plot include that the empirical variogram is approximately linear in the N–S direction, reaching a value of approximately 26; a sill exists around 13 in the E–W direction. There is a closer agreement at small lags in these two directions. This suggests a possible lack of isotropy.

The test was applied using $\mathbf{\Lambda} = \{(1, 0), (0, 1)\}$ and 4×4 subblocks (the subblocks are as described in Section 10.6), to estimate $\mathbf{\Sigma_R}$ for each of the 480 time points separately. The summary of the results is as follows: 26.3 percent of the p-values are less than 0.01, 41.0 percent are less than 0.05, and 51.7 percent are less than 0.1. Thus, the isotropic assumption does not appear to be very reasonable for these data. By comparing the actual sample variogram values, it was seen that slightly more than 77 percent of these sample variogram values in the N–S direction are larger than those in the E–W direction. This provides further evidence that the true variogram value in the N–S direction is typically larger than that in the E–W direction.

For irregularly spaced observations, consider the following experiment. First, the covariance model is as follows. Consider realizations from a zero-mean, second-order stationary Gaussian random field. Each random field was either isotropic or geometrically anisotropic with the following covariance structure:

$$C(r;m) = \begin{cases} \sigma^2\left(1 - \dfrac{3r}{2m} + \dfrac{r^3}{2m^3}\right) & \text{if } 0 \leq r \leq m \\ 0 \quad \text{otherwise,} \end{cases}$$

where $r = \sqrt{\mathbf{h}^{\mathrm{T}}\mathbf{B}\mathbf{h}}$, and $\mathbf{B}$ is a 2×2 positive definite matrix as defined in Section 5.1. The covariance model is the spherical model discussed in Section 3.3. The parameter σ^2 is a scale parameter which is set equal to 1.0. The parameter m defines the range and strength of dependence. The range, m, is taken to be 2, 5, and 8, denoting weak, moderate, and relatively strong spatial dependence, respectively.

The following five $\mathbf{B}$ matrices were examined for their effects on the testing:

$$\mathbf{B1} = \begin{bmatrix} 1 & 0 \\ 0 & 1 \end{bmatrix}, \quad \mathbf{B2} = \begin{bmatrix} 1 & 0 \\ 0 & 2 \end{bmatrix}, \quad \mathbf{B3} = \begin{bmatrix} 1 & 0 \\ 0 & 4 \end{bmatrix},$$

$$\mathbf{B4} = \begin{bmatrix} 1.5 & -0.5 \\ -0.5 & 1.5 \end{bmatrix}, \quad \mathbf{B5} = \begin{bmatrix} 2.5 & -1.5 \\ -1.5 & 2.5 \end{bmatrix}.$$

The matrix $\mathbf{B1}$ yields an isotropic variogram, so this can used to evaluate the size of the test. The matrices $\mathbf{B2}$, $\mathbf{B3}$, $\mathbf{B4}$, and $\mathbf{B5}$ yield geometrically anisotropic random fields as discussed in Section 5.1. More specifically, the main anisotropic axes are aligned with the (x, y) axes for $\mathbf{B2}$ and $\mathbf{B3}$ (i.e., the rotation matrix $\mathbf{R} = \mathbf{I}$), but are oriented at the 45 and 135 degree angles with respect to the x axis for $\mathbf{B4}$ and $\mathbf{B5}$.

In the notation of Section 5.1, the rotation matrix is

$$\mathbf{R} = \begin{bmatrix} \cos(\pi/4) & \sin(\pi/4) \\ -\sin(\pi/4) & \cos(\pi/4) \end{bmatrix}$$

for matrices $\mathbf{B4}$ and $\mathbf{B5}$. In addition, the anisotropy ratio, the ratio of the lengths of the main axes, is $\sqrt{2} : 1$ for $\mathbf{B2}$ and $\mathbf{B4}$ but $2 : 1$ for $\mathbf{B3}$ and $\mathbf{B5}$. In the notation of section 5.1,

$$\mathbf{T} = \begin{bmatrix} 1 & 0 \\ 0 & \sqrt{2} \end{bmatrix}, \quad \text{and} \quad \mathbf{T} = \begin{bmatrix} 1 & 0 \\ 0 & 2 \end{bmatrix},$$

in the two cases.

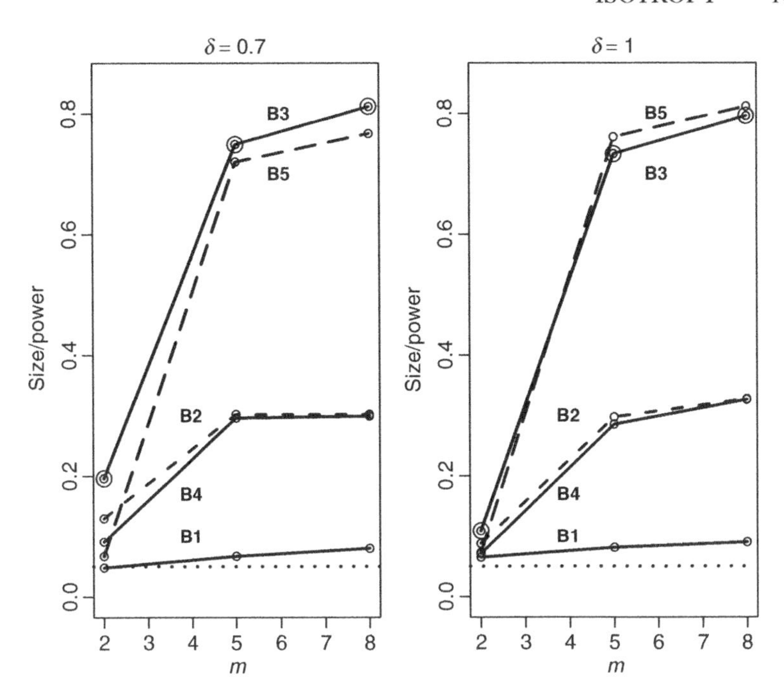

Figure 5.6 *Assessing isotropy in the unequally spaced locations setting. Given are the proportion of rejections at* $\alpha = 0.05$ *using the test statistic* $\mathrm{TS}_{\mathrm{IR},n}$ *in Section 5.6.4. The **B1** curve gives the sizes under isotropy, while **B2–B5** give the powers under anisotropic alternatives. The parameter m determines the range of correlation.*

The number of points for each realization was generated according to a Poisson distribution with an intensity parameter of $\nu = 1$ (per unit square) on a 20×20 square field. In other words, the expected number of points is 400. Given the outcome of the number of points, point locations were generated independently from a uniform distribution on the 20×20 square. Given these locations, the observations were then generated from a Gaussian process with the appropriate covariance structure. One thousand realizations were simulated for each choice of $\mathbf{B}$ and m. The lag set $\mathbf{\Lambda}$ is taken to be $\{(1, 0), (0, 1), (1, 1), (-1, 1)\}$. Two separate smoothing bandwidths in the variogram estimator, namely $\delta = 0.7$ and $\delta = 1.0$, were employed on each realization.

From Figure 5.6 we can see the following results:

i. All testing sizes are between approximately 0.05 and 0.10 (in fact between 0.048 and 0.081). This shows reasonably good agreement

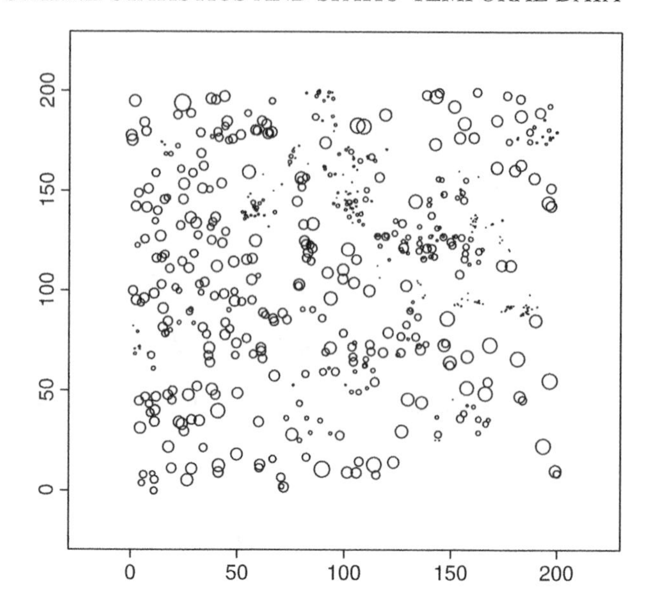

Figure 5.7 The longleaf pine data. The center of each circle gives the location of each of the 584 trees. The radius of each circle is proportional to the diameter of the tree at breast height.

with the nominal values of 0.05, although there is a somewhat inflated type-I error.

ii. As m increases, the power of all tests increase. This makes sense, as larger m corresponds to heavier correlation. The test should be more powerful in this situation.

iii. The power of the test is better for **B3** than **B2**, and better for **B5** than for **B4**. This is to be expected: in both cases the degree of anisotropy is stronger, and thus more easily detected.

iv. The power depends on bandwidth, in general. Results here are slightly better for $\delta = 0.7$ than for $\delta = 1.0$. The former gives approximately one square difference per window. Results for $\delta = 0.4$ are roughly comparable to those for $\delta = 1.0$.

We have observed the behavior of isotropy assessments in the case of equally spaced observations, and for unequally spaced locations determined by a homogeneous Poisson process. Useful as these results are, there are some situations where locations are not plausibly modeled by a homogeneous Poisson process. For example, consider the locations of longleaf pine trees given in Figure 5.7. Using the methodology in

Chapter 7, we find that the value of the standardized Donnelly's test statistic to determine random locations is very significant, suggesting that the locations are clustered beyond what we should expect under a homogeneous Poisson process. Not only are the locations clustered, but a visual inspection suggests that the process generating the tree locations may not be stationary. In particular, there appears to be a higher intensity of trees at the top (i.e., in the east), than at the bottom (in the west). It is desirable to have an assessment of isotropy for these types of data locations. This is the topic of the following section.

5.7 Assessment of isotropy for general sampling designs

5.7.1 A stochastic sampling design

Let $\{Z(\mathbf{s}) : \mathbf{s} \in R^2\}$ be a spatial random field. In order to proceed we need to describe a mechanism by which locations are determined. Assume that the locations at which $Z(\mathbf{s})$ is generated are random and follow a stochastic sampling design. Specifically, consider a general class of stochastic designs as given in, for instance, Lahiri and Zhu (2006). Specifically, the locations are determined as follows.

As in the previous section, let A be an open connected subset of $(0, 1]^2$. Then the sampling domain A_n is:

$$A_n = \lambda_n A,$$

where $\{\lambda_n\}$ is a sequence of positive numbers such that $\lambda_n = o(n^{-\epsilon})$ for some $\epsilon > 0$. In other words, λ_n plays the role of a scaling factor while keeping the shape of A_n the same for different values of n. Let $f(x)$ be a probability density function on A, and let $\mathbf{X}_1, \ldots, \mathbf{X}_n$ be a sequence of independent and identically distributed (bivariate) random vectors with density $f(x)$. The assumption is that the sampling sites $\mathbf{s}_1, \ldots, \mathbf{s}_n$ are obtained from a realization $\mathbf{x}_1, \ldots, \mathbf{x}_n$ of the random vectors $\mathbf{X}_1, \ldots, \mathbf{X}_n$, following the relation, for $i = 1, \ldots, n$,

$$\mathbf{s}_i = \lambda_n \mathbf{x}_i.$$

The sample $\{Z(\mathbf{s}_1), \ldots, Z(\mathbf{s}_n)\}$ is observed along with the sampling sites. This is the so-called stochastic design as opposed to the common equally spaced assumption on the sampling sites. Note that this formulation allows for a large variety of sampling locations. For example, if $f(\mathbf{x})$

is the uniform density, then the sampling locations follow a homogeneous Poisson process considered previously. For nonuniform densities, however, we have locations according to an inhomogeneous Poisson process, as described in Chapter 7, with intensity function given by $f(\mathbf{x})$.

5.7.2 Covariogram estimation and asymptotic properties

The testing procedure, which is described in Section 5.7.3, requires estimation of the covariance function $C(\mathbf{h})$ for a given set of lag vectors $\mathbf{h}_1, \ldots, \mathbf{h}_K \in R^2$. For a given lag vector $\mathbf{h}$, $C(\mathbf{h})$ is estimated using a nonparametric kernel method similar to Hall and Patil (1994). This is analogous to the kernel estimator for the variogram defined in Section 3.1.2. Let $w(\mathbf{h})$ denote a kernel function, and let h be a bandwidth. Let $\overline{Z} = n^{-1} \sum Z(\mathbf{s}_i)$, $Z_{ij} = [Z(\mathbf{s}_i) - \overline{Z}][Z(\mathbf{s}_j) - \overline{Z}]$ and $\mathbf{s}_{ij} = \mathbf{s}_i - \mathbf{s}_j$. Then, $C(\mathbf{h})$ is estimated by

$$\widehat{C}(\mathbf{h}) = \left\{ \sum_{i,j} w[(\mathbf{h} - \mathbf{s}_{ij})/h] \right\}^{-1} \left\{ \sum_{i,j} w[(\mathbf{h} - \mathbf{s}_{ij})/h] Z_{ij} \right\}.$$

Under regularity conditions, Hall and Patil (1994) showed that, for any given $c > 0$,

$$\lambda_n^2 \int_{\|\mathbf{h}\| \leq c} \{\widehat{C}(\mathbf{h}) - C(\mathbf{h})\}^2 \, d\mathbf{h} \to \int_{|t| \leq c} Y^2(\mathbf{h}) \, d\mathbf{h}$$

in distribution, where $Y(\mathbf{h})$ is a nonstationary Gaussian random field. For testing purposes, however, the *joint* distribution of $\widehat{C}(\mathbf{h}_1), \ldots, \widehat{C}(\mathbf{h}_K)$ is required.

For a given set of lag vectors $\mathbf{h}_1, \ldots, \mathbf{h}_K \in R^2$, let $G = [C(\mathbf{h}_1), \ldots, C(\mathbf{h}_K)]$ denote the vector of true values of the covariogram and define

$$\widehat{G}_n = [\widehat{C}(\mathbf{h}_1), \ldots, \widehat{C}(\mathbf{h}_K)]$$

to be the nonparametric estimates. We have the following result.

Theorem 5.7.1 *Under mild regularity conditions, we have*

$$\lambda_n(\widehat{G}_n - G) \to normal(\mathbf{0}, \mathbf{V})$$

in distribution as $n \to \infty$, where $\mathbf{V}$ *is the* $K \times K$ *covariance matrix of* $\widehat{G}_n$ *with the* (k, k')*th element given by*

$$V_{kk'} = c(f) \int E\{[Z(\mathbf{0})Z(\mathbf{h}_k) - C(\mathbf{h}_k)][Z(\boldsymbol{u})Z(\boldsymbol{u} + \mathbf{h}_{k'}) - C(\mathbf{h}_{k'})]\} d\boldsymbol{u}, \tag{5.9}$$

where $c(f) = (\int f^2)^{-2} \int f^4$.

We have stated Theorem 5.7.1 specifically assuming that the lag vectors $\mathbf{h}_1, \ldots, \mathbf{h}_K \in R^2$. However, a more general result can be stated for $\mathbf{h}_1, \ldots, \mathbf{h}_K \in R^d$, $d \geq 2$, as well. In fact, for general dimension d, the result will stay the same except that the scaling factor of $(\widehat{G}_n - G)$ changes to $\lambda_n^{d/2}$.

5.7.3 Testing for spatial isotropy

Recall that, given the observed sample from a random field $\{Z(\mathbf{s}) : \mathbf{s} \in R^2\}$, we want to test

$$H_0 : C(\mathbf{h}) = C(\|\mathbf{h}\|) \quad \text{for all } \mathbf{h};$$

that is, the covariance depends only on the length of the lag vector $\|\mathbf{h}\|$ between the two locations and not on the orientation.

To test H_0, the null hypothesis of isotropy, we again utilize Lu and Zimmerman's formulation, but in a general setting where sample locations are generated from a density $f(\cdot)$. We propose to compare values of $C(\mathbf{h})$ for $\mathbf{h} \in \mathbf{\Lambda}$, where $\mathbf{\Lambda} = \{\mathbf{h}_1, \ldots, \mathbf{h}_K\}$ is a set of user-chosen spatial lags. In general, we again test for

$$\mathbf{A}G = 0,$$

where $G = \{C(\mathbf{h}) : \mathbf{h} \in \mathbf{\Lambda}\}$ and $\mathbf{A}$ is a known matrix. For example, as in the equally spaced locations case, we may take $\mathbf{\Lambda} = \{(0, 1), (1, 0), (1, 2), (2, 1)\}$ and test for

$$C\{(0, 1)\} - C\{(1, 0)\} = 0 \quad \text{and} \quad C\{(1, 2)\} - C\{(2, 1)\} = 0.$$

In this case,

$$A = \begin{bmatrix} 1 & -1 & 0 & 0 \\ 0 & 0 & 1 & -1 \end{bmatrix}.$$

Using Theorem 5.7.1, we have that

$$\lambda_n \mathbf{A}(\widehat{G}_n - G) \to \mathrm{N}(\mathbf{0}, \mathbf{AVA}^{\mathrm{T}}),$$

in distribution. We propose to use the test statistic

$$T_n = (\mathbf{A}\widehat{G}_n)^{\mathrm{T}}(\mathbf{A}\widehat{\mathbf{V}}_0\mathbf{A}^{\mathrm{T}})^{-1}(\mathbf{A}\widehat{G}_n),$$

where $\widehat{\mathbf{V}}_0$ is an estimate of $\lambda_n^{-2}\mathbf{V}$.

Utilizing the asymptotic formula for $\mathbf{V}$ to construct $\widehat{\mathbf{V}}_0$ is not a simple matter, since the density $f(\cdot)$, and the inflation factor λ_n, are generally unknown. For this reason, we consider bootstrap methods.

There is a large literature in model-free bootstrapping and subsampling for temporally and spatially correlated observations. Künsch (1989) and Liu and Singh (1992) proposed the block bootstrap for stationary time series, while Carlstein (1986) justified subsampling for variance estimation in this setting. Subsampling methodology for equally spaced spatial data has been developed in Possolo (1991), Politis and Romano (1994), Sherman (1996), and Lahiri *et al.* (1999). Politis and Sherman (2001), and Politis *et al.* (1998) considered unequally spaced data with locations generated by stationary processes. Many of these methods are described in Chapter 10. In the current setting, with locations generated by a general $f(\cdot)$, these methods may not be reliable.

Instead, consider a particular spatial bootstrap, developed by Lahiri and Zhu (2006) to construct $\widehat{\mathbf{V}}_0$. The grid-based block bootstrap (GBBB) allows for nonuniformity of the spatial sampling design and nonstationarity of the underlying spatial process. Specifically, the GBBB algorithm is carried out on the data set B times. For $b = 1, \ldots, B$, let $\widehat{G}_{n,b}$ denote the estimate of G based on the bth bootstrapped data set. Then $\widehat{\mathbf{V}}_0$ is set equal to the sample covariance of the bootstrap replicates $\{\widehat{G}_{n,1}, \ldots, \widehat{G}_{n,B}\}$. The effectiveness of this procedure does depend on the choice of block size; the effects of this choice are examined in the numerical experiment in the following section and in Section 10.6.

Given the consistency of $\lambda_n^2\widehat{\mathbf{V}}_0$ for $\mathbf{V}$ we have, under the null hypothesis, that T_n has a large sample χ^2_ν distribution, where ν denotes the row rank of the matrix $\mathbf{A}$. Hence we can construct p-values using χ^2_ν critical values.

5.7.4 Numerical results for general spatial designs

In this section, we present a study to assess the performance of the proposed test. Realizations from a zero-mean second-order stationary Gaussian random field are generated using a specified covariance function. We consider two different covariance functions:

$$C_1(\mathbf{h}) = \exp(-r);$$
$$C_2(\mathbf{h}) = \exp(-r/8 - r^2/8),$$

with $r = (\mathbf{h}^{\mathrm{T}}\mathbf{B}\mathbf{h})^{1/2}$, where $\mathbf{B}$ is a 2×2 positive definite rotation matrix inducing geometric anisotropy in the random field, as discussed in Section 5.1. The graphs of $C_1(r)$ and $C_2(r)$ for different values of r in $(0, 8)$ are given in Figure 5.8.

The matrix $\mathbf{B}$ is determined by alignment of the main anisotropic axes compared to the x–y axes (counter clockwise rotation angle) and the anisotropic ratio (the ratio of lengths of the main axes). Let $\mathbf{I}_k$ denote the $k \times k$ identity matrix, and let $\mathbf{J}_k$ denote the $k \times k$ matrix with all entries equal to one. Consider the following seven different choices of $\mathbf{B}$:

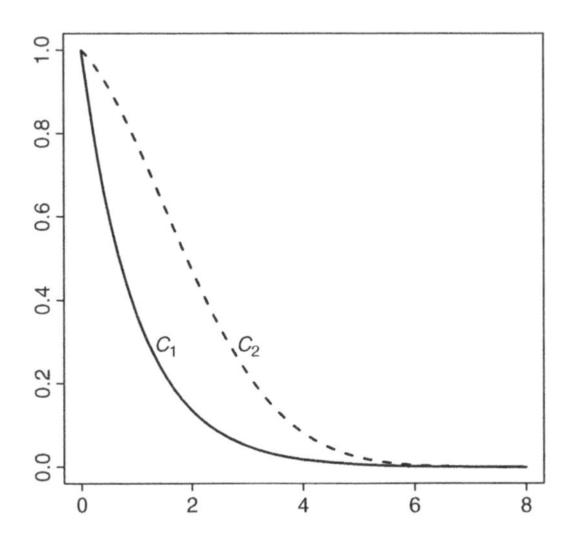

Figure 5.8 Two covariance functions used in the numerical experiment for assessing isotropy from nonuniform spatial locations. $C_1(\mathbf{h}) = \exp(-r)$ *and* $C_2(\mathbf{h}) = \exp(-r/8 - r^2/8)$.

C1: Isotropic (rotation = 0, anisotropic ratio = 1). $\mathbf{B} = \mathbf{I}_2$.

C2: Rotation $= 0$, anisotropic ratio = $\sqrt{2}$. $\mathbf{B} = \text{diag}(1, 2)$.

C3: Rotation $= 3\pi/4$, anisotropic ratio $=\sqrt{2}$. $\mathbf{B} = 2\mathbf{I}_2 - 0.5\mathbf{J}_2$.

C4: Rotation $= 0$, anisotropic ratio = 2. $\mathbf{B} = \text{diag}(1, 4)$.

C5: Rotation $= 3\pi/4$, anisotropic ratio = 2. $\mathbf{B} = 4\mathbf{I}_2 - 1.5\mathbf{J}_2$.

C6: Rotation $= 0$, anisotropic ratio = 3. $\mathbf{B} = \text{diag}(1, 9)$.

C7: Rotation $= 3\pi/4$, anisotropic ratio = 3. $\mathbf{B} = 9\mathbf{I}_2 - 4\mathbf{J}_2$.

The sample locations, $\mathbf{s} = (x, y)$, are generated according to the density $f_{xy} = f(x)f(y)$. We consider two different choices of design density $f(\cdot)$:

Uniform: $f(\cdot) = U(0, 1)$;

Truncated normal mixture: $f(\cdot) = 0.5TN\left(\mu = \frac{1}{2}, \sigma = \frac{1}{2}, \ell = 0, u = 1\right)$ $+ 0.5TN\left(\mu = \frac{3}{4}, \sigma = \sqrt{2}, \ell = 0, u = 1\right)$,

where $TN(\mu, \sigma, \ell, u)$ denotes a normal density with mean μ and standard deviation σ truncated in $[\ell, u]$. We use two combinations of sample sizes and scaling factors: (i) $n = 350, \lambda = 6$; and (ii) $n = 600, \lambda = 8$. These two situations allow for approximately the same number of observations per unit area.

For each of the above combinations, a test statistic is constructed based on three sets of spatial lags:

$$\begin{aligned} \mathbf{\Lambda}_1 &= \{(1, 0), (0, 1)\}; \\ \mathbf{\Lambda}_2 &= \left\{(1, 0), (0, 1), (1, 1)/\sqrt{2}, (1, -1)/\sqrt{2}\right\}; \qquad (5.10) \\ \mathbf{\Lambda}_3 &= \left\{(1, 0), (0, 1), (1, 1)/\sqrt{2}, (1, -1)/\sqrt{2}, (2, 1)/\sqrt{5}, (1, 2)/\sqrt{5}, \right. \\ &\qquad \left. (-2, 1)\sqrt{5}, (-1, 2)/\sqrt{5}\right\}, \end{aligned}$$

and compare $(1, 0)$ with $(0, 1)$; $(1, 1)/\sqrt{2}$ with $(1, -1)/\sqrt{2}$; $(2, 1)/\sqrt{5}$ with $(-1, 2)/\sqrt{5}$; and $(1, 2)/\sqrt{5}$ with $(-2, 1)/\sqrt{5}$.

We employ the product kernel

$$w[(s_1, s_2) - (x, y); (h_x, h_y)] = w_0[(s_1 - x)/h_x]w_0[(s_2 - y)/h_y]$$

to estimate the covariance function, where $w_0(\cdot)$ denotes the Epanechnikov kernel, and use bandwidth $h_x = \kappa\, std(D_x)N^{-1/5}$, and

$h_y = \kappa\, std(D_y) N^{-1/5}$ where *std* (**x**) is the standard deviation of **x**, D_x and D_y denote the differences between x- and y-coordinates, respectively, N denotes the total number of such differences, and κ is a scaling constant. The results displayed here correspond to $\kappa = 1$. We used block length $b_n = 2$ for performing the spatial resampling, and generated $B = 200$ bootstrap resamples for each generated random field. Although these choices of bandwidth and block length are not necessarily optimal, they give reasonable results, as will be seen. The problems of finding the optimal bandwidth in covariance function estimation and finding the optimal block length in the GBBB spatial bootstrap, are open problems at this time. We discuss block length determination for various data structures in Chapter 10.

The proportions of rejections at a nominal level of 0.05 for our test based on 500 runs for each setting are given in Table 5.2, for the cases C1–C7, for the sample sizes $n = 350, \lambda = 6$, and $n = 600, \lambda = 8$, and uniform and truncated normal design densities. We display the results for only the covariance function $C_1(\cdot)$; the results for $C_2(\cdot)$ show slightly larger power in the case of $C_2(\cdot)$ being the true covariance. This suggests that the test is not overly sensitive to the particular form of the underlying covariance function.

Several key observations can be made from the results. To start with, the level of the test, regardless of the lag set being used, remains similar and is reasonably close to the nominal level.

The power of the test using only two lags ($\mathbf{\Lambda}_1$) is relatively good, when there is no rotation present (cases C2, C4, C6), but suffers badly when there is rotation of the main anisotropic axis (cases C3, C5, and C7). This problem does not occur for the tests using four ($\mathbf{\Lambda}_2$) or eight ($\mathbf{\Lambda}_3$) spatial lags. For these last two lag sets, the powers increase as the anisotropic ratio increases. In fact, the powers for tests using four ($\mathbf{\Lambda}_2$) and eight ($\mathbf{\Lambda}_3$) lags are comparable to the test based on $\mathbf{\Lambda}_1$, even when there is no rotation present. In addition, the power of all tests increases significantly as the sample size increases, except for the lag set $\mathbf{\Lambda}_1$ in cases C3, C5, and C7, when this choice has essentially no power beyond the size of the test. These conclusions hold in both the cases of uniform and truncated normal spatial design densities. This suggests that the performance of the test is not overly sensitive to the specific distribution of spatial locations.

5.7.5 Effect of bandwidth and block size choice

To assess the impact of bandwidth choice numerically, consider the truncated normal design with $n = 600, \lambda_n = 8$, and $h_x = \kappa\, std(D_x) N^{-1/5}$,

Table 5.2 Results of the numerical experiment described in Section 5.7.4.

	Uniform design					
	$n = 350, \lambda = 6$			$n = 600, \lambda = 8$		
	$\mathbf{\Lambda}_1$	$\mathbf{\Lambda}_2$	$\mathbf{\Lambda}_3$	$\mathbf{\Lambda}_1$	$\mathbf{\Lambda}_2$	$\mathbf{\Lambda}_3$
C1	0.044	0.066	0.054	0.044	0.068	0.048
C2	0.102	0.118	0.104	0.164	0.156	0.132
C3	0.052	0.136	0.130	0.080	0.222	0.250
C4	0.340	0.244	0.176	0.466	0.372	0.358
C5	0.058	0.328	0.264	0.068	0.506	0.454
C6	0.518	0.386	0.346	0.714	0.570	0.516
C7	0.056	0.438	0.386	0.056	0.606	0.596
	Truncated normal design					
	$n = 350, \lambda = 6$			$n = 600, \lambda = 8$		
	$\mathbf{\Lambda}_1$	$\mathbf{\Lambda}_2$	$\mathbf{\Lambda}_3$	$\mathbf{\Lambda}_1$	$\mathbf{\Lambda}_2$	$\mathbf{\Lambda}_3$
C1	0.053	0.065	0.060	0.036	0.056	0.051
C2	0.128	0.110	0.138	0.174	0.154	0.138
C3	0.052	0.140	0.134	0.080	0.234	0.204
C4	0.294	0.248	0.236	0.444	0.306	0.294
C5	0.068	0.326	0.304	0.048	0.470	0.406
C6	0.486	0.406	0.428	0.632	0.534	0.524
C7	0.074	0.482	0.438	0.076	0.606	0.582

Displayed are the proportions of rejections of the tests based on lag sets $\mathbf{\Lambda}_1$, $\mathbf{\Lambda}_2$, and $\mathbf{\Lambda}_3$ for the cases C1–C7 for sample sizes $n = 350, \lambda = 6$, and $n = 600, \lambda = 8$, and uniform and truncated normal design densities. All values are computed based on the nominal level of 0.05.

and $h_y = \kappa\, std(D_y)N^{-1/5}$, where D_x and D_y denote the differences between the x-coordinates and the y-coordinates, respectively, N denotes the total number of such differences, and consider four scaling constants, namely $\kappa = 0.5, 1.0, 1.5, 2.0$. The block size is fixed at $b_n = 2$. For each value of κ, the power of the three test statistics is computed for the cases C1 (isotropic) and C5. The results are displayed in Table 5.3. It is seen that the level of the test remains close to the nominal level over the range

Table 5.3 The effects of bandwidth in the numerical experiment described in Section 5.7.5.

	Case: C1			Case: C5		
κ	Λ_1	Λ_2	Λ_3	Λ_1	Λ_2	Λ_3
0.5	0.036	0.064	0.086	0.016	0.282	0.322
1.0	0.036	0.056	0.051	0.048	0.470	0.406
1.5	0.044	0.066	0.066	0.060	0.550	0.464
2.0	0.050	0.066	0.076	0.036	0.574	0.484

Displayed are the proportions of rejections for different bandwidth choices corresponding to different values of κ for the cases C1 and C5 under the truncated normal design with $n = 600, \lambda = 8$. All values are computed based on the nominal level of 0.05.

Table 5.4 The effects of block length in the numerical experiment described in Section 5.7.5.

	Case: C1			Case: C5			Case: C7		
b_n	Λ_1	Λ_2	Λ_3	Λ_1	Λ_2	Λ_3	Λ_1	Λ_2	Λ_3
1	0.026	0.044	0.054	0.044	0.444	0.396	0.036	0.686	0.648
2	0.036	0.056	0.051	0.048	0.470	0.406	0.076	0.656	0.608
4	0.116	0.180	0.206	0.102	0.574	0.620	0.092	0.746	0.764

Displayed are the proportions of rejections for different block length choices (b_n) for the cases C1, C5, and C7 under the truncated normal design with $n = 600, \lambda = 8$. All values are computed based on the nominal level of 0.05.

of bandwidths considered. For the case C5, the power of the tests remains similar for $\kappa = 1, 1.5$, and 2, but decreases for $\kappa = 0.5$. This suggests that excessive undersmoothing is not a favorable option in this case.

To assess the effect of block length choice, fix $\kappa = 1$ and consider different values of $b_n = 1, 2$, and 4. The power of the three test statistics is computed for the cases C1, C5, and C7, for each value of b_n. The results are displayed in Table 5.4. Notice that for $b_n = 1$ and 2, the level of the test remains close to the nominal level for the case C1, and the power of the tests remains similar for C5 and for C7. However, for $b_n = 4$, the level of the test increases significantly. This suggests that choosing an overly large block length is detrimental to the level of the test.

5.8 An assessment of isotropy for the longleaf pine sizes

We test for isotropy in the data set on longleaf pine trees in an old-growth forest in the southern United States. This data set contains the diameters of 584 longleaf pine trees at breast height, along with their locations in a 200×200 domain. A bubble plot of the data set is displayed in Figure 5.7. As discussed, standard methods for the analysis of point patterns strongly demonstrate that the locations of the trees do not follow a Poisson process. For example, this can be seen using the K-function or nearest neighbor distances (these methods are discussed in Chapter 7). Using either measure, strong clustering is clearly indicated. Further, inspection suggests that the density generating the locations appears to be somewhat nonuniform, with a higher intensity of trees at the top (in the east) of the grid than at the bottom (in the west).

The methodology can be applied to assess the assumption of isotropy for these data. Consider three sets of spatial lags:

$$\begin{aligned}
\mathbf{\Lambda}_1 &= \{(1,0),(0,1)\};\\
\mathbf{\Lambda}_2 &= \{(1,0),(0,1),(1,1),(1,-1)\};\\
\mathbf{\Lambda}_3 &= \{(1,0),(0,1),(1,1),(1,-1),(2,1),(1,2),(-2,1),(-1,2)\},
\end{aligned}$$

and we compare $(1,0)$ with $(0,1)$; $(1,1)$ with $(1,-1)$; $(2,1)$ with $(-1,2)$; and $(1,2)$ with $(-2,1)$. We employ the product kernel,

$$w\{(s_1,s_2)-(x,y);(h_x,h_y)\} = w_0\{(s_1-x)/h_x\}w_0\{(s_2-y)/h_y\},$$

to estimate the covariance function, where $w_0(\cdot)$ denotes the Epanechnikov kernel, and use bandwidths $h_x = std(D_x)N^{-1/5}$, and $h_y = std(D_y)N^{-1/5}$, with D_x and D_y denoting the differences between x- and y-coordinates, respectively, and N denoting total number of such differences. We consider different values of block length, namely $b_n = 10, 20$, and 40 for performing the spatial resampling, and for each block length, we generate 200 resamples. The p-values of our test are between 0.16 and 0.44 across all $3 \times 3 = 9$ combinations of block lengths and spatial lag sets. This shows that we cannot reject the null hypothesis of isotropy for this data set, and that this conclusion holds across a variety of reasonable testing parameters. Thus, in terms of the size of the trees, there is no significant differential covariance in different directions.

To assess the effect of bandwidth on the test results, different bandwidths of the form $h_x = k_x \cdot std(D_x)N^{-1/5}$, and $h_y = k_y \cdot std(D_y)N^{-1/5}$ have also been considered, where k_x and k_y range from 0.5 to 2.0. The p-values using each of these bandwidth combinations give an indication of their sensitivity to the bandwidth. It has been found that the results do depend on the bandwidth combination used in all three cases. The conclusion reached, however, of an acceptance of isotropy, remains over a broad range of bandwidths. Further, the dependence on bandwidth is greatly reduced when a larger number of lags are used.

In summary, it is important to account for anisotropy in covariance functions when present. Not accounting for anisotropy leads to inefficient predictions. Graphical diagnostics, like the rose plot, are useful to help assess the appropriateness of an isotropic covariance function. Although useful, graphical diagnostics are generally not conclusive and are open to interpretation. For this reason, it is worth considering more formal procedures to assess isotropy. Both parametric methods and nonparametric methods have been employed to assess isotropy. In the nonparametric setting, observations may be equally spaced, randomly irregularly spaced, or irregularly spaced with varying intensities. In the cases where isotropy is deemed to be inappropriate, geometric anisotropy is often appropriate. This allows for isotropic fitting on revised coordinate axes. If not, then separate covariance models need to be fitted in separate directions, as has been discussed in Section 5.3.

6

Space–time data

There is a growing interest in data that are indexed in both space and time. Many fields including geoscience, meteorology, and ecology generate data that have both spatial and temporal components. For example, the wind-speed data introduced in Section 5.6.5 were used to assess spatial isotropy at each of 480 time points. This analysis views the data spatially, but does not account for the temporal aspect inherent in the actual data. Some of the concepts in considering spatio-temporal observations are natural generalizations of those developed for purely spatial data. Many concepts, however, are new to the spatio-temporal context. For example, how should the temporal and spatial correlation models interact with each other? Can or should they interact with each other? We consider this particular question, in detail, in Sections 6.6 and 6.7.

6.1 Space–time observations

Generalizing from the purely spatial setting, let $\{Z(\mathbf{s}, t) : \mathbf{s} \in \mathbb{R}^d, t \in \mathbb{R}\}$ denote a strictly stationary space–time random field. In most cases, we consider the time index, t, to be a subset of $\mathbb{Z}^+$, the set of positive integers. The spatial locations are typically either a subset of $\mathbb{Z}^2$, the two-dimensional integer lattice, or of $\mathbb{R}^2$. In the former case, the observations are on a grid, while in the latter, the spatial locations are said to be

Spatial Statistics and Spatio-Temporal Data: Covariance Functions and Directional Properties Michael Sherman

irregularly spaced. The wind-speed data are of the former type, lying on a 17×17 grid with grid spacing equal to approximately 210 km. The temporal spacing is 6 hours between adjacent observation times.

In general, we observe data, $\mathbf{Z}_{n,T} := Z(\mathbf{s}_i, t_j)$ for $i = 1, \ldots, n$, and $j = 1, \ldots, T$. In the wind-speed data, we have $n = 17^2 = 289$, and $T = 480$. Using these data, we seek to address the same type of questions considered for purely spatial data. Namely, what is the relationship between wind speeds at different locations and how can we use the available observations to predict wind speeds at unsampled locations. Here there are substantial data in both space and time. Often, there is a large number of temporal observations, with relatively few spatial locations. For example, the Irish wind data, analyzed in Section 6.7.1, has $n = 11$ spatial locations where wind speed is measured daily over an 18-year period. When there is this wealth of temporal information, but limited spatial information, less general conclusions can be made concerning spatial relationships.

6.2 Spatio-temporal stationarity and spatio-temporal prediction

Let $\mu(\mathbf{s}, t)$ denote the mean of $Z(\mathbf{s}, t)$, and define the covariance between two space–time variables as:

$$Cov[Z(\mathbf{s}, u), Z(\mathbf{t}, v)] = E\{[Z(\mathbf{s}, u) - \mu(\mathbf{s}, u)][Z(\mathbf{t}, v) - \mu(\mathbf{t}, v)]\}.$$

Second-order stationarity (SOS) requires that for any locations, $\mathbf{s}$, and times, t, it holds that:

i. $E[Z(\mathbf{s}, t)] = \mu$ and

ii. $Cov[Z(\mathbf{s} + \mathbf{h}, t + u), Z(\mathbf{s}, t)] = Cov[Z(\mathbf{h}, u), Z(\mathbf{0}, 0)] =: C(\mathbf{h}, u)$, for all spatial shifts, $\mathbf{h}$, and temporal shifts, u.

As in the purely spatial case, some form of stationarity is necessary to enable estimation of the underlying space–time covariance function.

In Chapter 2, we have given several reasons why the variogram function is a superior tool to the covariance function. The space–time variogram function is given by $\gamma(\mathbf{h}, u) := Var[Z(\mathbf{s} + \mathbf{h}, t + u) - Z(\mathbf{s}, t)]$. Despite this representation, we formulate most space–time models in terms of the covariance function (when it exists, i.e., second-order stationarity holds), although for estimation purposes we often consider the corresponding space–time variogram function. The relationship between the covariance and variogram functions is discussed in Chapter 2. Direct computations show that for SOS space–time processes, it holds that $\gamma(\mathbf{h}, u) = C(\mathbf{0}, 0) - C(\mathbf{h}, u)$.

We next address optimal prediction using the covariance function. The derivation of the optimal space–time predictor is completely analogous to the derivation in Section 2.2 for purely spatial observations. In particular, the predictor of $Z(\mathbf{s}_0, t_0)$ at an unsampled space–time combination based on nT space–time observations is given by:

$$\mu(\mathbf{s}_0, t_0) + \boldsymbol{\nu}^{\mathrm{T}}\boldsymbol{\Sigma}^{-1}(\mathbf{Z}_{n,T} - \mu_{n,T}),$$

where $\mu_{n,T} = E[\mathbf{Z}_{n,T}]$, $\boldsymbol{\Sigma}$ is the $nT \times nT$ matrix $Var[\mathbf{Z}_{n,T}]$, and $\boldsymbol{\nu}$ is the $nT \times 1$ vector of covariances, $Cov[Z(\mathbf{s}_0, t_0), \mathbf{Z}_{n,T}]$. In the case of simple kriging, this is the optimal predictor, while in the ordinary kriging formulation,

$$\mu(\mathbf{s}_0, t_0) = (\mathbf{1}^{\mathrm{T}}\boldsymbol{\Sigma}^{-1}\mathbf{1})^{-1}\mathbf{1}^{\mathrm{T}}\boldsymbol{\Sigma}^{-1}\mathbf{Z}_{n,T},$$

the GLS estimator of the mean. This is the exact same form as the kriging predictor in Chapter 2.

It is seen that, as in the purely spatial setting, knowledge of the covariance function, or of the variogram function, is essential for prediction. As usual, these functions are in general unknown, and thus need to be estimated. This is the topic of the next section.

6.3 Empirical estimation of the variogram, covariance models, and estimation

The initial goal is empirical estimation of space–time variograms and covariances. The semivariogram moment estimator is defined completely analogously to that in the purely spatial setting in Chapter 3 as:

$$\widehat{\gamma}(\mathbf{h}, u) = \frac{1}{2|N(\mathbf{h}, u)|} \sum_{N(\mathbf{h},u)} \{[Z(\mathbf{s}_i, t_i) - Z(\mathbf{s}_j, t_j)]^2\}, \tag{6.1}$$

where $N(\mathbf{h}, u) = \{[(\mathbf{s}_i, t_i), (\mathbf{s}_j, t_j)] : \mathbf{s}_i - \mathbf{s}_j = \mathbf{h} \text{ and } t_i - t_j = u\}$.

The covariance estimator is usually defined as:

$$\widehat{C}(\mathbf{h}, u) = \frac{1}{|N(\mathbf{h}, u)|} \sum_{N(\mathbf{h},u)} \left[Z(\mathbf{s}_i, t_i) - \overline{Z}\right]\left[Z(\mathbf{s}_j, t_j) - \overline{Z}\right]. \tag{6.2}$$

As in the purely spatial case, the semivariogram estimator is unbiased at observed space–time lags, while the covariance estimator is slightly biased.

As in the spatial setting, any covariance function needs to be nonnegative definite, and any variogram function needs to be nonpositive definite. Specifically, this requires that for any constants a_i, $i = 1, \ldots, k$, and set of space–time coordinates, $(\mathbf{s}_i, t_i)$, $i = 1, \ldots, k$, it holds for the covariance that

$$Var\left[\sum_{i=1}^{k} a_i Z(\mathbf{s}_i, t_i)\right] = \sum_{i=1}^{k}\sum_{j=1}^{k} a_i a_j Cov[Z(\mathbf{s}_i, t_i), Z(\mathbf{s}_j, t_j)] \geq 0.$$

Analogously, we require for the variogram that

$$\sum_{i=1}^{k}\sum_{j=1}^{k} a_i a_j \gamma[(\mathbf{s}_i, t_i) - (\mathbf{s}_j, t_j)] \leq 0,$$

for all a_i such that $\sum_{i=1}^{k} a_i = 0$. As in the purely spatial case, the moment estimators of the covariance and variogram functions do not satisfy these properties, in general. For this reason, we consider models that are guaranteed to be valid ones. Further, to carry out optimal prediction we require $Cov[Z(\mathbf{s}_0, t_0), \mathbf{Z}_{n,T}]$, that is, covariances at unobserved space–time lags. Estimates of these covariances, through direct empirical estimation, are not available. Again, it is desirable to have space–time models.

6.3.1 Space–time symmetry and separability

Although the definition of SOS in the space–time context is a straightforward analogue to that of SOS in the temporal setting in Chapter 1, there are some fundamental differences between the spatial and space–time settings. In particular, in the spatial setting, $C(\mathbf{h}) = C(-\mathbf{h})$, by definition of the covariance function. It is readily seen that this is not the case in the spatio-temporal setting at each time. Specifically, it does not hold, in general, that $C(\mathbf{h}, u) = C(-\mathbf{h}, u)$, for all $\mathbf{h}$ and u. Nor is it the case, in general, that $C(\mathbf{h}, u) = C(\mathbf{h}, -u)$.

Covariance functions for which $C(\mathbf{h}, u) = C(-\mathbf{h}, u)$ [and thus $C(\mathbf{h}, u) = C(\mathbf{h}, -u)$] holds, for all $\mathbf{h}$ and u, are called *fully symmetric*. Among the class of fully symmetric covariances, a covariance function is *separable* if $C(\mathbf{h}, u)/C(\mathbf{h}, 0) = C(\mathbf{0}, u)/C(\mathbf{0}, 0)$, for all $\mathbf{h}$ and u. Noting that $C(\mathbf{0}, 0)$ is a constant, separability is often written more simply as $C(\mathbf{h}, u) = C(\mathbf{h}, 0)C(\mathbf{0}, u)$. If this condition holds, we see that the space–time covariance can be factored (separated) into the product of a purely spatial covariance and a purely temporal covariance. It is straightforward to show that a separable covariance must be fully symmetric, but full symmetry does not imply separability. An important subcategorization of all space–time models is those with separable covariance functions and those with nonseparable covariance functions.

6.4 Spatio-temporal covariance models

From Chapter 3, a linear combination (with positive coefficients) of valid covariance functions is, itself, a valid covariance function. It is also the case that the product of valid covariance functions is itself valid. Using this, we have that any valid spatial covariance function and valid temporal covariance function can be multiplied together to give a valid space–time covariance function. If, for example, the spatial correlation is given by a member of the exponential family, with scale parameter σ^2, and decay parameter ν_s, and the temporal correlation is an autoregressive process of order 1 with positive AR(1) parameter $\rho = \exp(-\nu_t)$, then a nonnegative definite space–time covariance is given by:

$$C(\mathbf{h}, u, \theta) = \sigma^2 \exp(-\nu_s \|\mathbf{h}\|) \exp(-\nu_t |u|),$$

with $\theta = (\sigma^2 > 0,\ \nu_s > 0$ and $\nu_t > 0)$. Note that this space–time covariance function is clearly separable. The spatial covariance as defined here is isotropic; of course we could have an anisotropic spatial covariance as well.

As can be seen in Figure 6.1(a), the contour plot of a separable covariance shows contours that are parallel to each other. This means that, for any two given spatial lags, $\mathbf{h}_1$ and $\mathbf{h}_2$, we have that $C(\mathbf{h}_1, u, \theta) = C_1 \exp(-\nu_t |u|)$ and $C(\mathbf{h}_2, u, \theta) = C_2 \exp(-\nu_t |u|)$, say. It is clear that the two temporal covariances have the same shape and differ only by a constant scale factor.

While this model is relatively parsimonious and is nicely interpretable, there are many physical phenomena which do not satisfy the separability. For example, the wind-speed data [see, e.g., Figure 8 in Cressie and Huang (1999)] suggests a lack of separability. This can be seen through a plot of the moment estimator over the space–time lag grid. For this reason, it is appropriate to consider nonseparable space–time covariances.

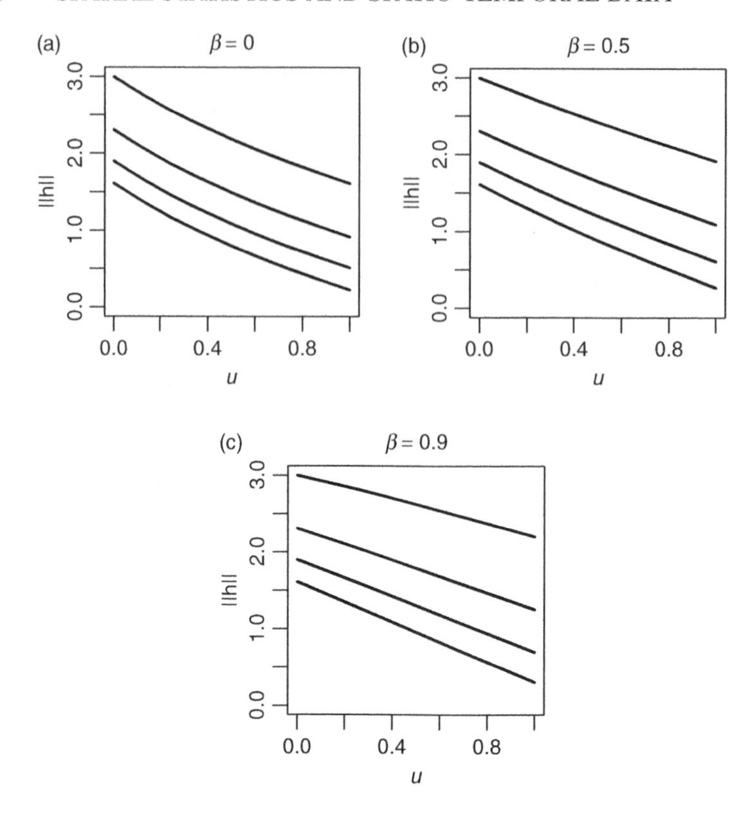

Figure 6.1 The contours of the space–time covariance in Equation 6.3 for three values of space–time interaction parameter β.

6.4.1 Nonseparable space–time covariance models

In the 1990s, there began a strong effort to allow for nonseparability in spatio-temporal covariance functions. Jones and Zhang (1997), through partial differential equations, developed nonseparable models in terms of integral equations. Cressie and Huang (1999) developed nonseparable covariances through a suitable class of corresponding spectral densities. They are able to give closed forms for the covariance functions, and give seven examples of such families of covariance functions. Gneiting (2002) developed a larger class of nonseparable covariances in the space–time (and not frequency) domain. Stein (2005) presented a class of nonseparable and asymmetric covariance functions, and characterized them in terms of their smoothness. One of Gneiting's examples of a nonseparable model is:

$$C(\mathbf{h}, u, \theta) = \frac{\sigma^2}{(|u|^{2\gamma} + 1)^{\tau}} \exp\left[\frac{-c\,\|\mathbf{h}\|^{2\gamma}}{(|u|^{2\gamma} + 1)^{\beta\gamma}}\right]. \tag{6.3}$$

Here, τ determines the smoothness of the temporal correlation, while $\gamma \in (0, 1]$ determines the smoothness of the spatial correlation, c determines the strength of the spatial correlation, and $\beta \in [0, 1)$ determines the strength of space/time interaction. In this parameterization, $\gamma = 1$ corresponds to the Gaussian covariance function, while $\gamma = \frac{1}{2}$ corresponds to the exponential distribution. Lesser values correspond to less smoothness, while larger values correspond to more smoothness. The spatial correlation is of the form of the stable family, described in Section 3.3.

Note that, if $\beta = 0$, the space–time covariance is the product of two factors, one a function of space, and the other a function of time, and thus the covariance is separable, in this case. Figure 6.1(a) displays the contours of this covariance function when $\beta = 0$. Note that the contours are all parallel to each other. Also displayed are contours for the values of (b) $\beta = 0.5$ and (c) $\beta = 0.9$. Note that the separation in the contours increases as the temporal separation, u, increases. Further, the contours are slightly convex when $\beta = 0$, close to linear for $\beta = 0.5$, and somewhat concave for $\beta = 0.9$. Often, this flexibility is very desirable to model a larger class of space–time phenomena than are possible under an assumption of separability.

All fitting methods discussed in Chapter 3 can be applied here. For example, WLS chooses the model parameters that minimize:

$$\sum_{i=1}^{m_s} \sum_{j=1}^{m_t} N(\mathbf{h}_i, u_j) \left[\frac{\widehat{\gamma}(\mathbf{h}_i, u_j)}{\gamma_i(\mathbf{h}_i, u_j, \theta)} - 1\right]^2 .$$

For Gaussian observations, likelihood methods can be applied in the usual manner as in Chapter 4, if the number of spatial and/or temporal observations is not too numerous. For larger data sets, approximate likelihood methods like pseudo-likelihood and MCML can be used, as also discussed in Chapter 4.

The assumption of either full symmetry, or separability greatly simplifies the interpretation of space–time covariance functions, and the modeling of these covariances. If, however, these properties do not hold, then predictions based on these covariance models can be badly wrong. Thus, an important part of model building in the space–time setting requires an assessment of full symmetry and separability. These assessments, of symmetry and separability, are considered in Section 6.6.

6.5 Space–time models

In Chapter 4, we formulated spatial models for continuous (Gaussian), and for binary variables. In this section, we discuss corresponding space–time models. Some space–time models are natural extensions of temporal models, while other models try to explicitly model space–time effects and interactions. Next, a brief survey of several types of space–time models.

i. VAR(1): recall the AR(1) process, discussed in Section 1.2.1. One way to view space–time data is to consider a multivariate time series, where the order of the multivariate observations is the number of spatial locations. Formally, let $\mathbf{Z}_t = [Z(\mathbf{s}_1, t), \ldots, Z(\mathbf{s}_n, t)]$ denote the vector of n responses across all spatial locations, at time t. Then, a multivariate ARMA(p, q) process is given by

$$\mathbf{Z}_t - \mathbf{a}_1\mathbf{Z}_{t-1} - \cdots - \mathbf{a}_p\mathbf{Z}_{t-p} = \epsilon_t + \mathbf{b}_1\epsilon_{t-1} + \cdots + \mathbf{b}_q\epsilon_{t-q}, \tag{6.4}$$

where $\mathbf{a}_1, \ldots, \mathbf{a}_p$ and $\mathbf{b}_1, \ldots, \mathbf{b}_q$ are $n \times n$ matrices, and the 'white noise' process, ϵ_t, is such that $Var(\epsilon_t) = \mathbf{\Sigma}$, and for $t \neq s$, $Cov(\epsilon_t, \epsilon_s) = 0$.

A special case is a vector AR(1) process which is given by

$$\mathbf{Z}_t = \mathbf{a}\mathbf{Z}_{t-1} + \epsilon_t.$$

Here the coefficient matrix, $\mathbf{a}$, determines the dependency between $\mathbf{Z}_t$ and $\mathbf{Z}_{t-1}$.

This formulation can be appropriate when the number of spatial locations is relatively small. For example, the Irish wind data ($n = 11$, see Section 6.7.3) can be modeled using this approach. One benefit to this formulation is that the variance of all empirical covariance estimates has a closed form solution [see, e.g., Priestley (1981)]. This is useful to compare different covariance estimates to assess separability (see Section 6.7).

ii. A STARMA (space–time autoregressive moving average) process truly combines the simultaneous spatial regression model in Chapter 4 with the temporal ARMA class of models, by allowing a term $\mathbf{a}_0\mathbf{Z}_t$ on the left hand side of Equation 6.4; that is, each spatial location depends on each other at the same time. To see the desirability of this, consider the case where there is one fixed time, that is, no temporal component. Then the STARMA formulation models correlation between observations at all spatial locations, while the multivariate ARMA(p, q) formulation gives complete independence.

iii. Dynamic spatial models have found a wide variety of applications. Examples of spatially dynamic models are found in modeling wetlands [Boumans and Sklar (1991), Costanza *et al.* (1990), and Kadlec and Hammer (1988)], oceanic plankton models [Show (1979)], modeling the growth of coral reefs [Maguire and Porter (1977)], and modeling fire ecosystems [Kessell (1977)]. Here, the general form is:

$$X(t+1) = F[X(t), Y(t)],$$

where X is the primary process of interest and Y denotes a correlated auxiliary process. The form shows that this is mainly a time series-based model. Note that if $F(x, y) = \mu + B(x - \mu) + \beta^{\mathrm{T}} Y(t)$, then we have a VAR(1). In this case, this is actually a multivariate AR(1) process and not a truly space–time process as is the STARMA model.

iv. A general model is given by:

$$Z(\mathbf{s}, t) = m(\mathbf{s}) + \mu(t) + e(\mathbf{s}, t),$$

where $m(\mathbf{s})$ models a purely spatial mean function, $\mu(t)$ is a purely temporal mean function, and $e(\mathbf{s}, t)$ denotes an intrinsically stationary space–time error process. This model is sufficiently flexible to model a large range of spatio-temporal data sets.

v. Subba Rao (2008) fits separate time series models at each spatial location. The time series coefficients are assumed to vary smoothly over space. The fitted model then allows for kriging from this (possibly nonstationary) space–time model.

6.6 Parametric methods of assessing full symmetry and space–time separability

Separable, and fully symmetric models were introduced and discussed in Section 6.4. The moment estimate of the wind-speed data suggests that a separable model may be inappropriate for these data. This conclusion, based on a plot of the space–time covariance contours, cannot be firm. It is desirable to have a more objective measure of the appropriateness of a separable or fully symmetric covariance model, in the spirit of the tests for isotropy in Chapter 5. As another example, Gneiting (2002) suggests that the fitted value of $\widehat{\beta} = 0.61$ in the covariance model in Equation 6.3 indicates a nonseparable covariance structure. Given that the parameter space for the space–time interaction parameter, β, is $[0, 1)$, it seems reasonable that $\beta \neq 0$. Is there, however, a clear indication that β is not equal to 0 in the population? We have no way to know without some knowledge of the distribution of the nonseparability parameter estimator $\widehat{\beta}$.

One approach is to carry out parametric testing, assuming that a given space–time covariance model is correct. For example, recall that in Gneiting's model in Equation 6.3, if $\beta = 0$ then this covariance function factors into a product of a spatial covariance and a temporal covariance, and is thus separable. Using this, one approach is to assess the appropriateness of separability using the assumed model. In particular, if we assume a *Gaussian* space–time field as the model, then $l(\mathbf{Z}; \mu, \mathbf{\Sigma}) = \ln[L(\mathbf{Z}; \mu, \mathbf{\Sigma})]$ denotes the log of the likelihood given and discussed in Section 4.1. In this case, we can compute:

$$\chi^2 := 2\{l[\mathbf{Z}; \widehat{\mu}, \widehat{\Sigma}(\widehat{\beta})] - l[\mathbf{Z}; \widehat{\mu}(0), \widehat{\Sigma}(0)]\},$$

where $l[\mathbf{Z}; \widehat{\mu}, \widehat{\Sigma}(\widehat{\beta})]$ denotes the unrestricted loglikelihood, and $l[\mathbf{Z}; \widehat{\mu}(0), \widehat{\Sigma}(0)]$ denotes the loglikelihood under the assumed separable model, with $\beta = 0$. For large sample sizes this can be compared to a chisquared distribution with 1 d.f. The validity of this procedure is asymptotic (to make the chisquared the correct reference distribution), depends on the Gaussianity of the observations, and depends on the chosen class of covariance models being the correct one.

Mitchell *et al.* (2006) proposed a likelihood ratio test for separability of covariance models in the context of multivariate repeated measures assuming the multivariate normality of observations. Mitchell *et al.* (2005) implemented this test in the spatio-temporal context. Scaccia and Martin (2002) and Scaccia and Martin (2005) presented a spectral method to test the symmetry and separability for spatial lattice processes.

6.7 Nonparametric methods of assessing full symmetry and space–time separability

Fuentes (2006) proposed a nonparametric test for separability of a spatio-temporal process based on a spectral method, in the frequency domain. First, separability is formulated in the spectral domain and the coherency function is seen to be independent of frequency under separability. Then, significant departures of the empirical coherence from constancy across frequencies signifies a lack of separability. In the following, we consider space–time covariances in the space–time domain.

As in Li *et al.* (2007), we consider a framework to assess the assumptions of full symmetry and separability simultaneously, based on the asymptotic joint normality of sample space–time covariance estimators. Let $\boldsymbol{\Lambda}$ denote a set of user-chosen space–time lags, and let m denote the cardinality of $\boldsymbol{\Lambda}$. Let D be the domain of observations, and let $\widehat{C}(\mathbf{h}, u)$ denote an estimator of $C(\mathbf{h}, u)$ over D. Let $\mathbf{G} = \{C(\mathbf{h}, u), (\mathbf{h}, u) \in \boldsymbol{\Lambda}\}$, and let $\widehat{\mathbf{G}} = \{\widehat{C}(\mathbf{h}, u), (\mathbf{h}, u) \in \boldsymbol{\Lambda}\}$ denote the estimator of $\mathbf{G}$ over D. The appropriately standardized $\widehat{\mathbf{G}}$ has an asymptotic multivariate normal distribution for a random field with a fixed spatial domain and an increasing temporal domain. There are no requirements on the marginal or joint distributions generating the observations other than mild moment and mixing conditions on the strictly stationary random field. In this sense, the approach is nonparametric.

A very general form of hypothesis applied to many assumptions made for space-time covariances is

$$H_0 : \mathbf{Af(G)} = \mathbf{0}, \tag{6.5}$$

where $\mathbf{A}$ is a contrast matrix of row rank q, and $\mathbf{f} = (f_1, \ldots, f_r)^{\mathrm{T}}$ are real-valued functions which are differentiable at $\mathbf{G}$. For example, the null hypotheses of full symmetry and separability exactly follow this form. We used this approach in Section 5.6 to test for isotropy, but in that case the function $\mathbf{f}(\cdot)$ was simply the identity function in all cases.

According to the definitions of full symmetry and separability, the null hypothesis for full symmetry, denoted by H_0^1, and the null hypothesis for separability, denoted by H_0^2, are given as:

$$H_0^1\colon C(\mathbf{h}, u) - C(\mathbf{h}, -u) = 0, \ (\mathbf{h}, u) \in \boldsymbol{\Lambda},$$

$$H_0^2\colon C(\mathbf{h}, u)/C(\mathbf{h}, 0) - C(\mathbf{0}, u)/C(\mathbf{0}, 0) = 0, \ (\mathbf{h}, u) \in \boldsymbol{\Lambda}.$$

Observe that the hypothesis H_0^1 is a contrast of covariances, while H_0^2 is a contrast of ratios of covariances. Thus, H_0^1 can be rewritten in the form of $\mathbf{A}_1\mathbf{G} = \mathbf{0}$, while H_0^2 can be rewritten as $\mathbf{A}_2\mathbf{f}(\mathbf{G}) = \mathbf{0}$ for a specified group of space–time lags, $\mathbf{\Lambda}$, and contrast matrices $\mathbf{A}_1$ and $\mathbf{A}_2$, where $\mathbf{f}$ takes pairwise ratios of elements in $\mathbf{G}$. For example, if $\Lambda = \{(\mathbf{0}, 0), (\mathbf{h}_1, u_1), (\mathbf{h}_1, -u_1), (\mathbf{h}_2, u_2), (\mathbf{h}_2, -u_2), (\mathbf{h}_1, 0), (\mathbf{0}, u_1), (\mathbf{h}_2, 0), (\mathbf{0}, u_2)\}$, that is, $\mathbf{G} = \{C(\mathbf{0}, 0), C(\mathbf{h}_1, u_1), C(\mathbf{h}_1, -u_1), C(\mathbf{h}_2, u_2), C(\mathbf{h}_2, -u_2), C(\mathbf{h}_1, 0), C(\mathbf{0}, u_1), C(\mathbf{h}_2, 0), C(\mathbf{0}, u_2)\}^{\mathrm{T}}$, then

$$\mathbf{A}_1 = \begin{bmatrix} 0 & 1 & -1 & 0 & 0 & 0 & 0 & 0 & 0 \\ 0 & 0 & 0 & 1 & -1 & 0 & 0 & 0 & 0 \end{bmatrix},$$

and

$$\mathbf{f}(\mathbf{G}) = \left[\frac{C(\mathbf{h}_1, u_1)}{C(\mathbf{h}_1, 0)}, \frac{C(\mathbf{h}_1, -u_1)}{C(\mathbf{h}_1, 0)}, \frac{C(\mathbf{h}_2, u_2)}{C(\mathbf{h}_2, 0)}, \frac{C(\mathbf{h}_2, -u_2)}{C(\mathbf{h}_2, 0)}, \frac{C(\mathbf{0}, u_1)}{C(\mathbf{0}, 0)}, \frac{C(\mathbf{0}, u_2)}{C(\mathbf{0}, 0)} \right]^{\mathrm{T}},$$

with

$$\mathbf{A}_2 = \begin{bmatrix} 1 & 0 & 0 & 0 & -1 & 0 \\ 0 & 1 & 0 & 0 & -1 & 0 \\ 0 & 0 & 1 & 0 & 0 & -1 \\ 0 & 0 & 0 & 1 & 0 & -1 \end{bmatrix}.$$

Under their corresponding null hypotheses, it holds that $\mathbf{A}_1\mathbf{G} = \mathbf{0}$ and $\mathbf{A}_2\mathbf{f}(\mathbf{G}) = \mathbf{0}$. The two hypotheses, H_0^1 and H_0^2, are two special cases of the general hypothesis in Equation 6.5. it is seen that the assessment of separability is somewhat more difficult than that of full symmetry. In many cases, however, the latter test is not practically necessary. In particular, once H_0^1 is rejected, H_0^2 is automatically rejected as well, as the latter hypothesis is a subset of the former.

Let $\widehat{C}_n(\mathbf{h}, u)$ denote a sample-based estimator of $C(\mathbf{h}, u)$ based on observations in a sequence of increasing index sets D_n. For example, this can be the empirical estimator in Section 6.3, Equation 6.2. Let $\widehat{\mathbf{G}}_n = \{\widehat{C}_n(\mathbf{h}, u) : (\mathbf{h}, u) \in \mathbf{\Lambda}\}$ denote the estimator of $\mathbf{G}$ computed over D_n. Decompose D_n into $D_n = \mathcal{F} \times \mathcal{I}_n$, where $\mathcal{F} \subset \mathbb{R}^{d_1}$, $\mathcal{I}_n \subset \mathbb{R}^{d_2}$. Suppose $\mathcal{F}$ is a fixed space, in the sense that finitely many observations are located within this space, and $\mathcal{I}_n$ is an increasing space. Account for the shape of the space–time domain in which data is observed as in Chapter 5. Let $\mathcal{A}$ denote the interior of a closed surface contained in a d_2-dimensional cube with edge length 1, and let $\mathcal{A}_n$ denote the inflation of $\mathcal{A}$ by a factor n. Define $\mathcal{I}_n = \mathbb{Z}^{d_2} \cap \mathcal{A}_n$ if the observations are regularly spaced, and $\mathcal{I}_n = \mathcal{A}_n$ otherwise. This formulation allows for a wide variety of spatial domains as discussed in Chapter 5.

In many situations, the observations are taken from a fixed space $S \subset \mathbb{R}^d$ at regularly spaced times $T_n = \{1, \ldots, n\}$. For example, the Irish wind data, in Section 6.7.1, are of this form. In this particular case, $d_1 = d$, $d_2 = 1$, and we define the mixing coefficient (e.g., Ibragimov and Linnik (1971), p. 306):

$$\alpha(u) = \sup_{A,B} \{|P(A \cap B) - P(A)P(B)|, A \in \mathfrak{F}_{-\infty}^{0}, B \in \mathfrak{F}_{u}^{\infty}\},$$

where $\mathfrak{F}_{-\infty}^{0}$ is the σ-algebra generated by the past time process until $t = 0$, and $\mathfrak{F}_{u}^{\infty}$ is the σ-algebra generated by the future time process from $t = u$. This is the temporal mixing coefficient analogous to the spatial mixing coefficient discussed in Section 5.6.1. As in the spatial setting, assume the mixing coefficient, $\alpha(u)$, satisfies a strong mixing condition:

$$\alpha(u) = \mathrm{O}(u^{-\epsilon}) \quad \text{for some } \epsilon > 0. \tag{6.6}$$

There are many classes of temporal processes that satisfy such a mixing condition. For example, an autoregressive times series of order one satisfies (6.6), with Gaussian, double exponential, or even Cauchy errors, as shown by Gastwirth and Rubin (1975). For Gaussian errors $\alpha(u) \leq C\,|\rho|^u$, for a constant C, where ρ denotes the AR(1) autoregressive parameter. For other Gaussian processes, from Doukhan (1994), it holds that if $Cov[Z(t), Z(t+u)] = O(u^{-\beta})$, then $\alpha(u) = O(u^{1-\beta})$. Thus, (6.6) holds when $\beta > 1$, which in turn implies that the temporal covariances are summable.

Let $S(\mathbf{h}) = \{\mathbf{s} : \mathbf{s} \in S, \mathbf{s} + \mathbf{h} \in S\}$ denote the set of locations such that through translation by $\mathbf{h}$ remain in S, and let $|S(\mathbf{h})|$ denote the number of elements in $S(\mathbf{h})$. For simplicity, assume the mean of Z is known and equal to 0. If this assumption is removed, then let $\widehat{C}_n^*(\mathbf{h}, u)$ and $\widehat{\mathbf{G}}_n^*$ denote the mean-corrected estimators of $C(\mathbf{h}, u)$ and $\mathbf{G}$, respectively. It is straightforward to show that $\widehat{\mathbf{G}}_n^*$ and $\widehat{\mathbf{G}}_n$ have the same large sample properties. The estimator of C under the zero mean assumption is defined s:

$$\widehat{C}_n(\mathbf{h}, u) = \frac{1}{|S(\mathbf{h})||T_n|} \sum_{\mathbf{s} \in S(\mathbf{h})} \sum_{t=1}^{n-u} Z(\mathbf{s}, t) Z(\mathbf{s} + \mathbf{h}, t + u),$$

and assume the moment condition for $\widehat{C}_n(\mathbf{h}, u)$:

$$\sup_n E\left\{\left|\sqrt{|T_n|}\left[\widehat{C}_n(\mathbf{h}, u) - C(\mathbf{h}, u)\right]\right|^{2+\delta}\right\} \leq C_\delta$$

$$\text{for some } \delta > 0, C_\delta < \infty. \tag{6.7}$$

Theorem 6.7.1 *Let $\{Z(\mathbf{s}, t), \mathbf{s} \in \mathbb{R}^d, t \in \mathbb{Z}\}$ be a strictly stationary spatio-temporal random field observed in $D_n = S \times T_n$, where $S \subset \mathbb{R}^d$ and $T_n = \{1, \ldots, n\}$. Assume*

$$\sum_{t \in \mathbb{Z}} |Cov[Z(\mathbf{0}, 0) Z(\mathbf{h}_1, u_1), Z(\mathbf{s}, t) Z(\mathbf{s} + \mathbf{h}_2, t + u_2)]| < \infty, \tag{6.8}$$

for all $\mathbf{h}_1, \mathbf{h}_2 \in S, \mathbf{s} \in S(\mathbf{h}_2)$ and all finite u_1, u_2.

Then $\mathbf{\Sigma} = \lim_{n \to \infty} |T_n| Cov(\widehat{\mathbf{G}}_n, \widehat{\mathbf{G}}_n)$ exists, the (i, j)th element of which is

$$A_{ij} \sum_{\mathbf{s}_1 \in S(\mathbf{h}_i)} \sum_{\mathbf{s}_2 \in S(\mathbf{h}_j)} \sum_{t \in \mathbb{Z}} Cov[Z(\mathbf{s}_1, 0) Z(\mathbf{s}_1 + \mathbf{h}_i, u_i),$$

$$Z(\mathbf{s}_2, t) Z(\mathbf{s}_2 + \mathbf{h}_j, t + u_j)].$$

where $A_{ij} := 1/[|S(\mathbf{h}_i)||S(\mathbf{h}_j)|]$. Further, if it holds that $\mathbf{\Sigma}$ is positive definite, and that conditions (6.6) and (6.7) hold, then

$$\sqrt{|T_n|}(\widehat{\mathbf{G}}_n - \mathbf{G}) \xrightarrow{d} \mathrm{N}_m(\mathbf{0}, \mathbf{\Sigma}) \text{ as } n \to \infty.$$

The conditions assumed for Theorem 6.7.1 to hold are relatively mild ones. The assumption on the temporal correlation given by (6.6) holds for a large class of temporal processes, for instance, AR(1) processes with normal, double exponential or Cauchy errors from Gastwirth and Rubin (1975). These examples allow for the mixing coefficient to not account for cardinality of the two sets separated by time u, unlike in the spatial situation in Chapter 5. The moment condition, (6.7), is only slightly stronger than the existence of an asymptotic variance of the covariance estimator, given in the theorem by Σ.

In this theorem, the observations are allowed to be either regularly spaced or irregularly spaced in S. However, even for irregularly spaced observations, only the covariances of observed spatial lags are considered, due to the assumed limited number of observations in S. Note that, in this section, we require the observations to be taken at the same spatial locations over time, which is often the case. For example, monitoring stations usually are observed in the same locations over time. This is the case for the Irish wind data, and for the Pacific Ocean wind data that are analyzed in Section 6.7.1.

The large sample distribution of $\widehat{\mathbf{G}}_n$ in other space–time regimes can be stated in a similar fashion. For example, regimes can be regularly spaced observations with an increasing spatio-temporal domain; spatially irregularly spaced observations with an increasing spatio-temporal domain; and irregularly spaced observations with an increasing spatio-temporal domain. Under appropriate conditions, the asymptotic joint normality of $\widehat{\mathbf{G}}_n$ holds in all these types of data structures under various mixing and moment conditions. This guarantees the large sample validity of the tests over a wide variety of data structures under mild assumptions on the domain shape and the strength of dependence in the underlying space–time random field.

Replacing $\mathbf{G}$ with an estimator $\widehat{\mathbf{G}}_n$ in (6.5), we obtain a contrast vector for testing H_0 as the estimated left hand side of (6.5), $\mathcal{C} = \mathbf{Af}(\widehat{\mathbf{G}}_n)$. Specifically, the contrasts for testing H_0^1 and H_0^2 are given by

$$\mathcal{C}_1 = \widehat{C}_n(\mathbf{h}, u) - \widehat{C}_n(\mathbf{h}, -u), \ (\mathbf{h}, u) \in \mathbf{\Lambda}, \quad \text{and}$$

$$\mathcal{C}_2 = \widehat{C}_n(\mathbf{h}, u)/\widehat{C}_n(\mathbf{h}, 0) - \widehat{C}_n(\mathbf{0}, u)/\widehat{C}_n(\mathbf{0}, 0), \ (\mathbf{h}, u) \in \mathbf{\Lambda}.$$

Apparently, $\mathcal{C}_1$ and $\mathcal{C}_2$ can be rewritten into the form of $\mathbf{A}_1\widehat{\mathbf{G}}_n$ and $\mathbf{A}_2\mathbf{f}(\widehat{\mathbf{G}}_n)$, respectively.

It is then straightforward to obtain the large sample distribution of the test statistics based on the asymptotic joint normality of $\widehat{\mathbf{G}}_n$. By the multivariate delta theorem [e.g., Mardia *et al.* (1979), p. 52], it holds that,

$$\sqrt{|T_n|}[\mathbf{f}(\widehat{\mathbf{G}}_n) - \mathbf{f}(\mathbf{G})] \xrightarrow{d} \mathrm{N}_r(\mathbf{0}, \mathbf{B}^{\mathrm{T}}\mathbf{\Sigma}\mathbf{B}), \tag{6.9}$$

where $\mathbf{B}_{ij} = \partial f_j/\partial \mathbf{G}_i, i = 1, \ldots, m, j = 1, \ldots, r$. The test statistic (*TS*) is formed based on the contrasts of $\mathbf{f}(\widehat{\mathbf{G}}_n)$, and the distribution of *TS* is obtained under the null hypothesis as

$$TS = |T_n|[\mathbf{Af}(\widehat{\mathbf{G}}_n)]^{\mathrm{T}}(\mathbf{AB}^{\mathrm{T}}\mathbf{\Sigma}\mathbf{BA}^{\mathrm{T}})^{-1}[\mathbf{Af}(\widehat{\mathbf{G}}_n)] \xrightarrow{d} \chi^2_q, \tag{6.10}$$

for a matrix $\mathbf{A}$ with row rank q. As discussed in Chapter 5, the idea of employing a quadratic form to assess the discrepancy between two vectors emerged in Lu and Zimmerman (2001). This idea lends itself naturally to testing general hypotheses of the form (6.5) for spatio-temporal random fields.

By (6.10), it holds that under H_0^1, the contrast C_1 yields the test statistic for full symmetry,

$$TS1 = |T_n|(\mathbf{A}_1\widehat{\mathbf{G}}_n)^{\mathrm{T}}(\mathbf{A}_1\mathbf{\Sigma}\mathbf{A}_1^{\mathrm{T}})^{-1}(\mathbf{A}_1\widehat{\mathbf{G}}_n) \xrightarrow{d} \chi^2_{q_1},$$

and under H_0^2, the contrast C_2 yields the test statistic for separability,

$$TS2 = |T_n|\left\{\mathbf{A}_2\left[\mathbf{f}(\widehat{\mathbf{G}}_n)\right]\right\}^{\mathrm{T}}(\mathbf{A}_2\mathbf{B}^{\mathrm{T}}\mathbf{\Sigma}\mathbf{B}\mathbf{A}_2^{\mathrm{T}})^{-1}\left\{\mathbf{A}_2\left[\mathbf{f}(\widehat{\mathbf{G}}_n)\right]\right\} \xrightarrow{d} \chi^2_{q_2},$$

for matrices $\mathbf{A}_1$ and $\mathbf{A}_2$ with row ranks q_1 and q_2, respectively. The matrix $\mathbf{B}$ can be estimated empirically by replacing $\mathbf{G}$ with $\widehat{\mathbf{G}}_n$ in the test statistic. For the other types of domains, D_n, the distribution of the test statistic can be derived in an analogous manner with the normalizing sequence $|\mathcal{I}_n|$ replacing $|T_n|$, and noting the corresponding change in $\mathbf{\Sigma}$. The covariance matrix $\mathbf{\Sigma}$ defined in Theorem 6.7.1 is usually unknown, and thus needs to be estimated. While empirical estimation is, in principle, possible, due to the large number of elements in $\mathbf{\Sigma}$, subsampling techniques (Chapter 10) seem to perform better. Specifically, if the data are observed over a fixed spatial domain S and an increasing time domain, we can form overlapping $S \times l(n)$ subblocks using a moving subblock window along time. Much research has addressed the choice of the optimal block length, $l(n)$, in the sense of minimizing the mean squared error (MSE) of estimators in a variety of contexts [see, e.g., Chapter 10 or Lahiri (2003)]. Carlstein (1986) suggested choosing the block length under a parametric model. The development in Section 10.4.1 gives a formula based on nonoverlapping blocks. Here, however, one is developed using all overlapping blocks. This reduces the variance of the variance estimator by a factor of $\frac{2}{3}$ [Künsch (1989)], and enables the use of slightly longer subblocks by a factor of $\left(\frac{3}{2}\right)^{1/3}$. The block length for a series of length n is then

$$l(n) = \left(\frac{2\rho}{1-\rho^2}\right)^{2/3}\left(\frac{3n}{2}\right)^{1/3},$$

for an AR(1) process with parameter ρ. The AR(1) parameter ρ can be estimated by $\widehat{\rho}_n = \widehat{C}_n(\mathbf{0}, 1)/\widehat{C}_n(\mathbf{0}, 0)$. The large sample justification for this approach is based on an assumption that the statistic of interest is the sample mean, and that the temporal correlation follows an AR(1) process with parameter ρ. Although not optimal in the current context, this procedure often works well in practice. For more fully model-free approaches to determine block length, see the methods in, for instance Lahiri (2003).

In Section 6.7, examples of the contrast matrices were given that were used in testing the appropriate linear hypothesis. The choices are not unique though, as one can choose a subset of rows from $\mathbf{A}_1$ or $\mathbf{A}_2$ to obtain new test statistics for their corresponding tests. For example, we can pick only the first two rows of $\mathbf{A}_2$ to form a new test statistic whose asymptotic distribution follows χ_2^2. Although these tests will have approximately the same sizes, the power will depend on the specific choice of contrasts chosen and on the underlying space–time covariance structure.

Although it seems not possible to make absolute recommendations on choice of contrast matrices, some general recommendations can be made. Generally, it is preferable to use lags combining small spatial and temporal lags because, typically, covariance estimators of smaller lags are obtained over more observations than larger lags and they play a more important role in making predictions over the random field (as seen in Chapter 3). Given sufficient data, however, one can also include a larger variety of lags in terms of both space and time to ensure assessment of the characteristics of the whole space–time random field. In addition, contingent on the understanding of the physical process, one should definitely take the features of the physical process into consideration while choosing testing lags. For example, if the random field relates to the wind or precipitation, it is more appropriate to use the dominant wind direction as a guide to choose space–time lags with strong correlation, since the wind plays an important role in governing the structure of this type of random field. We use these ideas in the following section.

We apply the testing procedures to the Irish wind data and the Pacific Ocean wind data to illustrate how the methodology aids in choosing appropriate covariance models for the data.

6.7.1 Irish wind data

The Irish wind data described in Haslett and Raftery (1989) consists of time series of daily average wind speed at 11 meteorological stations in Ireland, during the time period 1961–1978. In order to normalize the data, as in Haslett and Raftery (1989) and Gneiting (2002), we perform a square-root transformation, and subtract seasonal effects and the spatially varying mean from the wind speed to obtain velocity measures before carrying out testing.

Stein (2005) noted an apparent asymmetric property of the covariance function by viewing plots of the sample space–time variogram as given in Equation 6.1. Gneiting *et al.* (2007) explored the validity of the assumptions of full symmetry and separability of the covariance function. Based on visual inspection, they fitted a separable, a fully symmetric, and a general stationary covariance model on training data from 1961–1970. They also assessed the prediction performance using these three fitted models on the test period of 1971–1978. Their exploration suggests that the data do not support the assumptions of full symmetry and separability. However, their parameter estimate $\widehat{\lambda} = 0.0573$ seems reasonably close to $\lambda = 0$. This could suggest that full symmetry does hold. Section 6.7.3 defines the covariance parameter λ, and gives its interpretation.

To formally test the full symmetry and separability using the data, we choose 5 pairs of stations among the $\binom{11}{2} = 55$ pairs, and choose time lags $u = 1$ and 2 days. This latter choice is due to the fact that correlations for the velocity measures decay rapidly in time. An apparently natural choice of the station pairs is the 5 pairs with the shortest spatial distance $\|\mathbf{h}\|$ among the 55 pairs. However, the prevailing westerly wind suggests choosing the 5 pairs of stations with the smallest ratio, h_2/h_1, where h_1 and h_2 are the east–west component and the north–south component of the spatial lag $\mathbf{h}$, respectively. In order to show the effects of testing lags so as to provide guidance on how to choose them, we compare the test based on the five station pairs with the smallest ratio to the tests based on the five pairs with the largest ratio and with the shortest $\|\mathbf{h}\|$, respectively. For each station pair, we choose time lags $u = 1$ and 2 days with the west station leading the east station, as the wind propagates from west to east.

The testing set $\mathbf{\Lambda}$ contains 10 lags of the combination of 5 spatial lags, $\mathbf{h}$, and 2 temporal lags, us, and the other lags introduced by these 10 lags in the test. Thus, $m = r = 20$ for the full symmetry test, and $m = 18$, $r = 12$ for the separability test, whereas $q = 10$ for both tests. The test statistics and p-values for the full symmetry and separability tests based on the five station pairs with the smallest ratio are: *TS1* = 262.7, p-value = 0; *TS2* = 445.2, p-value = 0. Therefore, the test results thoroughly reject the assumptions of full symmetry and separability. This provides a theoretical basis for the statement of asymmetry in Stein (2005), and gives support to the suggested model of Gneiting *et al.* (2007) which allows for a non-fully symmetric (and thus nonseparable) covariance function. The test based on the five station pairs with the largest ratio gives $TS1 = 20.2$, $TS2 = 103.6$. Although these are both quite significant, it is seen that the strength of the results is diminished. The same holds for the test based on the five pairs with the smallest $\|\mathbf{h}\|$, which gives $TS1 = 132.9$, $TS2 = 202.5$. It is seen that the test based on the smallest ratio detects further departure from full symmetry and separability by taking the wind direction into account. Note that in all cases the test statistic for *TS1* is smaller than that for *TS2*. This is appropriate, as there is more evidence against the stronger hypothesis of separability. A further difference between the two hypotheses is as follows: if the east station is leading the west station in time for each pair, the value of *TS2* drops appreciably. The value of *TS1*, however, remains the same because H_0^1 is invariant under time reversal. This implies that the power of the separability test is weakened by this latter choice of test lags.

6.7.2 Pacific Ocean wind data

These data consist of the east–west component of the wind velocity vector from a region over the tropical western Pacific Ocean for the period from November, 1992 to February, 1993. The wind velocities are given every 6 hours, on a 17×17 grid with grid interval of about 210 km. The reader is referred to Wikle and Cressie (1999) for a more detailed description of these data. Cressie and Huang (1999) graphically showed an apparent nonseparability of the spatio-temporal covariance through examination of the empirical space–time variogram plot. Based on this, they fitted several nonseparable covariance models to the data. To assess the necessity of fitting a nonseparable model, we perform the tests for full symmetry and separability to these wind data.

For each grid location, the time-averaged mean is removed to create the zero-mean data set, as was done in Wikle and Cressie (1999). Three sets of east–west spatial lags are chosen, each consisting of three distinct $\|\mathbf{h}\|$s, and two sets of time lags, each consisting of five distinct us. Since neither east wind nor west wind is clearly predominant, u is simply chosen with the east grid location leading the west grid location in time. The degrees of freedom are $q = 15$ for all the tests, and the testing results are shown in Table 6.1. It is seen from this table that the tests provide strong evidence against separability for all testing lags, and against full symmetry for small spatial lags. This corroborates the necessity of Cressie and Huang's nonseparable models. This table also clearly illustrates that different lags can lead to quite different p-values for the test. For example,

Table 6.1 Pacific Ocean wind data: testing full symmetry and separability.

		Full symmetry		Separability	
$\|\mathbf{h}\|$	u	*TS1*	p-value	*TS2*	p-value
1,2,3	u_1	152.7	0.000	459.7	0.000
1,5,10		85.6	0.000	205.5	0.000
10,11,12		27.6	0.025	40.4	0.000
1,2,3	u_2	96.8	0.000	458.1	0.000
1,5,10		87.5	0.000	130.3	0.000
10,11,12		24.7	0.054	49.7	0.000

Units of $\|\mathbf{h}\|$: grid interval; units of u: 6 hours. All p-values of '0.000' are less than 1.4×10^{-5}. u_1 denotes the time lags $1, 3, 5, 7, 9$, while u_2 denotes the time lags $1, 2, 3, 4, 5$.

it is not easy to detect the asymmetric property of the covariance if we use large spatial lag $\|\mathbf{h}\|$, rather than small $\|\mathbf{h}\|$. This is in accordance with the empirical spatio-temporal variogram in Cressie and Huang (1999). If u is chosen with the opposite leading direction, that is, the west grid location leading the east grid location in time, then the results are very similar to Table 6.1. The results change dramatically, however, if $\mathbf{h}$ is chosen in the north–south direction.

We have seen a firm rejection of full symmetry and separability in the data sets in Sections 6.7.1 and 6.7.2. More can be learned concerning the reliability of these conclusions by assessing the size and power of the testing procedures. In order to do this, we consider two simulations. In the first, the test for full symmetry and separability for the data structure in the Irish wind data is assessed. In the second, the test for separability for data on a grid as in the Pacific Ocean data over a range of grid sizes, temporal lengths, and temporal correlations is considered. Each situation is analyzed over 1000 simulated space–time data sets.

6.7.3 Numerical experiments based on the Irish wind data

To assess the size and power of the tests, it is necessary to choose an appropriate underlying space–time covariance model from which to generate observations. We employ the correlation model fitted to the Irish wind data in Gneiting *et al.* (2007) to simulate a space–time random field, $\mathbf{Z}$, of size $11 \times |T_n|$ at the 11 stations. Let $\mathbf{\Sigma}_0$ denote the spatio-temporal correlation matrix of the vectorized $\mathbf{Z}$, and let $\mathbf{Z}_0$ denote a vector of independent, zero-mean standard normal variables. In this case, $\mathbf{\Sigma}_0$ is of dimension $11n \times 11n$ and $\mathbf{Z}_0$ is of length $11n$, where $n = |T_n| = 3650$. An apparently natural way to simulate $\mathbf{Z}$ is via $\mathbf{\Sigma}_0^{1/2}\mathbf{Z}_0$, where $\mathbf{\Sigma}_0^{1/2}$ can be computed via an eigenvalue decomposition. Although possible in principle, it is not feasible, in practice, to generate $\mathbf{Z}$ in this manner due to the large dimension of the matrix $\mathbf{\Sigma}_0$. One method to bypass this difficulty is to split the temporal component of the whole random field, $\mathbf{Z}$, into $\mathbf{Z}_1, \mathbf{Z}_2, \ldots, \mathbf{Z}_{\lceil n/k \rceil}$, each of size $11 \times k$. The value of k should be such that temporal correlation exists only between the nearest neighbors of $\mathbf{Z}_i$, $i = 1, 2, \ldots, \lceil n/k \rceil$. An appropriate k can be obtained empirically. In these data, it appears to be justified by the negligible temporal correlation when the time lag exceeds a certain value k. In this situation, $k = 15$ is appropriate. Then the explicit form of the conditional distribution of two jointly normal random vectors allows for generating the data consecutively. Specifically, an initial 11×15 Gaussian random field $\mathbf{Z}_1$ is generated, and then given $\mathbf{Z}_1 = \mathbf{z}_1$ we generate $\mathbf{Z}_2|(\mathbf{Z}_1 = \mathbf{z}_1)$, $\mathbf{Z}_3|(\mathbf{Z}_2 = \mathbf{z}_2)$ and so on. This methods frees the user from the restriction of the dimension of $\mathbf{\Sigma}_0$, enabling generation of the space–time random field of a size that is the same size as the training data set on which the model is fitted. The training period is 3650 days, so the space–time random field of size 11×3650 is chosen.

Gneiting *et al.* (2007) employed the following spatio-temporal covariance model:

$$C(\mathbf{h}, u) = A\left\{\exp\left[-\frac{c\|\mathbf{h}\|}{(1 + a|u|^{2\alpha})^{\beta/2}}\right] + \frac{\nu}{1-\nu}\delta_{\mathbf{h}=0}\right\} + \lambda\left(1 - \frac{1}{2v}|h_1 - vu|\right)_+, \tag{6.11}$$

where $A = [(1-\nu)(1-\lambda)]/[1 + a|u|^{2\alpha}]$, $(\cdot)_+ = \max(\cdot, 0)$, and the constants a and c are nonnegative scale parameters of time and space, respectively. The smoothness parameter α, the space–time interaction parameter β, the nugget parameter ν, and the symmetry parameter λ all take values in [0, 1]. The vector $\mathbf{h} = (h_1, h_2)^{\mathrm{T}}$ is comprised of an east–west component h_1, and north–south component h_2. The scalar $v \in \mathbb{R}$ is an east–west velocity. When $\lambda = 0$, this model simplifies to a fully symmetric model. Further assuming $\beta = 0$, this model reduces to a separable model. When these two parameters are nonzero, λ and β control the degree of nonsymmetry and nonseparability, respectively.

It is particularly interesting to see the size at the null values ($\lambda = 0$ and $\beta = 0$), and the power at the observed values of the empirical spatio-temporal model with $\lambda = 0.0573$ and $\beta = 0.681$. When using the pairs with the smallest ratio, we find that the empirical sizes with respect to the nominal level 0.05 are 0.074 and 0.084 for the tests of fully symmetry and separability. The powers at the empirical values fitted from the data are 0.88 and 1.00. It makes sense that we have greater power against the more restrictive hypothesis of separability, but we have strong confidence in the rejection of full symmetry as well. By comparison, when we use the five stations with the smallest $\|\mathbf{h}\|$, we have empirical sizes of 0.083 and 0.091, and powers of 0.48 and 0.87, respectively. It is seen that the sizes are further from nominal and the powers are lower. These facts demonstrate that the physically motivated choice of space–time lags increases the reliability of the test. In all four tests, the sizes of the tests are between 2 and 4% above the nominal level.

6.7.4 Numerical experiments on the test for separability for data on a grid

This experiment is focused on evaluating the separability test for data on equally spaced grids, such as the Pacific Ocean wind data. The Pacific Ocean wind data is collected over a moderately sized spatial grid. However, it is informative to study small spatial grid sizes, which presents a more challenging setting for the test. In addition to evaluating the test in terms of size and power, we assess the effect of estimation of the matrix $\mathbf{\Sigma}$ by its subsampling-based estimator, $\widehat{\mathbf{\Sigma}}$, the grid size, the temporal size $|T_n|$, and the temporal correlation on the test, respectively. To implement all these tasks, consider the first-order vector autoregressive model, VAR(1), as presented in Section 6.5 [see also de Luna and Genton (2005)]. This model allows for explicit computation of the asymptotic covariance matrix $\mathbf{\Sigma}$ [Priestley (1981), p. 693]. In the VAR(1) process, $\mathbf{Z}_t = \mathbf{R}\mathbf{Z}_{t-1} + \boldsymbol{\epsilon}_t$, where $\mathbf{Z}_t = [Z(\mathbf{s}_1, t), Z(\mathbf{s}_2, t), \ldots, Z(\mathbf{s}_K, t)]^{\mathrm{T}}$, K denotes the cardinality of S, $\boldsymbol{\epsilon}_t$ is a Gaussian multivariate white noise process with a spatially stationary and isotropic exponential correlation function, and $\mathbf{R}$ is a matrix of coefficients which determines the dependency between $\mathbf{Z}_t$ and $\mathbf{Z}_{t-1}$.

First, a choice of spatial covariance is necessary. We take an exponential covariance with $C(\mathbf{h},0)=\exp\left(-\frac{\|\mathbf{h}\|}{\phi}\right)$, $\phi=3.48$, and $\mathbf{R}=\rho\mathbf{I}$, where $\mathbf{I}$ denotes the identity matrix. This produces random fields with a separable space–time covariance. By varying grid size, temporal length $|T_n|$, and temporal correlation parameter ρ in the experiment, we can assess their effect on the size of the test. We choose two lags ($\|\mathbf{h}\|=1$, $u=1$) and ($\|\mathbf{h}\|=1, u=2$) in the test, so the degrees of freedom of the test statistic are $q=2$. The nominal level is set to be 0.05. The summary of results is as follows:

i. The grid size does not have a large effect on the size of the test, and the test size is around 0.05 even with $|T_n|=200$, while the temporal correlation parameter, ρ, brings the size upward slightly as correlation increases.

ii. Estimation of the covariance matrix $\mathbf{\Sigma}$ by $\widehat{\mathbf{\Sigma}}$ has a moderate effect on the size of the test when $|T_n|=200$, but little effect when $|T_n|=1000$.

To assess power in this setting, the coefficients in $\mathbf{R}$ are as follows: for each $(\mathbf{s}_i, t)$, the coefficient is ρ for $(\mathbf{s}_i, t-1)$, whereas it is 0.05 for $\{(\mathbf{s}_j, t-1) : \|\mathbf{s}_j - \mathbf{s}_i\| = 1\}$, and 0 for the remaining $(\mathbf{s}, t-1)$s. This produces nonseparable space–time covariances. The summary of results is as follows:

i. The power is approximately constant but increases mildly as ρ increases.

ii. The power increases dramatically as temporal size $|T_n|$ increases.

iii. In assessing power of the test, estimation of the covariance matrix $\mathbf{\Sigma}$ by $\widehat{\mathbf{\Sigma}}$ has little effect, especially when $|T_n|$ reaches 1000.

The overall conclusion is that the tests for symmetry and separability are reasonably reliable for time series of length 200, and quite reliable for time series of length 1000.

6.7.5 Taylor's hypothesis

Taylor's hypothesis [Taylor (1938)] addresses the relationship between the purely spatial and the purely temporal covariance by examining if there exists a velocity vector $\mathbf{v} \in \mathbb{R}^d$ such that $C(\mathbf{0}, u) = C(\mathbf{v}u, 0)$ for all u. Several researchers have discussed various restrictions for a specific covariance function to satisfy Taylor's hypothesis.

Specifically, Gupta and Waymire (1987) and Cox and Isham (1988) studied the approximate validity of the hypothesis for various space–time covariance functions. Among them is the geometrically anisotropic covariance model,

$$C(\mathbf{h}, u) = \sigma^2 \exp[-(a^2\|\mathbf{h}\|^2 + b^2u^2)^{1/2}],$$

where a and b are positive constants. Basic calculations show that Taylor's hypothesis holds for any $\mathbf{v} \in \mathbb{R}^d$ having $\|\mathbf{v}\| = b/a$. In other words, given a covariance model of this class we would expect, using a test for Taylor's hypothesis, to see large p-values on a circle of radius $\|\mathbf{v}\| = b/a$, with smaller p-values for vectors of smaller or larger length.

If the candidate vector $\mathbf{v}$ is known, then, analogously to the hypotheses in Section 6.7, we can write the null hypothesis as H_0^3: $C(\mathbf{0}, u) - C(\mathbf{v}u, 0) = 0$ and the contrast as $\mathcal{C}_3 = \widehat{C}_n(\mathbf{0}, u) - \widehat{C}_n(\mathbf{v}u, 0)$. Given Λ including the involved lags, we can find a matrix, say $\mathbf{A}_3$, to make $\mathbf{A}_3\mathbf{G} = \mathbf{0}$ under the null hypothesis. For example, if

$$\Lambda = \{(\mathbf{0}, u_1), (\mathbf{0}, u_2), (\mathbf{0}, u_3), (\mathbf{v}u_1, 0), (\mathbf{v}u_2, 0), (\mathbf{v}u_3, 0)\},$$

that is,

$$\mathbf{G} = [C(\mathbf{0}, u_1), C(\mathbf{0}, u_2), C(\mathbf{0}, u_3), C(\mathbf{v}u_1, 0), C(\mathbf{v}u_2, 0), C(\mathbf{v}u_3, 0)]^{\mathrm{T}},$$

then

$$\mathbf{A}_3 = \begin{bmatrix} 1 & 0 & 0 & -1 & 0 & 0 \\ 0 & 1 & 0 & 0 & -1 & 0 \\ 0 & 0 & 1 & 0 & 0 & -1 \end{bmatrix}. \tag{6.12}$$

Under the null hypotheses, we have $\mathbf{A}_3\mathbf{G} = \mathbf{0}$. The rest is completely analogous to the previous tests for full symmetry and separability. However, as defined, Taylor's hypothesis states the possible *existence* of such a vector $\mathbf{v}$. Typically, the possible vector $\mathbf{v}$ satisfying the hypothesis is unknown. It is informative to then use the test in an exploratory fashion by viewing the testing results over a range of vectors $\mathbf{v}$. Such a possible set of vectors is depicted in Figure 6.2. The test can be performed for each $\mathbf{v}$ in the set, to obtain a field of p-values corresponding to the vector field. Then, determination of the vector $\mathbf{v}$ (if any) satisfying Taylor's relationship can be made by examining the p-value for each vector. Large p-values indicate that Taylor's hypothesis is admitted with vectors corresponding to these large p-values.

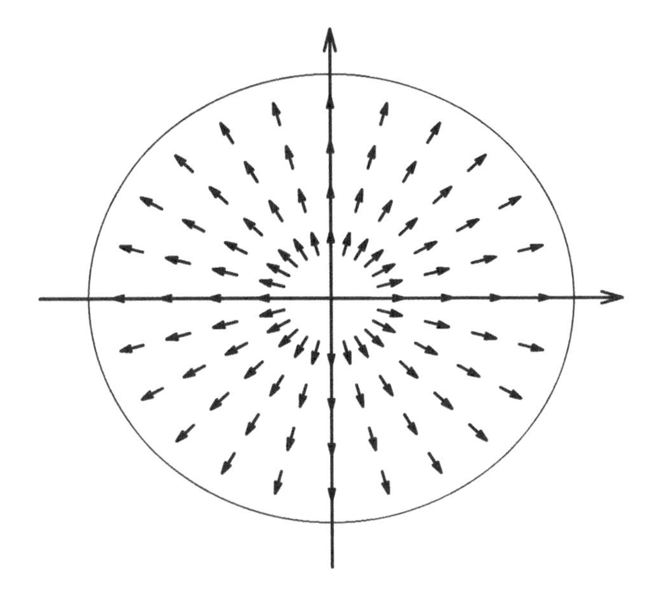

Figure 6.2 A depiction of lags used to test for Taylor's hypothesis.

This approach was carried out in Li *et al.* (2009) on NEXRAD reflectivity data of the type described in Section 9.1. The spatial observations are the radar reflectivity on a $4\,\text{km} \times 4\,\text{km}$ spatial grid in the southeastern United States. Between May 2, 2002 and May 5, 2002, this reflectivity was measured every 15 minutes. The hypothesis test for space lags of 4 km and a time lag of 15 minutes suggests that there are vectors $\mathbf{v}$ such that Taylor's hypothesis may well hold. In other words, the hypothesis tests for Taylor's hypothesis yield nonsignificant p-values. For larger space and/or time lags, however, testing strongly indicates that Taylor's hypothesis does not hold.

6.8 Nonstationary space–time covariance models

All spatio-temporal covariance models discussed until now have been stationary. This is often a reasonable assumption after accounting for a suitable mean function, $m(\mathbf{s}) + \mu(t)$, or more generally, $\mu(\mathbf{s}, t)$, as described in Section 6.5. Often, however, it is more appropriate to model nonstationary behavior through the covariance function. In the purely spatial situation this has been discussed in Section 3.7. For nonstationary covariance models in the spatio-temporal setting there are many application papers, for instance, Calder (2007) and Fuentes *et al.* (2008). More research into nonstationary space–time covariance models would clearly expand the opportunity for application.

7

Spatial point patterns

In this chapter, we introduce and discuss some fundamental properties of spatial point patterns. A spatial point pattern consists of locations of events in a given region $D \subset R^2$. The number of events in any subset $B \subseteq D$ will be denoted by $n(B)$, while $|B|$ will denote the area of this subset. For example, Figure 7.1 displays the locations of $n(D) = 71$ pine trees in Sweden on a $D = 96\,\text{m} \times 100\,\text{m}$ rectangular domain with $|D| = 9600\,\text{m}^2$. This point pattern is taken to be the (random) outcome of a point process. It is typically certain characteristics of this underlying point process that we hope to model. For example, is there any spatial connection between the location of a given tree and that of any other trees? It may be that the existence of a tree suggests a favorable environment, so it becomes likely that another tree will be nearby. This type of behavior is described as 'clustering.' On the other hand, a tree may use up moisture and sunlight, and thus the presence of a tree may make it less likely that another tree will be nearby. This suggests an 'inhibition' induced by existing trees. In fact, both clustering and inhibition could be present at different distances. Of course, it may be the case that neither clustering nor inhibition is present, at any distance, and that the presence of a tree has no effect on the existence of other trees nearby. This corresponds to 'complete spatial randomness.' In order to formalize these notions, we need to develop some basic properties of an underlying point process.

Spatial Statistics and Spatio-Temporal Data: Covariance Functions and Directional Properties Michael Sherman

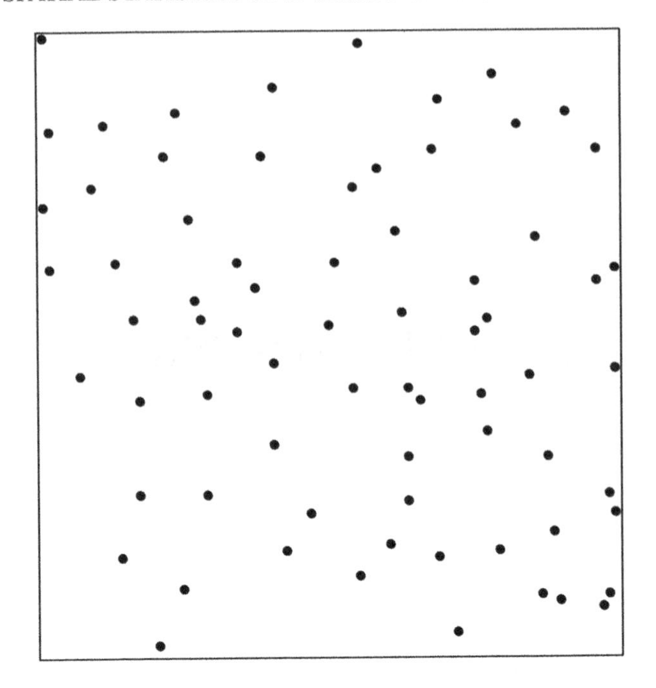

Figure 7.1 Locations of 71 Swedish pine trees.

7.1 The Poisson process and spatial randomness

The most basic characteristic of a point process is the first-order intensity. This gives, at each location **s**, the expected number of events in a very small neighborhood around the location **s**. Formally, the first order intensity function is:

$$\lambda(\mathbf{s}) = \lim_{|d\mathbf{s}|\to 0} \frac{E[N(d\mathbf{s})]}{|d\mathbf{s}|}.$$

For a (first-order) stationary process we have $\lambda(\mathbf{s}) = \lambda$, for all **s**. This means that we expect no more or less events in any two regions of equal area. In this case, we can estimate the first-order intensity by $\widehat{\lambda} = n(D)/|D|$. For example, if the pine trees come from a first-order stationary process, then the first-order intensity is estimated by $\widehat{\lambda} = 71/[(96\,\mathrm{m})(100\,\mathrm{m})] = 0.00739/\mathrm{m}^2$.

The most fundamental question is: which, if any, of the types of behaviors discussed in the first paragraph, 'clustering,' 'inhibition' or 'complete randomness' holds. Complete spatial randomness is given by the Poisson process; namely, that for some positive intensity, λ, and any disjoint areas $B_1, \ldots, B_k$ in D, it holds that

$$P[n(B_i) = n_i, i = 1, \ldots, k] = \prod_{i=1}^{k} \frac{(\lambda|B_i|)^{n_i} \exp(-\lambda|B_i|)}{n_i!}.$$

This definition implies the following basic properties:

i. The Poisson process is stationary. This is seen by observing that, for any region, B, the expected number of events in this region, $\lambda|B|$, depends only on the size of the region. Note, further, that the entire joint distribution of the $n(B_i)$s depends only on the areas, and not on their exact locations. This property is 'strict stationarity' for point processes.

ii. The number of events in any region has a Poisson distribution.

iii. The numbers of events in disjoint regions are independent random variables.

A fundamental question for any point pattern is whether or not it is compatible with complete spatial randomness (CSR). If CSR holds, then there is no further modeling necessary to describe the underlying point process, as all joint distributions are given by the definition of the Poisson process. Further, a rejection of CSR would hopefully point to a reasonable alternative, for example clustering or inhibition models.

An apparently natural way to gauge the appropriateness of CSR is to compute all the distances, u_{ij}, between all pairs of events, $i \neq j = 1, \ldots, n$, and compare the distribution of these to what is expected under CSR. In turns out that this doesn't work too well. Tests based on these interevent distances are not very powerful in detecting either clustering or inhibition, when present. Any clustering or inhibition is often focused on events that are *close to* other events. This suggests observing and studying distances between events and their *closest* neighbor event. Let v_i, $i = 1, \ldots, n$, denote the distances from each event to its nearest neighbor event.

It is useful to have simple summary statistics of these n nearest neighbor distances in order to assess characteristics of the underlying point process. The simplest of such statistics are $V_1 = \min_i v_i$ and, more generally, $V_k = v_{(k)}$, the kth smallest of the v_i. For large n, the distribution under CSR of

$$TS_k := \frac{n(n-1)\pi V_k^2}{|D|}$$

is χ^2_{2k}.

For the Swedish pines data we have $n = 71$, $V_1 = 2.236$ m, and $|D| = 9600$ m^2, giving $TS_1 = 8.13$. The associated p-value is $P\left[\chi^2_2 > 8.13\right] = 0.017$. This suggests (rather strongly) that the minimum distance between events, V_1, is too large, and is suggestive of inhibition.

One difficulty with V_1 is that for a moderate to a large number of events it becomes unstable. To see this, consider $n = 150$ events on the unit square. Say, for example, that all distances are rounded to two decimal places. Then $V_1 = 0$ or $V_1 \geq 0.01$. Then $TS_1 = 0$ or $TS_1 \geq 7.0$. The former suggests clustering and the latter suggests inhibition. These two opposite conclusions seem quite inappropriate for two outcomes that could be arbitrarily close. One way to make the arithmetic more stable is to use V_k, but then the choice of k becomes somewhat arbitrary.

A more satisfying way to stabilize the arithmetic, and to use a broader measure is to use *all* the nearest neighbor distances, v_i, $i = 1, \ldots, n$. Specifically, let

$$\widehat{G}(v) = \frac{\#(v_i \leq v)}{n}$$

denote the empirical distribution function of the nearest neighbor distances. Under the null hypothesis of CSR we have (ignoring edge effects) that the number of points, $n(B)$, in any region B, is Poisson distributed with mean $\lambda|B|$. Thus, the probability that no events are within v of another given event is equal to $\exp(-\lambda\pi v^2)$, so that

$$G(v) = 1 - P(V > v) = 1 - \exp(-\lambda\pi v^2).$$

Estimating λ by $\widehat{\lambda}$ allows for direct graphical comparison between the empirical G, $\widehat{G}$, and the theoretical G under CSR. This can be done through the distribution functions or through the densities. Figure 7.2 shows the two distribution functions for the Swedish pines data set, and Figure 7.3 shows a comparison of the histogram of v_is and the (estimated) density of nearest neighbor distances under CSR,

$$g(v) = 2\pi v\widehat{\lambda}exp[-\pi\widehat{\lambda}v^2].$$

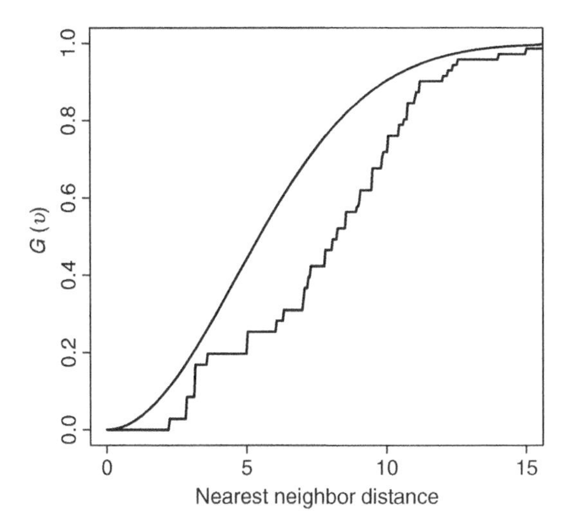

Figure 7.2 The distribution function of observed nearest neighbor distances, and the theoretical distribution under CSR for the Swedish pines.

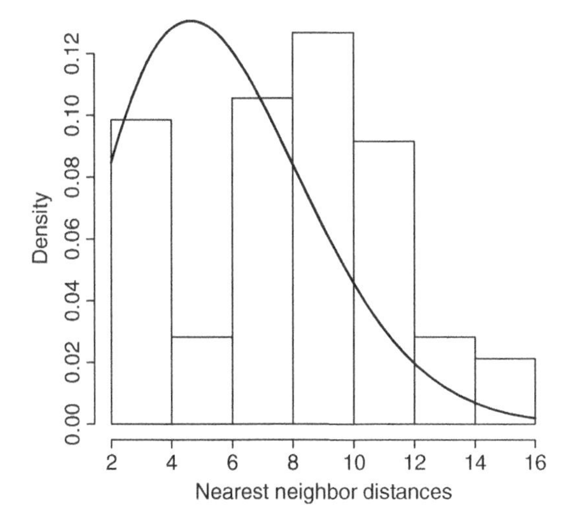

Figure 7.3 Histogram of the observed nearest neighbor distances and the theoretical density under CSR for the Swedish pines.

From the distribution functions, we see that $\widehat{G}(v)$ is below $G(v)$ for small v, which indicates regularity or inhibition. The density comparison suggests that the mode of the data histogram is between 8 m and 10 m, while the mode of G is between 4 and 6 m. To make a formal test based on the nearest neighbor distances we require the distribution of some function of the nearest neighbor distances. A natural function is simply the sample average of the nearest neighbor event distances. Donnelly (1978), for moderate to large number of events, n, and not too irregular domains, D, gives that

$$\mu_{n,D} := E(\overline{v}) = \frac{1}{2}\left(\frac{|D|}{n}\right)^{1/2} + \left(0.051 + \frac{0.042}{n^{1/2}}\right)\frac{l(D)}{n},$$

$$Var_{n,D} := Var(\overline{v}) = \frac{0.07|D|}{n^2} + \frac{0.037|D|^{1/2}l(D)}{n^{5/2}},$$

where $l(D)$ denotes the length of the boundary of the region D. Further, for a moderate to large number of events, he argues that

$$Z_{n,D} := (\overline{v} - \mu_{n,D})/Var_{n,D}^{1/2}$$

has an approximate standard normal distribution. For the Swedish pines data set, the summaries are $\overline{v} = 7.908$ m, $\mu_{n,D} = 6.123$ m, and

$$Var_{n,D}^{1/2} = 0.408\,\text{m}.$$

This gives a (one-sided) p-value of $P(Z_{n,D} \geq 4.375) = 6.07 \times 10^{-6}$, which leads to a very strong rejection of CSR. All graphics and test statistics have led to the conclusion that the Swedish pines data are not compatible with CSR, and that there is inhibition present. Figure 7.6 (d) shows $n = 71$ points from a Poisson process. The original data set is shown in (a). To many eyes, the Swedish pines appear to be more compatible with CSR than the point pattern from the CSR model, which appears somewhat clustered. Thus, to the eye, the inhibited process is often taken as random while the truly random one is taken as clustered. This demonstrates how our eye can be a rather poor judge of appropriate model dynamics, and clearly points to the need for the more objective measures we have discussed.

Closely related to the nearest neighbor distribution function, $G(\cdot)$, is the K-function. The K-function, $K(y)$, is heuristically defined by

$$\frac{1}{\lambda}E(\#\text{ events within } y \text{ of an arbitrary event}),$$

where λ is the first-order intensity. Under CSR,

$$\begin{aligned}&E(\#\text{ points within } y \text{ of an arbitrary event})\\&\quad = E(\#\text{ events in a circle of radius } y) = \lambda\pi y^2.\end{aligned}$$

Thus, under CSR, $K(y) = \pi y^2$. This definition of the K-function is more directly interpretable than is the nearest neighbor distribution function, $G(\cdot)$, as the number of additional events within an arbitrary single event. Further, we can often compute K-functions under alternatives to CSR, which makes this function very useful for model fitting. Computing G under alternatives to CSR is typically quite difficult. The K-function is (crudely) estimated by

$$\widehat{K}(y) = \frac{1}{\widehat{\lambda}} \frac{[\#\text{ pairs } (i, j) \text{ with } d(i, j) \leq y]}{n}.$$

This is often adequate for small y; for larger y, various weighting functions reduce the negative bias of this estimator due to edge effects.

Under CSR, $Var[\widehat{K}(y)] \simeq Cy^2$, for some constant C. The practical importance of this is that the fluctuations around the empirical K-function increase as the distance y increases. For this reason, a more informative comparison of the empirical to a hypothetical K-function is given by

$$\widehat{L}(y) = \sqrt{\frac{\widehat{K}(y)}{\pi}},$$

which estimates y under CSR.

Figure 7.4 plots the empirical L-function for the Swedish pines point pattern. We see that the empirical L-function lies below the theoretical L-function over small distances. While this suggests inhibition, we note that the comparison between this plot and Figure 7.2 suggests that the nearest neighbor distances are more powerful to detect departures from CSR than is the L-function. A useful and flexible alternative method of detecting departures from CSR is the paired correlation function [Illian *et al.* (2008)]. The cumulative nature of the empirical nearest neighbor distribution function and empirical K-function can mask features at different distances. The paired correlation function enables detection of departures from CSR at different spatial distances, although empirical estimation depends on a smoothing parameter.

Due to the thorough rejection of CSR in the point pattern, we now desire to describe a point process model which is more compatible with the observed events. As all indicators point to inhibition, we first consider inhibition models. Clustered models are discussed in the following section.

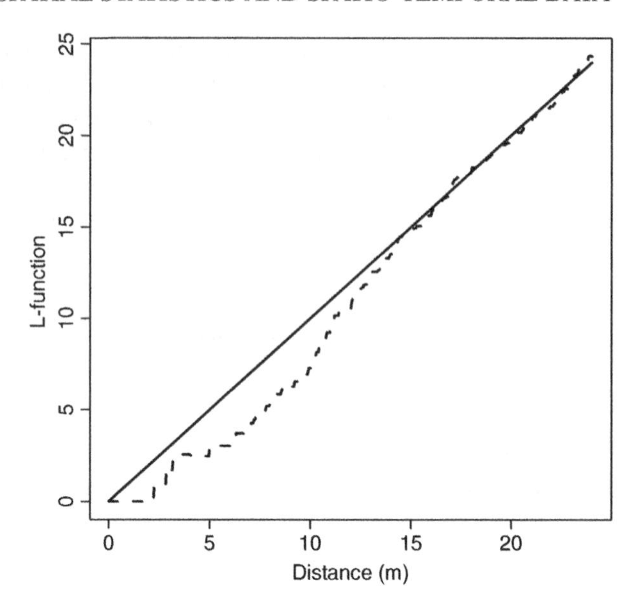

Figure 7.4 *The theoretical (solid) and empirical (dashed) L-functions for the Swedish pines.*

7.2 Inhibition models

A *simple inhibition model* evolves sequentially:

i. Event $\mathbf{s}_1$ is uniformly distributed over D according to CSR;

ii. Given $\mathbf{s}_1, \ldots, \mathbf{s}_{i-1}$, $\mathbf{s}_i$ is uniformly distributed on $\mathbf{t} \in D$ such that $d(\mathbf{t}, \mathbf{s}_j) \geq \delta$, $j = 1, \ldots, i-1$

This gives n events on D, none within δ of any other event. As δ increases, the observed pattern becomes more regular. Although often useful, this model is too crude to model the Swedish pines data, where there seems to be inhibition at short distances, but this seems to diminish the number of nearby trees, not to make such neighbors nonexistent. We seek a model that will discourage events from being too close to each other, while allowing for events to be arbitrarily close. Towards this end, call two locations, $\mathbf{s}_1$ and $\mathbf{s}_2$, *neighbors*, if they are separated by distance less than some constant δ. Consider the joint density of n events,

$f(\mathbf{s}_1, \ldots, \mathbf{s}_n)$. Roughly, this is the likelihood of observing events at these n locations simultaneously. A more flexible model than that of simple inhibition is the Strauss model, for which

$$f(\mathbf{s}_1, \ldots, \mathbf{s}_n) = \alpha \beta^n \gamma^s,$$

for $\beta > 0$ and $0 \leq \gamma \leq 1$. In the joint density, s denotes the number of pairs of neighbors in D, β is the overall intensity, γ denotes the interaction strength between neighbors, and α is a normalizing constant so that the density integrates to one. The value $\gamma = 1$ corresponds to no dependence on the number of neighbors, that is, to a Poisson process. At the other extreme value, $\gamma = 0$, corresponds to a simple inhibition process (also known as a 'hard-core Strauss process'). Intermediate values of γ express the degree of inhibition present.

To fit a Strauss model to the Swedish pines data we need to choose a reasonable value of δ, the range of inhibition. An often reasonable ad hoc approach is to find where $\widehat{K}(t) - \pi t^2$ reaches a local minimum, where $\widehat{K}(t)$ denotes the empirical K-function. For these data, there is a pronounced local minimum at $t = 6.96$ m, and a global minimum at $t = 9.84$ m. Thus, we consider $\delta = 7$ and $\delta = 10$ as candidate models. To assess which is more appropriate, $B = 99$ data sets were generated from each of the two models. Figure 7.5 gives the histograms of mean nearest neighbor distances from the patterns from the two models. The average mean nearest neighbor distance from data sets from the $\delta = 7$ model is 7.668 ($SD = 0.0468$), while the $\delta = 10$ model average mean nearest neighbor distance is 7.172 ($SD = 0.0583$). The two simulation (two-sided) p-values are 0.62 and 0.22, respectively. Thus, overall, the simulations with the radius of inhibition of $\delta = 7$ are somewhat more compatible with the observed $\overline{v} = 7.908$, but both models are reasonably compatible.

For the $\delta = 7$ model, the parameter estimates are $\widehat{\beta} = 0.017$ and $\widehat{\gamma} = 0.268$. For comparison's sake, the $\delta = 10$ model has $\widehat{\gamma} = 0.461$ corresponding to less inhibition, and to the lower mean nearest neighbor distances in the simulation. Figure 7.6 shows one data set from each of the fitted models ($\delta = 7$, $\gamma = 0.268$, and $\delta = 10$, $\gamma = 0.461$), along with the original data set. The two simulated point patterns from the two models seem broadly compatible with each other, and with the original point pattern. CSR, however, is incompatible with the other three.

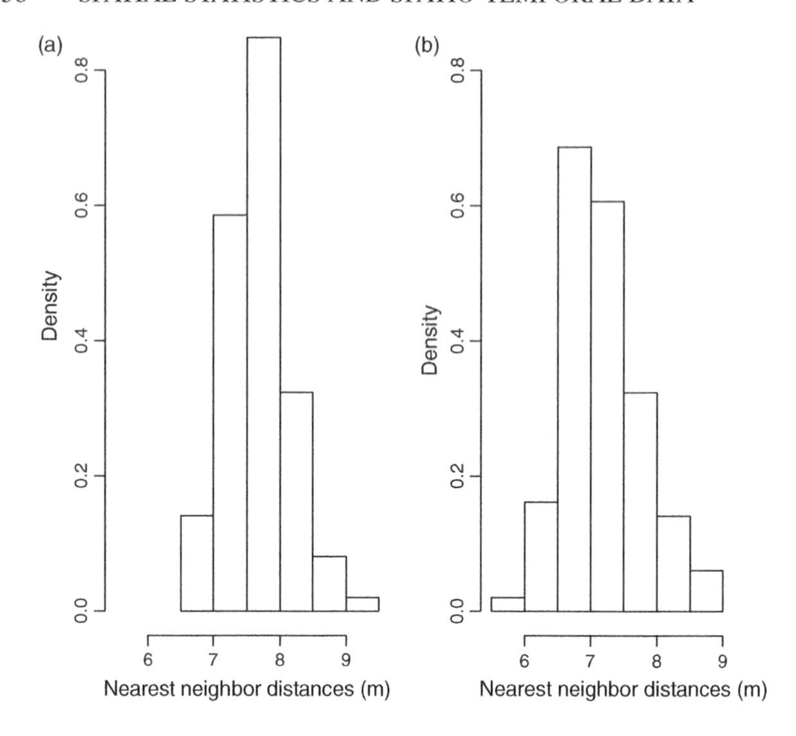

Figure 7.5 Distributions of mean nearest neighbor distances from two Strauss models for the Swedish pine tree locations. (a) $\delta = 7$; (b) $\delta = 10$.

7.3 Clustered models

Figure 7.7 (a) gives the standardized locations of $n = 62$ redwoods. In other words, all locations are rescaled to be located on the unit square $D = [0, 1]^2$. The distribution of locations is apparently quite different from the Swedish pines. Naively, we might assume they are clustered. We are aware, however, from the analysis of the Swedish pines, that the eye tends to see more clustering than is actually present. For this reason we desire a more objective analysis.

Figure 7.8 shows the empirical nearest neighbor distribution function with the theoretical function under CSR. It is seen that the majority of the empirical curve is above that of the theoretical, indicating clustering. In particular, the median nearest neighbor distance is approximately 0.03 for

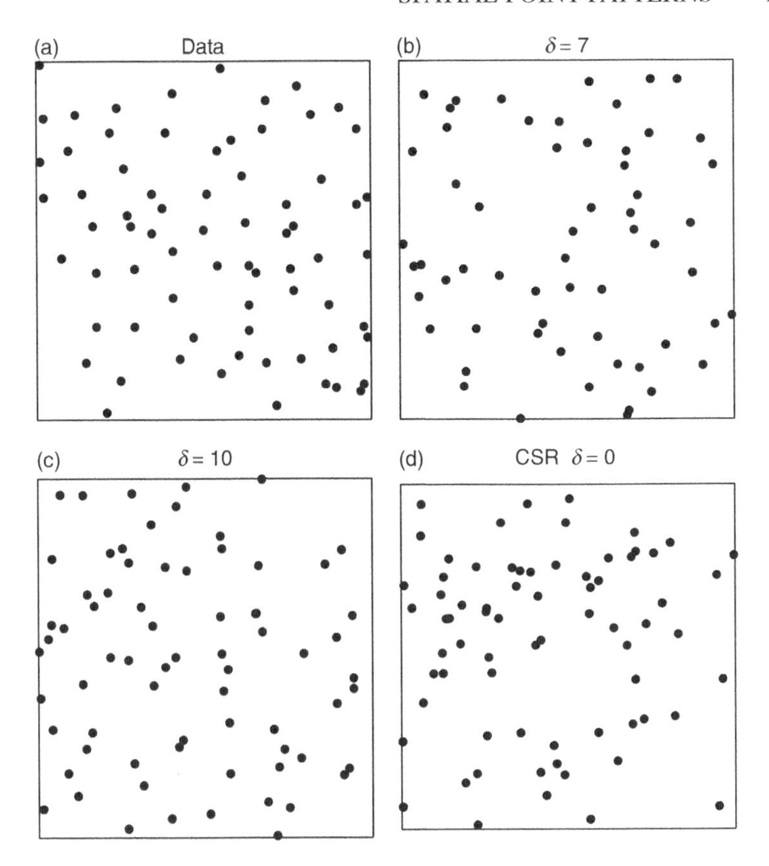

Figure 7.6 Four data sets: (a) original point pattern of Swedish pines; (b) data from Strauss process with $\gamma = 7$; (c) data from Strauss process with $\gamma = 10$; (d) data from a Poisson process.

the empirical, while the median is approximately 0.06 for the theoretical. The standardized Donnelly statistic is

$$Z_{n,D} = (\overline{v} - \mu_{n,D})/Var_{n,D}^{1/2} = (0.0385 - 0.06713)/0.00481 = -5.95,$$

and $P[Z_{n,D} \leq -5.95] = 1.34 \times 10^{-9}$. Thus we do thoroughly reject the null hypothesis of CSR. The large negative value of the test statistic suggests clustering. In order to find a reasonable alternative to CSR we need to discuss some clustering models. We discuss the dynamics arrived at through a Poisson cluster process model; in particular, we now describe the Neyman–Scott process.

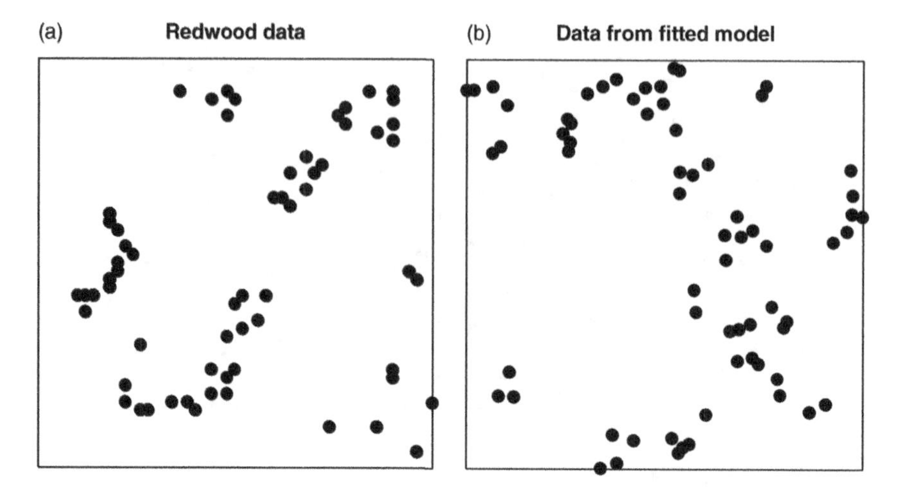

Figure 7.7 Redwood point pattern, and data from the fitted model. (a) Original point pattern of redwoods. (b) One data set from the best-fitting Thomas process.

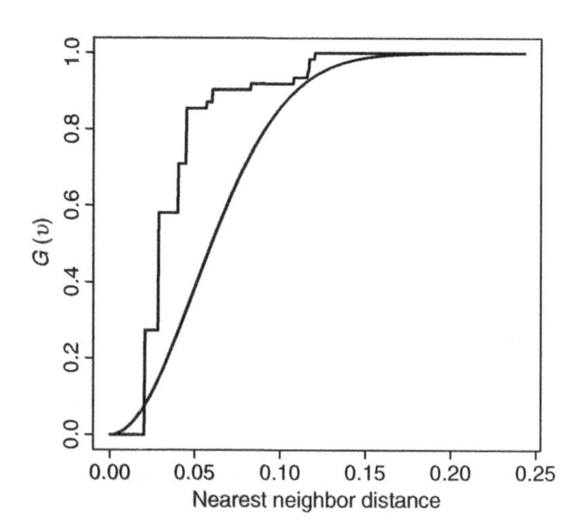

Figure 7.8 Distribution function of observed nearest neighbor distances, and theoretical distribution under CSR for the redwoods. The empirical above the theoretical indicates clustering.

A Neyman–Scott process [Neyman and Scott (1958)] has the following components:

i. 'Parent events' are CSR, with first order intensity ρ.
ii. Each parent has S 'offspring.' The distribution of S is given by a density function, $p_s, s = 0, 1, \ldots$
iii. Offspring are located i.i.d. around their parent, according to a bivariate density function $f(\cdot)$.

The resulting point pattern is the locations of the offspring. Note that the variance in the density $f(\cdot)$ determines the degree of clustering in the process. A special case of this process is the 'Thomas process,' for which, in Step (ii), S has a Poisson distribution, and, in Step (iii), the bivariate density $f(\cdot)$ is normal, isotropic, and with independent directional components.

Notice that the Poisson cluster process is first-order stationary as, for any region A, we have

$$E(\#\text{events in } A) = E[E(\#\text{events in } A|\#\text{ parent events})]$$
$$= E[\#\text{parent events} E(S)] = E(\#\text{parent events})E(S) = \rho|A|E(S).$$

Thus, the Poisson cluster process is first-order stationary with intensity $\lambda = \rho E(S)$. This is simply the intensity of the parents multiplied by the expected number of offspring per parent.

The first order intensity corresponds to the mean function of a spatial process. A natural question is: what corresponds to the covariance function? In the initial section of Chapter 7, we introduced the K-function. In some sense this gives the second-order properties of a point process. A more fundamental answer, however, is the second-order intensity, given by:

$$\lambda_2(\mathbf{s}, \mathbf{t}) = \lim_{|d\mathbf{s}| \to 0} \frac{E[N(d\mathbf{s})N(d\mathbf{t})]}{|d\mathbf{s}||d\mathbf{t}|}.$$

This is roughly interpreted as the chance of observing an event at both locations $\mathbf{s}$ and $\mathbf{t}$. Under CSR, we have $\lambda_2(\mathbf{s}, \mathbf{t}) = \lambda^2$, whenever $\mathbf{s} \neq \mathbf{t}$. Under second-order stationarity, $\lambda_2(\mathbf{s}, \mathbf{t}) = \lambda_2(\mathbf{s} - \mathbf{t}, \mathbf{0})$, for all $\mathbf{s} \neq \mathbf{t}$. Under isotropy, we have that $\lambda_2(\mathbf{s} - \mathbf{t}, \mathbf{0}) = \lambda_2(d)$, whenever $d(\mathbf{s}, \mathbf{t}) = d$. We assume that isotropy holds for all models discussed in this chapter. Isotropy assessment for point process models based on observed point patterns will be discussed in Chapter 8.

It can be shown that under stationarity and isotropy, the second-order intensity and the K-function are related via

$$K(y) = \frac{2\pi}{\lambda^2} \int_0^y r\lambda_2(r)dr.$$

This is a more rigorous definition of the K-function than given previously, and can be used to fit clustered models. Under CSR, we have $\lambda_2(d) = \lambda^2$, for any d, and thus $K(y) = \pi y^2$, as it must. Further, for the Neyman–Scott Process, at any distance d, we have that $\lambda_2(d) = \lambda^2 + \rho E[S(S-1)]$ $f_2(d)$, where the first term gives the contribution of two offspring from different parents, and the second gives the contribution from two offspring from the same parent, separated by distance d. The density $f_2(d)$ is the density of the distance between two offspring from the same parent, $f_2(d) = \int_{R^2} f(\mathbf{w}) f(\mathbf{v} - \mathbf{w})\mathbf{dw}$, where $\|\mathbf{v}\| = d$. Thus,

$$K(y) = \pi y^2 + \frac{E[S(S-1)]F_2(y)}{\rho[E(S)]^2}$$

where $F_2(y)$ is the distribution function of the distance between two events in the same cluster. For example, if the distribution of offspring events around each parent is bivariate normal with independent components, and common variances, σ^2, then d^2 is distributed as $2\sigma^2\chi^2_2$, so that

$$f_2(d) = \frac{d}{2\sigma^2} \exp\left(\frac{-d^2}{4\sigma^2}\right),$$

and $F_2(y) = 1 - \exp\left[\frac{-y^2}{4\sigma^2}\right]$. If the number of offspring, S, has a Poisson distribution, then $E[S(S-1)] = [E(S)]^2$. We see that the Thomas process has K-function $K(y, \theta) = \pi y^2 + (1/\rho)[1 - \exp(-y^2/4\sigma^2)]$ where $\theta = (\rho, \sigma^2)$.

To find a fitted model using the K-function, consider the discrepancy measure:

$$D(\theta) = \int_0^{y_0} \{[\widehat{K}(y)]^c - [K(y, \theta)]^c\}^2 dy,$$

where y_0 gives the range of interest, and c is chosen to make the variance of the estimated K-function approximately constant. An empirically reasonable value for the range of integration is $y_0 = 0.25k$ for a domain of a $k \times k$ square. Parameters computed by minimizing $D(\theta)$ are often called minimum contrast estimators; see, for instance, Moller and Waagepetersen (2004) or Illian *et al.* (2008) for more on minimum contrast estimators, and on likelihood-based inference for spatial point processes.

To carry out this procedure, good initial values are useful. Here, note that for the Thomas process, $K(y, \theta) - \pi y^2 \simeq (1/\rho)$ for large y. Further, when $y = 4\sigma$, then $K(y, \theta) - \pi y^2 \simeq (1/\rho)$. This suggests setting

$$1/\widehat{\rho}_0 = \max_y[\widehat{K}(y) - \pi y^2] \text{ and } 4\widehat{\sigma}_0 = \arg\max_y[\widehat{K}(y) - \pi y^2].$$

For the redwood data, this gives $(1/\rho_0) = 0.044$, and so $\widehat{\rho}_0 = 22.7$, and $\arg\max_y[\widehat{K}(y) - \pi y^2] = 0.16$, and thus $\widehat{\sigma}_0 = (0.16/4) = 0.04$. Minimizing $D(\theta)$ with these initial values, and $c = 0.25$, gives the solutions $\widehat{\rho} = 23.6$ and $\widehat{\sigma} = 0.047$ (with $y_0 = 0.25$). This gives approximately 24 clusters, each with S offspring, where S has a Poisson distribution with $\lambda = 62/23.6 = 2.63$. To judge the compatibility of this fitted model with the data, we compute the mean nearest neighbor distances over 99 simulations from the fitted model. Figure 7.9(a) shows the distribution of the mean nearest neighbor distances from these simulations. We see that the mean nearest neighbor distances range from 0.040 to 0.078 over 99 simulations. Unfortunately, these are not compatible with the observed 0.0393.

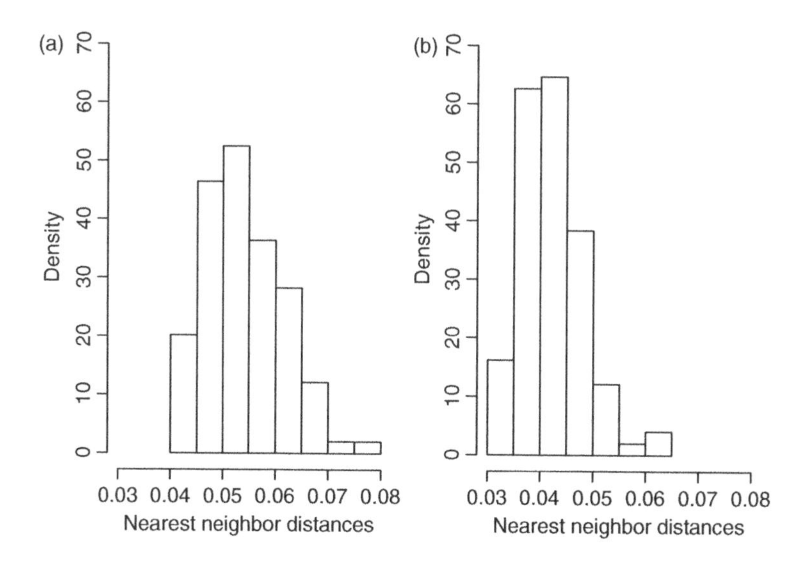

Figure 7.9 Distributions of mean nearest neighbor distances from two Thomas process models for the redwood trees. (a) c = 0.25; *(b)* c = 0.5.

Before rejecting the Poisson cluster process class of models, it is natural to gauge how sensitive the fitted model is to the choice of c in $D(\theta)$. Another reasonable value is $c = 0.5$; this value stabilizes the variance of $\widehat{K}(y)$ under CSR. Using this power of c (with the same initial values), we find $\widehat{\rho} = 26.9$ and $\widehat{\sigma} = 0.034$. The larger estimate of the number of clusters, ρ, corresponds to a tighter distribution of less offspring around the parents. Mean nearest neighbor distances from this fitted model range from 0.032 to 0.061 over 99 simulations from this fitted model. Figure 7.9(b) shows the distribution of the mean nearest neighbor distances from the simulations from this model. The two-sided simulation p-value $= 0.70$ for the model fitted with power $c = 0.5$. It is somewhat surprising that the mean nearest neighbor distances from these simulations *are* compatible with those of the data set.

In viewing the data plot in Figure 7.7, our eye tends to see far fewer than 24 or 27 clusters. Once again, it seems the eye tends to group together events that really shouldn't be grouped together. Figure 7.7 also shows one realization from the fitted model with $\widehat{\rho} = 26.9$. It seems broadly similar to the original data set. It also seems to have far fewer than 27 clusters, despite having been generated from a model with this many clusters!

The two models for these data are not practically very different. However, the radically different formal conclusions arrived at, based on the choice of c, for this data set is somewhat unsatisfying. While it may be the case that other goodness of fit tests might show better agreement between the two choices of c, we find that the performed test based on mean nearest neighbor distances gives quite different conclusions on model appropriateness. For this reason, a data-based choice of this tuning parameter is useful.

Towards this end, consider the general criterion to be minimized:

$$\tilde{D}(\theta) = \int_0^{y_0} w(y)\{[\widehat{K}(y)] - [K(y,\theta)]\}^2 dy,$$

where $w(y)$ denotes a function equal to the reciprical of $Var[\widehat{K}(y)]$. The function $w(y)$, in general, depends on the underlying point process and is thus generally unknown. Guan and Sherman (2007) developed a method to nonparametrically estimate the function $w(y)$ in $\tilde{D}(\theta)$, and thus eliminate the difficult choice of the appropriate power c in the expression for $D(\theta)$. Further, they show that, under mild conditions, the model parameter estimates arrived at by minimizing either $D(\theta)$ or $\tilde{D}(\theta)$ [with estimated $w(y)$] are consistent for the true model parameters, and have an asymptotic normal distribution. The estimators minimizing $\tilde{D}(\theta)$ [with estimated $w(y)$], however, are asymptotically efficient, as the estimated $w(y)$ is consistent for the true reciprocal of $Var[\widehat{K}(y)]$ under a large class of stationary point process models including CSR, simple inhibition, Strauss process, and Poisson cluster processes. This method, however, is not appropriate for an observed point pattern of the size of the redwood data. The nonparametric estimation of $w(y)$ is accomplished via spatial resampling (discussed in Chapter 10), and requires at least 100–200 events to perform well. For point patterns with smaller numbers of events, it is always a good idea to perform model fitting over a range of the tuning parameters (like y_0 and c) to find a good consensus model, if possible.

We have focused on models for stationary point processes. There are many point patterns that cannot be adequately modeled by a member of the stationary class. A basic nonstationary model is an inhomogeneous point process. The process is nonstationary through a nonconstant first-order intensity function, $\lambda(\mathbf{s})$. Often this intensity is modeled as a function of covariates. This is discussed, for instance, in Diggle (2003) and Illian *et al.* (2008). Nonparametric estimation of a nonconstant intensity is addressed in Guan (2008). When fitting models, it is often desired to assess the goodness of fit through residuals. Residual analysis from fitted spatial point pattern models is developed in Baddeley *et al.* (2005).

8

Isotropy for spatial point patterns

All the models discussed in Sections 7.2 and 7.3 are isotropic, that is, there are no directional effects to the distribution of event locations. Often this assumption is made out of convenience and is not appropriate. Here we discuss approaches to assess isotropy for spatial point patterns.

As discussed in Chapter 7, spatial point process models are useful tools to model irregularly scattered point patterns that are frequently encountered in biological, ecological and epidemiological studies. Further examples include locations of biological cells in a tissue, or leukemia patients in a state. A spatial point process is *stationary* if its distribution is invariant under translations. It is further said to be *isotropic* if its distribution is invariant under rotations about the origin. Otherwise, it is said to be *anisotropic*. As in the case of quantitative data in Chapter 5, the assumption of isotropy is often made in practice due to simpler interpretation and ease of analysis.

In many applications, however, it is important to assess isotropy and to consequently perform a thorough directional analysis if isotropy is rejected. Consider the leukemia data described in Section 8.5. Although these data motivate this study, there are several further examples of

Spatial Statistics and Spatio-Temporal Data: Covariance Functions and Directional Properties Michael Sherman

directional analyses of spatial point patterns. See, for example, Howe and Primack (1975), Haase (2001), Phillips and Barnes (2002), Schenk and Mahall (2002), and Staelens *et al.* (2003).

The majority of statistical analyses of spatial point patterns assume isotropy without checking/testing for it. For example, the words 'anisotropic' and 'anisotropy' do not even appear in the subject indices of Diggle (2003) and Upton and Fingleton (1985), as noted by Stoyan (1991). The more recent text of Illian *et al.* (2008) does *define* isotropy. In practice, the goodness of fit of a fitted (isotropic) model has been sometimes assessed through graphical methods. For example, by analogy with the methods in Chapter 5, one can estimate second-order intensities in separate directions and draw a rose plot to see if this intensity is similar at the same distance in different directions. Unfortunately, as discussed in Chapter 5, these methods often have little power in detecting an inadequate fit due to anisotropy. Ohser and Stoyan (1981), Castelloe (1998), and Rosenberg (2004) are among the very few who have proposed methods to assess isotropy (and to detect anisotropy) for spatial point processes. These approaches, however, are either informal or limited to certain classes of models.

As noted in Chapter 5, anisotropy has been extensively studied in geostatistics (i.e., for numerical spatial data). It is well understood that misspecifying an isotropic model as anisotropic may result in inappropriate spatial modeling and/or less efficient spatial prediction. Numerous procedures to test for isotropy for quantitative observations, and to model anisotropy have been proposed, as discussed in Chapter 5.

We discuss a formal approach to test for isotropy in spatial point patterns that is not restricted to a specific class of data-generating point process models. The approach is based on the asymptotic joint normality of the sample second-order intensity function (SSIF), which can be used to compare this function at several lags in multiple directions. We use an L_2 consistent subsampling estimator for the asymptotic covariance matrix of the SSIF, and use it to construct a test statistic with a limiting χ^2 distribution. The approach requires only mild moment conditions and a weak dependence assumption for the underlying point process.

In Section 8.1, the definition of the SSIF is reintroduced. The test statistic is discussed in Section 8.2. We conduct a numerical experiment to investigate the performance of the proposed testing approach and apply it to the leukemia data sets in Section 8.5.

8.1 Some large sample results

Consider a two-dimensional spatial point process, N. Let D denote the region where observations are taken, $|D|$ denote the area of D, and $N(D)$ denote the random number of points in D. Let $d\mathbf{s}$ be an infinitesimal region which contains the point $\mathbf{s}$. As in Chapter 7, define the first- and second-order intensity functions as

$$\lambda(\mathbf{s}) \equiv \lim_{|d\mathbf{s}|\to 0} \left\{ \frac{\mathrm{E}[N(d\mathbf{s})]}{|d\mathbf{s}|} \right\},$$

$$\lambda_2(\mathbf{s}, \mathbf{t}) \equiv \lim_{|d\mathbf{s}|,|d\mathbf{t}|\to 0} \left\{ \frac{\mathrm{E}[N(d\mathbf{s}) \times N(d\mathbf{t})]}{|d\mathbf{s}| \times |d\mathbf{t}|} \right\}.$$

If the process N is stationary, then $\lambda(\mathbf{s}) = \lambda$ and $\lambda_2(\mathbf{s}, \mathbf{t}) = \lambda_2(\mathbf{u})$, where $\lambda_2(\mathbf{u})$ denotes $\lambda_2(\mathbf{0}, \mathbf{u})$ and $\mathbf{u} = \mathbf{s} - \mathbf{t}$, which is referred to, in analogy to the quantitative case, as the lag, throughout the remainder of this chapter. If N, further, is isotropic, then $\lambda_2(\mathbf{u}) = \lambda_2^o(\|\mathbf{u}\|)$, where λ_2^o is a function defined on $\mathbb{R}$, and $\|\mathbf{u}\|$ is the Euclidean distance from $\mathbf{u}$ to the origin. The simplest example of a stationary and isotropic point process is a homogeneous Poisson process for which $\lambda_2(\mathbf{s}, \mathbf{t}) = \lambda^2$. In this chapter, we consider more general spatial point patterns, like the Strauss process and Thomas process of Chapter 7. Note that, under isotropy, if $\|\mathbf{u}_i\| = \|\mathbf{u}_j\|$, then $\lambda_2(\mathbf{u}_i) = \lambda_2(\mathbf{u}_j)$. Thus, we can test for isotropy by comparing SSIFs at lags with the same length in different directions. Note that *strict* isotropy requires that all joint distributions are direction independent. Our notion of isotropy is denoted as *weak* isotropy. This is in accordance with the analogous notion for quantitative processes as given in Chapter 5.

From now on, assume that the process N is stationary, but not necessarily isotropic. To estimate $\lambda_2(\cdot)$, let h be a positive constant and let $w(\cdot)$ be a bounded, nonnegative, symmetric density function which takes positive values only on a finite support. Define $N^{(2)}(d\mathbf{s}_1, d\mathbf{s}_2) \equiv N(d\mathbf{s}_1)N(d\mathbf{s}_2)I(\mathbf{s}_1 \neq \mathbf{s}_2)$, where $I(\mathbf{s}_1 \neq \mathbf{s}_2) = 1$ if $\mathbf{s}_1 \neq \mathbf{s}_2$ and 0 otherwise. Define the sample second-order intensity function (SSIF) to be the following natural kernel estimator of $\lambda_2(\mathbf{u})$:

$$\widehat{\lambda}_2(\mathbf{u}) = \int_{\mathbf{s}_1 \in D} \int_{\mathbf{s}_2 \in D} \frac{w[(\mathbf{u} - \mathbf{s}_1 + \mathbf{s}_2)/h]}{|D \cap (D - \mathbf{s}_1 + \mathbf{s}_2)| \times h^2} N^{(2)}(d\mathbf{s}_1, d\mathbf{s}_2), \quad (8.1)$$

where $|D \cap (D - \mathbf{s}_1 + \mathbf{s}_2)|$ denotes the intersected area of D and $D - \mathbf{s}_1 + \mathbf{s}_2 \equiv \{\mathbf{y} : \mathbf{y} = \mathbf{z} - \mathbf{s}_1 + \mathbf{s}_2, \mathbf{z} \in D\}$, which serves as an edge-correction term in Equation 8.1.

8.2 A test for isotropy

The test is based on $\widehat{\lambda}_{2,n}(\mathbf{u})$, computed from events within domain D_n, evaluated at a set of lags representing multiple directions. Using direct evaluation, it holds that:

$$\mathrm{E}\big[\widehat{\lambda}_{2,n}(\mathbf{u})\big] = \int w(\mathbf{x})\lambda_2(\mathbf{u} - h_n\mathbf{s})d\mathbf{s}.$$

Consider two lags $\mathbf{u}_i$ and $\mathbf{u}_j$, where $\|\mathbf{u}_i\| = \|\mathbf{u}_j\|$. Since $w(\cdot)$ is a symmetric kernel function and $\lambda_2(\mathbf{u})$ is also symmetric under isotropy, it can be concluded that

$$\mathrm{E}\big[\widehat{\lambda}_{2,n}(\mathbf{u}_i)\big] = \mathrm{E}\big[\widehat{\lambda}_{2,n}(\mathbf{u}_j)\big],$$

under isotropy. Thus the null hypothesis can be expressed in terms of the expected SSIFs as

$$H_0 : \mathrm{E}\big[\widehat{\lambda}_{2,n}(\mathbf{u}_i)\big] = \mathrm{E}\big[\widehat{\lambda}_{2,n}(\mathbf{u}_j)\big], \quad \mathbf{u}_i \neq \mathbf{u}_j, \text{ with } \|\mathbf{u}_i\| = \|\mathbf{u}_j\|.$$

Let $\widehat{\mathbf{G}}_n$ denote the SSIFs at some user-chosen lags. We form a set of contrasts based on the above equations in H_0 such that $\mathbf{A}\mathrm{E}(\widehat{\mathbf{G}}_n) = \mathbf{0}$ for some fixed full row rank matrix $\mathbf{A}$. To test for isotropy, we test the hypothesis $H_0 : \mathbf{A}\mathrm{E}(\widehat{\mathbf{G}}_n) = \mathbf{0}$. A similar technique has been used in Chapters 5 and 6, testing for spatial isotropy in the setting of spatial and spatio-temporal observations.

In light of the asymptotic results analogous to those in Chapter 5, we define the following test statistic

$$TS_n \equiv |D_n| \times h_n^2 \times (\mathbf{A}\widehat{\mathbf{G}}_n)^{\mathrm{T}}(\mathbf{A}\widehat{\boldsymbol{\Sigma}}_n\mathbf{A}^{\mathrm{T}})^{-1}(\mathbf{A}\widehat{\mathbf{G}}_n),$$

where $\widehat{\boldsymbol{\Sigma}}_n$ represents a consistent estimator of $\boldsymbol{\Sigma}$. Let r denote the row rank of $\mathbf{A}$. It follows that

$$TS_n \xrightarrow{D} \chi_r^2$$

under isotropy as $n \to \infty$ by the multivariate Slutsky's theorem [Ferguson (1996)]. Therefore, an approximate size-α test for isotropy rejects H_0 if

TS_n is bigger than $\chi^2_{r,\alpha}$, the upper α percentage point of a χ^2 distribution with r degrees of freedom.

To obtain the test statistic, an estimate of the matrix $\mathbf{\Sigma}$ is needed. Using the large sample covariance matrix is unwieldy, so it is often better to consider a subsampling approach for variance estimation. Specifically, let $D_{l(n)}$ be a subshape that is congruent to D_n both in configuration and orientation but rescaled, where $l(n) = cn^{\alpha}$ for some positive constant c and $\alpha \in (0, 1)$. Define a shifted copy $D_{l(n)}(\mathbf{y}) \equiv \{\mathbf{s} + \mathbf{y} : \mathbf{s} \in D_{l(n)}\}$, where $\mathbf{y} \in D_n^{1-c} \equiv \{\mathbf{y} : D_{l(n)}(\mathbf{y}) \subset D_n\}$. Let $\widehat{\mathbf{G}}_{l(n)}(\mathbf{y})$ be the SSIFs on $D_{l(n)}(\mathbf{y})$, and $h_{l(n)}$ be the bandwidth used to obtain $\widehat{\mathbf{G}}_{l(n)}(\mathbf{y})$. The subsampling estimator (denoted by $\widehat{\mathbf{\Sigma}}_n$) is given by

$$\frac{1}{|D_n^{1-c}|\, f_n} \int_{D_n^{1-c}} |D_{l(n)}| h_{l(n)}^2 \big[\widehat{\mathbf{G}}_{l(n)}(\mathbf{y}) - \overline{\mathbf{G}}_{l(n)}\big]\big[\widehat{\mathbf{G}}_{l(n)}(\mathbf{y}) - \overline{\mathbf{G}}_{l(n)}\big]^{\mathrm{T}} d\mathbf{y},$$

where $\overline{\mathbf{G}}_{l(n)} \equiv \frac{1}{|D_n^{1-c}|} \int_{D_n^{1-c}} \widehat{\mathbf{G}}_{l(n)}(\mathbf{y}) d\mathbf{y}$ and $f_n = 1 - \frac{|D_{l(n)}|}{|D_n|}$ is a finite sample bias correction. Using the results on spatial resampling in Chapter 10, we have that $\widehat{\mathbf{\Sigma}}_n$ is a L_2 consistent estimator for $\mathbf{\Sigma}$ under some regularity conditions.

8.3 Practical issues

To use the formal testing method proposed in Section 8.2, the user needs to determine which spatial lags to compare, the smoothing bandwidth, and the subblock size for variance estimation. The results in Sherman (1996) suggest that the subblock size should be such that approximately $cn^{1/4}$ points are included in each subblock, where c is a constant and n now denotes the total number of events. Empirical experimentation finds that $c \approx 0.8$ generally performs well. For data-driven methods of selecting the value of c, see, for example, Hall and Jing (1996) and Politis and Sherman (2001).

Our simulation experience suggests that the choice of $\mathbf{u}$ and h affects the test size only slightly but has a greater influence on the power. Overall we have found that a good choice for $\|\mathbf{u}\|$ is $\frac{1}{3}$ to $\frac{1}{2}$ of the dependence range. This range can be roughly estimated by studying an isotropic SSIF plot. Unless strong a priori knowledge of a suspected anisotropic pattern exists, it is appropriate to choose evenly spaced directions of $\mathbf{u}$, say between 0 and 180 degrees. Specifically, for data sets with sample sizes comparable to the examples in Section 8.4, it is reasonable to compare no more than four directions at a time. The choice of the bandwidth h should be determined based on two criteria:

i. There should be sufficient sample size, say at least 150–200 pairs of points, used to calculate each second-order intensity function.
ii. No, or very few, overlapping pairs are used to calculate the estimated function in different directions.

In practice, the choice of h should be considered together with that of $\mathbf{u}$ in order to satisfy all the discussed criteria.

In addition to the general guidelines listed in the previous paragraph on the selection of the bandwidth, h, we now discuss a data-driven method. Assume that the second-order intensity function is twice continuously differentiable. Direct evaluation shows that:

$$\mathrm{E}[\widehat{\lambda}_{2,n}(\mathbf{u})] = \lambda_2(\mathbf{u}) + \left\{\int \left[\mathbf{s}^{\mathrm{T}}\lambda_2''(\mathbf{u})\mathbf{s}\right] w^2(\mathbf{s})d\mathbf{s}\right\} h_n^2/2 + o\left(h_n^2\right).$$

Combining the above result and the variance expression of $\widehat{\lambda}_{2,n}(\mathbf{u})$, we obtain the (asymptotically) optimal bandwidth choice by minimizing the mean squared error, yielding:

$$h_n = \left(\frac{2\lambda_2(\mathbf{u})\int w^2(\mathbf{s})d\mathbf{s}}{|D_n|\left\{\int \left[\mathbf{s}^{\mathrm{T}}\lambda_2''(\mathbf{u})\mathbf{s}\right] w^2(\mathbf{s})d\mathbf{s}\right\}^2}\right)^{\frac{1}{6}} \equiv c \times \frac{1}{|D_n|^{\frac{1}{6}}}.$$

In the spirit of the work of Hall and Jing (1996), this implies that, for a region that is smaller than D_n, but still relatively large, say D_m, the optimal bandwidth $h_m \approx c \times |D_m|^{-1/6}$ and thus $h_n \approx (|D_m|/|D_n|)^{1/6} \times h_m$. To obtain an estimate of h_m, the user can first estimate $\lambda_2(\mathbf{u})$ by $\widehat{\lambda}_{2,n}(\mathbf{u})$ using a pilot bandwidth h_n' on D_n, and then split D_n into n_m subblocks of size $|D_m|$. Let $\widehat{\lambda}_{2,m}^i(\mathbf{u}; h)$ denote the estimate of $\lambda_2(\mathbf{u})$ calculated on subblock D_m^i using a bandwidth h. Then, $\widehat{h}_m$ is defined as the minimizer of the empirical mean squared error:

$$MSE(h) \equiv \sum_{i=1}^{n_m} \left[\widehat{\lambda}_{2,m}^i(\mathbf{u}; h) - \widehat{\lambda}_{2,n}(\mathbf{u})\right]^2 / n_m,$$

and then define $\widehat{h}_n = (|D_m|/|D_n|)^{1/6} \times \widehat{h}_m$. To reduce the effect of different starting values of h_n' on the selection, this procedure can be iterated by setting h_n' equal to the newly obtained $\widehat{h}_n$, until satisfactory convergence is reached. If the optimal bandwidth for estimating the second-order intensity function at a set of lags (instead of one particular lag) is needed, which is the case in the current setting, one can simply set $\widehat{h}_m$ to be the one that minimizes the sum of the empirical mean squared errors for all the lags. It is a good idea to plot $MSE(h)$ as a function of h to guide the choice of $\widehat{h}_n$, since several local minima can occur. In addition, attempts should be made to satisfy the general guidelines on the selection of h given in the preceding paragraph. Finally, this procedure is computationally intensive and also requires a moderately large sample size. Thus, this method is used to determine $\widehat{h}$ in the data example in Section 8.5 but not in all numerical experiments in Section 8.4. The effect of the joint choice of h and $\|\mathbf{u}\|$ on the testing results is evaluated in the next section.

8.4 Numerical results

8.4.1 Poisson cluster processes

Simulated realizations from a Poisson cluster process (as discussed in Section 7.3) on a unit square are observed. Specifically, parent locations are generated from a homogeneous Poisson process, with intensity $\rho = 50$ or 100, and then the number of offspring per parent are determined by a Poisson random variable with mean, $\mu = E(S)$, equal to 8 or 4, respectively. The position of offspring relative to their parents is then determined by an i.i.d. normal distribution with p.d.f.

$$f(\mathbf{u}) = \frac{1}{2\pi\sigma^2 \times |\mathbf{B}|^{-1}} \exp(-\|\mathbf{B}\mathbf{u}\|/2\sigma^2). \tag{8.2}$$

Thus, each of these two setups leads to approximately 400 events for each realization. The parameter σ in Equation 8.2 determines the spread of each cluster. For the isotropic case, $2\sigma^2$ is the mean squared distance from an offspring to its parent. In the simulation, σ is set at 0.02 and 0.03. These values are similar to the estimates of σ in Cressie's analysis [Cressie (1993)] of longleaf pine data, and the analysis of the redwood seedlings (Section 7.3), respectively. The matrix $\mathbf{B}$, in Equation 8.2, is a 2×2 matrix defined as

$$\mathbf{B} = \begin{bmatrix} 1 & 0 \\ 0 & p \end{bmatrix} \begin{bmatrix} \cos\theta & \sin\theta \\ -\sin\theta & \cos\theta \end{bmatrix}, \quad p \geq 1.$$

The parameter p defines the degree of anisotropy, while θ defines the major direction of anisotropy. This is analogous to the types of anisotropy considered for quantitative spatial fields in Chapter 5. Note that, as p increases, the degree of anisotropy increases. We consider $p = \sqrt{2}$, and 2, and $\theta = 60$ degrees with respect to the x axis. When $\mathbf{B} = \mathbf{I}$ and $p = 1$ we have the (isotropic) Thomas process introduced in Section 7.3.

One thousand realizations for each combination of parameters are simulated. The lag $\|\mathbf{u}\|$ is set in the four directions of 0, 45, 90, and 135 degree angles with respect to the x axis. Three values of h, 0.01, 0.02, and 0.03, are considered to evaluate the effect of the bandwidth h on test performance. The contrast matrix $\mathbf{A}$ was chosen such that the 0 degrees direction was compared to each of the remaining directions. The resulting test statistic was then compared to a χ^2 distribution with three degrees of freedom. For estimation of the covariance matrix $\boldsymbol{\Sigma}$, 0.18×0.18 subsquares are used.

We slightly modified the subsampling estimator in Section 8.2 for better finite sample performance. Specifically, to approximate the integral in the subsampling estimator, we laid a fine grid over the domain D_n. A subregion window was moved across the domain and on each subregion, $D_{l(n)}(\mathbf{y}_i)$, and for each $\mathbf{u}$ we calculate $\tilde{\lambda}^i_{2,l(n)}(\mathbf{u})$:

$$\iint_{\mathbf{s}_1,\mathbf{s}_2 \in D_{l(n)}(\mathbf{y}_i)} \frac{w[(\mathbf{u} - \mathbf{s}_1 + \mathbf{s}_2)/h_{l(n)}]}{\sqrt{|D_{l(n)}(\mathbf{y}_i) \cap \{D_{l(n)}(\mathbf{y}_i) - \mathbf{s}_1 + \mathbf{s}_2\}| \times h_{l(n)}}} \times N^{(2)}(d\mathbf{s}_1, d\mathbf{s}_2).$$

for $i = 1, \ldots, k_n$. We then obtained the (j,k)th element of the subsampling estimator by:

$$\frac{1}{k_n f_n} \times \sum_{i=1}^{k_n} \left[\tilde{\lambda}^i_{2,l(n)}(\mathbf{u}_j) - \overline{\lambda}_{2,l(n)}(\mathbf{u}_j)\right] \times \left[\tilde{\lambda}^i_{2,l(n)}(\mathbf{u}_k) - \overline{\lambda}_{2,l(n)}(\mathbf{u}_k)\right],$$

where $\overline{\lambda}_{2,l(n)}(\mathbf{u})$ denotes the average of all $\tilde{\lambda}^i_{2,l(n)}(\mathbf{u})$, $i = 1, \ldots, k_n$. Table 8.1 shows the empirical powers at the nominal 5 percent level.

Although not shown, we have also considered the case of $p = 1$, which corresponds to isotropy. The sizes of the tests in all situations are between 0.03 and 0.08. This is true regardless of the choice of ρ, μ, σ, $\|\mathbf{u}\|$, and h. For a fixed choice of ρ, μ, and σ, we see from Table 8.1 that the powers increase as the anisotropic ratio increases (i.e., as p increases

Table 8.1 Cluster process: empirical sizes and powers at the 5 percent nominal level.

			h ($p=\sqrt{2}$)			h ($p=2$)		
σ	(ρ, μ)	$\|\mathbf{u}\|$	0.01	0.02	0.03	0.01	0.02	0.03
0.02	(100,4)	0.02	0.16	0.36	0.35	0.76	0.96	0.81
		0.03	0.27	0.41	0.45	0.92	0.97	0.92
		0.04	0.27	0.40	0.38	0.77	0.91	0.88
		0.05	0.19	0.27	0.27	0.34	0.56	0.66
	(50,8)	0.02	0.22	0.50	0.53	0.95	0.99	0.95
		0.03	0.48	0.59	0.59	0.99	0.99	0.98
		0.04	0.51	0.61	0.54	0.95	0.98	0.96
		0.05	0.34	0.42	0.41	0.61	0.78	0.83
0.03	(100,4)	0.03	0.09	0.13	0.19	0.37	0.68	0.82
		0.045	0.11	0.19	0.23	0.52	0.79	0.82
		0.06	0.13	0.18	0.22	0.31	0.56	0.63
		0.075	0.10	0.14	0.16	0.14	0.22	0.29
	(50,8)	0.03	0.11	0.22	0.37	0.68	0.94	0.96
		0.045	0.23	0.36	0.39	0.86	0.96	0.95
		0.06	0.20	0.32	0.36	0.63	0.80	0.84
		0.075	0.15	0.18	0.21	0.26	0.39	0.48

from $\sqrt{2}$ to 2). For a fixed choice of ρ, μ, and p, the powers tend to decrease as σ increases from 0.02 to 0.03. This is because the strength of clustering becomes weaker as the cluster spread increases, and thus the true SSIFs in different directions are more similar than under smaller σ values. As a consequence, the power to detect directional differences becomes weaker. For the same reason, the empirical powers increase as μ increases from four to eight, as a higher value of μ indicates a stronger clustering effect.

In addition, it can be seen that the powers are affected by the choice of $\|\mathbf{u}\|$ and h. Firstly, $\|\mathbf{u}\| = 1.5\sigma$ or 2σ, which roughly corresponds to $\frac{1}{3}$ to $\frac{1}{2}$ of the dependence range (4σ is treated as the dependence range as in Section 7.3), typically yields the best power. Secondly, $h = 0.01$ yields the lowest powers in almost all cases. This is mainly due to an insufficient sample size when calculating the SSIFs. Thirdly, the powers when $h = 0.02$ are comparable to and, in fact, in many cases higher than those when $h = 0.03$. This is possibly due to the increasing amount of overlapping pairs used to calculate different SSIFs when h increases, which may offset the benefit caused by an increasing sample size. Additional results suggest that good power can often be achieved when the number of pairs of points used to calculate each SSIF is 200 or more. These observations lead to the recommendation on the choice of h that we have given in Section 8.3.

8.4.2 Simple inhibition processes

We next consider realizations from a simple inhibition process as described in Section 7.2 on a 40×40 square region. For each realization, a Poisson process is generated and then thinned by the deletion of all pairs of events such that $\mathbf{u}^T\mathbf{B}\mathbf{u} \leq \delta^2$, where the choices of $\mathbf{B}$ are as defined in Section 5.6.5, and $\delta = 0.4$ or 0.6. We control the intensity for the initial Poisson process such that approximately 500 events are generated for each realization.

One thousand realizations are simulated for each choice of anisotropy matrix $\mathbf{B}$ in Section 5.6.5. The value of $\|\mathbf{u}\|$ is equal to 0.4, 0.6, and 0.8 for $\delta = 0.4$, and 0.6, 0.9, and 1.2 for $\delta = 0.6$. These values correspond to 1, 1.5, and 2 times δ, respectively. The same set $\mathbf{\Lambda}$ is used to form the two-, four-, and eight-lag tests as in the Poisson clustering case, and c is chosen as 0.8. The bandwidth h is selected by the data-driven method introduced in Section 8.3.

Table 8.2 reports the empirical sizes and powers at the 5 percent nominal level. The sizes are all close to the nominal. As in the Poisson clustering case, the two-lag test is the most powerful for **B2** and **B3** (no rotation), while the four-lag test is the most powerful for **B4** and **B5**. The power in general increases as the anisotropic ratio increases, and the best power is achieved when $\|\mathbf{u}\| = \delta$. When δ increases from 0.4 to 0.6, the power generally increases. Observe that the values of $\|\mathbf{u}\|$ are larger when $\delta = 0.6$. As a result, the directions of event pairs used to estimate $\lambda_2(\mathbf{u})$ are typically closer to the targeted directions when $\delta = 0.6$ than when $\delta = 0.4$. This in turn leads to a larger difference between the expected sample second-order intensity function values in the major anisotropic directions and consequently larger power when $\delta = 0.6$. When compared to the Poisson clustering case, it is seen that the powers are much lower. Overall, based on these experiments, the test can be broadly recommended for random or clustered point patterns, but not for inhibited patterns.

Table 8.2 Simulation results from one thousand realizations of simple inhibition processes.

δ	$\|\mathbf{u}\|$	Lags	**B1**	**B2**	**B3**	**B4**	**B5**
0.4	0.4	2	0.06	0.13	0.19	0.06	0.06
		4	0.06	0.10	0.14	0.17	0.13
		8	0.07	0.11	0.12	0.13	0.10
	0.6	2	0.06	0.08	0.11	0.05	0.05
		4	0.08	0.08	0.09	0.10	0.07
		8	0.06	0.06	0.09	0.08	0.07
	0.8	2	0.06	0.07	0.06	0.06	0.04
		4	0.07	0.09	0.07	0.08	0.07
		8	0.06	0.07	0.06	0.06	0.07
0.6	0.6	2	0.06	0.27	0.53	0.06	0.06
		4	0.07	0.19	0.41	0.21	0.38
		8	0.06	0.15	0.36	0.17	0.36
	0.9	2	0.06	0.11	0.17	0.05	0.05
		4	0.07	0.11	0.15	0.09	0.14
		8	0.07	0.07	0.12	0.07	0.11
	1.2	2	0.05	0.07	0.06	0.04	0.05
		4	0.06	0.08	0.07	0.06	0.07
		8	0.04	0.06	0.07	0.05	0.06

Each entry is the percentage of rejections at the 5 percent nominal level. δ denotes the minimum permissible distance between events.

8.5 An application to leukemia data

The data are the locations of leukemia cases diagnosed in the state of Texas between 1990 and 1998. For each case, a control matched on sex and date of birth is randomly selected from all births in Texas during that period. We study observations within a rectangular area in the city of Houston. The western side of this rectangle borders Highway 6, while its eastern, northern, and southern sides are all along Highway 8. The Houston Ship Channel, an area of major petroleum refining and petrochemical industries, sits to the east of the region. There are 545 cases and 566 controls observed in this region. Figure 8.1 plots the (shifted) locations of leukemia cases and controls in a (approximately) 25×22 field.

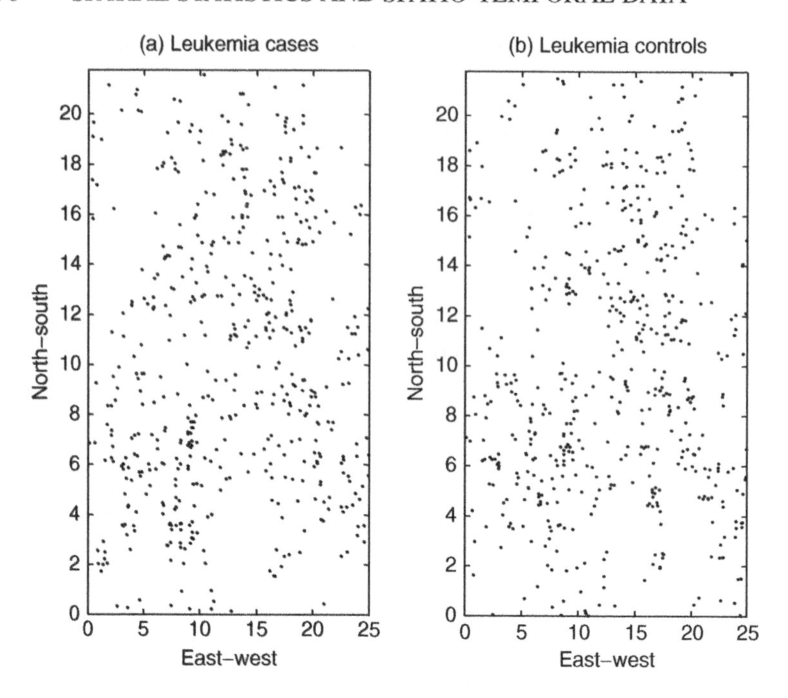

Figure 8.1 Locations of leukemia patients in eastern Texas. (a) Leukemia cases. (b) A control data set.

It is suspected that exposure to pollution, for example, agricultural pesticides in a rural area, emissions of chemicals, say Benzene, from traffic and industry in an urban area, may increase the risk of developing leukemia. If this is the case, we expect to see different clustering patterns in the cases than in the controls at certain scales. We address whether these data support such a conclusion.

For our study region, there are two main possible sources of pollution: local traffic and the ship channel to the east. Since roads in this area are mostly in the east–west or the north–south direction, and in particular, the ship channel is directly to the east of this region, we choose $\mathbf{\Lambda}$ to be

$$\left\{(0.6,0),(0,0.6),\left(0.6/\sqrt{2},0.6/\sqrt{2}\right),\left(-0.6/\sqrt{2},0.6/\sqrt{2}\right)\right\}.$$

The length 0.6 is selected for $\|\mathbf{u}\|$ due to the earlier recommendation that $\frac{1}{3}-\frac{1}{2}$ of the range is a good choice of lag $\|\mathbf{u}\|$. The SSIF for each of the cases and controls flattens out at a value of approximately 1.5 for these data. Isotropy of the case and control groups is assessed separately. For both cases and controls, the contrast matrix is $\mathbf{A} = (1, 1, -1, -1)$.

The bandwidths are chosen to be 0.26 for cases and 0.23 for controls using the data-based algorithm discussed in Section 8.3. The subblock size is approximately 4.0×3.5, which gives approximately 39 nonoverlapping replicates. The calculated test statistics are equal to 2.061 for the cases and 0.0674 for the controls, with respective p-values 0.1511 and 0.7952. Thus, we do not reject spatial isotropy for either the cases or the controls, although there is some weak support for anisotropy in the case locations.

We now compare the strengths of clustering between these two groups. Cuzick and Edwards (1990) introduced a testing approach based on k nearest neighbors that utilizes the assumption of isotropy. In light of the testing results (acceptance of isotropy for both cases and controls), their method can be applied to the leukemia data. In particular, we perform the test using the $k = 1$ nearest neighbor. This yields a test statistic equal to -0.5305 and a p-value equal to 0.7021. Thus, we conclude there is no evidence of a difference in spatial clustering between cases and controls. This may be due to the fact that the region being studied here is too small and thus exposure to some risk factors, for example, pollution, is relatively homogeneous over the entire region. Conversely, there could be a local effect that our global testing procedure does not detect. Of course, there may not be any differential effects due to the given levels of environmental factors.

9

Multivariate spatial and spatio-temporal models

In previous chapters all variables of interest were generally univariate. For example, for the wheat yields or cancer rates in Chapter 4, or the wind speeds in Chapters 5 and 6, there was one response variable, Z. This often suffices; however, there is an increasing wealth of multivariate spatial, and multivariate spatio-temporal data appearing. For example, in the state of California, 83 monitoring stations made hourly measurements of carbon monoxide (CO), nitric oxide (NO), and nitrogen dioxide (NO_2). It is of great interest to be able to predict the levels of these three pollutants at unsampled locations by incorporating *joint* information from the three variables at all locations and times. Another setting is in real estate. For example, the selling prices and rents of commercial real estate in Chicago, Dallas, and San Diego are studied in Gelfand *et al.* (2004), with the purpose of providing information for real estate finance and investment analysis. As another example, the racial distribution of residents in southern Louisiana has been studied in Sain and Cressie (2007) to determine whether locations of toxic facilities impact the spatial distribution of white people and minorities.

In all of these data sets, effective predictions of multivariate responses requires knowledge of the underlying covariance function describing variable, spatial, and temporal correlations.

Spatial Statistics and Spatio-Temporal Data: Covariance Functions and Directional Properties Michael Sherman

Consider a p-dimensional multivariate random field $\mathbf{Y}(x) = \{Y_1(x), \ldots, Y_p(x)\}^{\mathrm{T}}, x \in D$, for a region $D \subset \mathbb{R}^q, q \geq 1$. In the air pollution setting we have $p = 3$. A common model for multivariate observations naturally extends a general model for univariate observations as

$$\mathbf{Y}(x) = \boldsymbol{\mu}(x) + \mathbf{Z}(x) + \boldsymbol{\epsilon}(x). \tag{9.1}$$

The mean function, $\boldsymbol{\mu}(x)$, is commonly parameterized linearly by $\boldsymbol{\mu}(x) = \mathbf{A}(x)\boldsymbol{\beta}$, with $\mathbf{A}(x)$ a matrix of covariates depending on coordinates and/or other location-specific variables, $\boldsymbol{\beta}$ is an unknown vector of parameters, and $\boldsymbol{\epsilon}(x)$ denotes a white noise vector, that is, $\boldsymbol{\epsilon}(x) \sim F(0, \Omega)$, where F is a p-variate distribution function with diagonal covariance matrix Ω. For example, F could be the multivariate normal distribution given in Section 4.1, with independent components. Assume that the mean function $\boldsymbol{\mu}(x)$ in Equation 9.1 can be adequately modeled. The focus then becomes the covariance structure of the process $\mathbf{Z}(x)$. Assume that a second-order stationary multivariate random field, as in Chapter 2 (in the univariate setting), describes the variation in $\mathbf{Z}(x)$. That is, $E[\mathbf{Z}(x)] = \boldsymbol{\mu}$ and

$$\mathbf{C}(k) = Cov[\mathbf{Z}(x), \mathbf{Z}(x+k)] = E\{[\mathbf{Z}(x) - \mu][\mathbf{Z}(x+k) - \mu]^{\mathrm{T}}\}, \tag{9.2}$$

where $\mathbf{C}(k) = [C_{ij}(k)]$ is a $p \times p$ cross-covariance matrix for each lag vector $k \in \mathbb{R}^q$. This formulation allows for several settings of great interest. For example:

i. A univariate spatial random field with equally or unequally spaced locations, as described in Chapters 2–5. In this situation, $p = 1$, $x = \mathbf{s} = (s_1, s_2)$ in $\mathbb{R}^2$ for scalars s_i;

ii. A univariate spatio-temporal random field, as discussed in Chapter 6. In this case $p = 1, x = (\mathbf{s}, t)$ in $\mathbb{R}^{d+1}$ for a scalar t representing a time index;

iii. For general p, $x = \mathbf{s}$ in $\mathbb{R}^d$, we observe a multivariate spatial random field;

iv. For general p, $x = (\mathbf{s}, t)$ in $\mathbb{R}^{d+1}$, we observe a multivariate space–time random field.

In this chapter we focus on situations (iii) and (iv). To see the importance of multivariate correlation functions we first consider case (iii) and the cokriging setting.

9.1 Cokriging

In cokriging, one variable is typically considered as primary, and the other variables are considered secondary. That is, they are of use to aid in predicting the first variable but not of primary interest. Often the primary variable is under sampled but of high quality, while the secondary variable is sampled with higher spatial density, but is of lower quality. We consider this situation in rainfall estimation.

Precipitation is an important hydrologic model input, and is characterized by spatial and temporal variability. Often, point precipitation measurements at rain gauges have been used with hydrological models. Since rain gauges physically measure the depth of rainfall at the points of measurement, they generally provide good quality data. Although the measurements are of good quality, rain gauge networks are usually too sparse to capture the spatial variability of precipitation over an entire hydrologic system.

Consider data from Texas in 2003. In this data set, there are 311 official rain gauges to measure rainfall. Compared to rain gauges, weather radars provide precipitation data with much better spatial and temporal sampling frequencies. One such weather radar system is the Next Generation Weather Radar (NexRad), formally known as the Weather Surveillance Radar – 1988 Doppler (WSR-88D) of the United States. The NexRad system, in Texas in 2003, gave 50 151 estimates of rainfall at each location on a 4×4 km grid over the state. Although more plentiful, the radar estimates of precipitation are less reliable than those from gauges. There are several possible sources of error or data contamination in the radar estimates. These are discussed, for example, in Jayakrishnan *et al.* (2004). The radar estimates can be useful, however, where gauge measurements are sparse or nonexistent. In order to make use of the NexRad measurements for kriging-based prediction of rain fall, we need the univariate spatial covariances of rain gauges and NexRad measurements, as well as the cross-covariance between NexRad measurements and rain gauge measurements.

For notational simplicity, assume that all variables have been measured at all locations. This is the so-called isotopic (*not* isotropic!) situation for multivariate spatial observations. As in the rainfall setting, consider the situation where, given the $n \times p$ observations $Z_1(\mathbf{s}_1), \ldots, Z_p(\mathbf{s}_1), Z_1(\mathbf{s}_2), \ldots, Z_p(\mathbf{s}_n)$, we seek to predict one of the variables (say variable k) at a new location, $Z_k(\mathbf{s}_0)$, say. For example, this would be the rainfall at a location with no rain gauge. Analogous to the development in Section 2.2, we seek an unbiased linear estimator of the form:

$$\sum_{i=1}^{n}\sum_{j=1}^{p} \lambda_{ij} Z_j(\mathbf{s}_i)$$

that minimizes

$$E\left[\sum_{i=1}^{n}\sum_{j=1}^{p} \lambda_{ij} Z_j(\mathbf{s}_i) - Z_k(\mathbf{s}_0)\right]^2. \tag{9.3}$$

Unbiasedness is implied by $\sum_{i=1}^{n} \lambda_{ik} = 1$, while for $j \neq k$ we have $\sum_{i=1}^{n} \lambda_{ij} = 0$. We now seek to minimize Equation 9.3 subject to these p constraints. We now have p Lagrange multipliers, but otherwise the cokriging predictor is derived analogously to the kriging predictor when $p = 1$. There are, however, some interesting options in the multivariable setting.

To simplify, consider the case where there are two variables, Z and Y, and the first is of primary interest and measured at n locations, and the second, auxiliary variable at m locations. The predictor is now of the form $\widehat{Z}(\mathbf{s}_0) = \sum_{i=1}^{n} \lambda_{i1} Z(\mathbf{s}_i) + \sum_{i=1}^{m} \lambda_{i2} Y(\mathbf{s}_i)$.

Letting $\mathbf{Z} = [Z(\mathbf{s}_1), \ldots, Z(\mathbf{s}_n), Y(\mathbf{s}_1), \ldots, Y(\mathbf{s}_m), Z(\mathbf{s}_0)]^{\mathrm{T}}$ and letting

$$\tilde{\boldsymbol{\lambda}} = [\lambda_{11}, \ldots, \lambda_{n1}, \lambda_{12}, \lambda_{m2}, -1]^{\mathrm{T}},$$

we have:

$$Var[\widehat{Z}(\mathbf{s}_0) - Z(\mathbf{s}_0)] = \tilde{\boldsymbol{\lambda}}^{\mathrm{T}} \boldsymbol{\Sigma} \tilde{\boldsymbol{\lambda}} = F(\tilde{\boldsymbol{\lambda}}),$$

say. In the notation of Section 2.2, we can set $H = F - m_1\big(-\sum_{i=1}^{n} \lambda_{i1} + 1\big) + m_2 \sum_{i=1}^{m} \lambda_{i2}$. Differentiating H with respect to $\boldsymbol{\lambda}$ (and m_1, m_2) gives:

$$\sum_{i=1}^{n} \lambda_{i1} C(Z_i, Z_j) + \sum_{i=1}^{m} \lambda_{i2} C(Y_i, Z_j)$$
$$+ m_1/2 = C(Z_0, Z_j) \quad \text{for } j = 1, \ldots, n,$$

$$\sum_{i=1}^{n} \lambda_{i1} C(Z_i, Y_j) + \sum_{i=1}^{m} \lambda_{i2} C(Y_i, Y_j)$$
$$+ m_2/2 = C(Z_0, Y_j) \quad \text{for } j = 1, \ldots, m,$$

$$\sum_{i=1}^{n} \lambda_{i1} = 1,$$

$$\sum_{i=1}^{m} \lambda_{i2} = 0.$$

These $n + m + 2$ equations have $n + m + 2$ unknowns, and thus have a unique solution in $\boldsymbol{\lambda} = (\lambda_{11}, \ldots, \lambda_{n1}, \lambda_{12}, \ldots, \lambda_{m2}, m_1/2, m_2/2)^{\mathrm{T}}$. The cokriging variance based on this predictor is then obtained by substituting the necessary components from this solution in $\boldsymbol{\lambda}$ into $F(\tilde{\lambda})$, above. In this two-variable case, the variance is given by:

$$Var(Z_0) + m_1/2 - \sum_{i=1}^{n} \lambda_{i1} C(Z_i, Z_0) - \sum_{i=1}^{m} \lambda_{i2} C(Y_i, Z_0).$$

Note that we can have other constraints on the coefficients λ. For example, let μ_Z denote the mean of Z, and let μ_Y denote the mean of Y. Then consider the predictor:

$$\widehat{Z}(\mathbf{s}_0) = \sum_{i=1}^{n} \lambda_{i1} Z(\mathbf{s}_i) + \sum_{i=1}^{m} \lambda_{i2} [Y(\mathbf{s}_i) - \mu_Y + \mu_Z].$$

By taking the expectation of each side, we see that the constraint

$$\sum_{i=1}^{n} \lambda_{i1} + \sum_{i=1}^{m} \lambda_{i2} = 1$$

gives us an unbiased estimator. This is not formally a predictor, because μ_Z and μ_Y are unknown. These means, however, can be estimated by their respective sample means and this will not alter the unbiasedness condition. Now there is only one constraint as opposed to two constraints in the bivariate cokriging case. This is sometimes useful to make predictions that better conform to known bounds. For example, consider a case like the rainfall prediction where we know that Z is nonnegative. The constraint $\sum_{i=1}^{m} \lambda_{i2} = 0$ necessarily makes some of the $\lambda_{i2} < 0$. If these λ_{i2} tend to correspond to large Y values, then the predictions of Z can become negative. For positive variables there is a much smaller chance of negative predictions using the single constraint.

9.2 An alternative to cokriging

In the case where the primary variable is very sparse, it might be difficult, or even impossible, to obtain information on the covariance function $C(Z_i, Z_j)$ which is necessary for the cokriging system. In the rainfall setting, the primary observations, the gauge measurements, are so spread out that they are approximately independent of each other, and thus there is no reliable information on their covariance function.

To see an alternative, note that we have been assuming that $\mu(x) = \mu$ in Equation 9.1, namely that the mean vector is constant for each of the p variables. Alternatively, consider allowing for a nonconstant mean function. For example, to predict the rainfall consider the prediction:

$$\widehat{Z}(\mathbf{s}_0) = \widehat{\beta}_0 + \widehat{\beta}_1 Y(\mathbf{s}_0),$$

where the regression parameters are estimated from the locations where both the primary and secondary variables are measured. It may, however, be unreasonable to assume that the errors on the grid are independent. For this reason, it is useful to explore the covariance function of the errors in the regression model. We illustrate this procedure on the Texas rainfall data.

There are actually three sets of daily rainfall data: one is daily NexRad estimates over a 4×4 km grid over Texas; another one is gauge data that is of relatively good quality and contains 664 rain gauges; the last one is station data. These are also measurements from gauges, but are of higher quality. There are, however, only 60 weather stations. The NexRad data is Stage III WSR-88D precipitation data from the National Weather Service (NWS), and the gauge data is compiled at the National Climate Data Center (NCDC) of the NWS. Both NexRad data and station data give daily rainfalls from 7 a.m. to 7 a.m., but not all of the gauges are collected from 7 a.m. to 7 a.m. Only 311 out of 664 are collected at 7 a.m. and thus are appropriate for use in the real data analysis, as they are the only ones that are comparable with the station and NexRad data. Although the 60 stations are the most reliable, they are too sparse to capture the spatial features of the rainfall data. They can naturally be used as a 'gold standard' to assess the quality of the model predictions. The 311 gauges are dense enough to give some spatial information, so we use the 311 gauges to calibrate the NexRad data. Among the 60 stations, 14 of them are collocated with gauges, while the measurements at the remaining 46 station locations are not being used in the calibration, hence it is appropriate to use these 46 stations to assess the quality of model predictions.

9.2.1 Statistical model

Let $Z(\mathbf{s}_i, t)$ denote the gauge measurement, $i = 1, \ldots, 311$, $Y\left(\mathbf{s}'_j, t\right)$ denote radar NexRad measurements, $j = 1, \ldots, 50\,151$, and $t = 1, \ldots,$ 365, for both sets of measurements. For $\mathbf{s}'_j$ close to $\mathbf{s}_i$, we expect that the value of $Y\left(\mathbf{s}'_j, t\right)$ is close to the value of $Z(\mathbf{s}_i, t)$ as they are recorded on the same day. For $Y\left(\mathbf{s}'_j, t\right)$ with $\min_j \left\|s'_j - s_i\right\|$, define $Y\left(\mathbf{s}'_j, t\right) = Y(\mathbf{s}_i, t)$, that is, assume they are collocated. The NexRad measurements relatively often register spurious, nonzero rainfall measurements, and it is useful to try to eliminate these measurements via a threshold. There are four cases in the data:

i. $Z(\mathbf{s}_i, t) = 0$, $Y(\mathbf{s}_i, t) = 0$;
ii. $Z(\mathbf{s}_i, t) > 0$, $Y(\mathbf{s}_i, t) = 0$;
iii. $Z(\mathbf{s}_i, t) = 0$, $Y(\mathbf{s}_i, t) > 0$;
iv. $Z(\mathbf{s}_i, t) > 0$, $Y(\mathbf{s}_i, t) > 0$.

Cases (i) and (iv) are what we hope to see in the data, while cases (ii) and (iii) are not. The model calibrates the NexRad data Y using the gauge data Z. If we use $Y(\mathbf{s}_i, t) = 0$ as the classifier for $Z(\mathbf{s}_i, t) = 0$, we misclassify the $Z(\mathbf{s}_i, t)$s in cases (ii) and (iii). To reduce the misclassification error, it is useful to introduce a classifier threshold β_2 and assume: $Z(\mathbf{s}_i, t) = 0$, if $Y(\mathbf{s}_i, t) \leq \beta_2$; $Z(\mathbf{s}_i, t) > 0$, if $Y(\mathbf{s}_i, t) > \beta_2$. The introduction of β_2 can reduce the misclassification rate. For Y above the threshold, linear regression removes possible bias in the NexRad data. The linear relationship between the NexRad and gauge measurements may not be constant across time, and for this reason it is useful to consider a varying-coefficient model to allow the coefficients to vary over time.

Based on the above discussion, the following model is assumed:

$$Z(\mathbf{s}, t)^{1/k} = \{\beta_0(t) + \beta_1(t) Y(\mathbf{s}, t)^{1/k} + \xi[Y(\mathbf{s}, t), \boldsymbol{\beta}_t]\epsilon_t\} I[Y(\mathbf{s}, t) > \beta_2(t)] \qquad (9.4)$$

where $Z(\mathbf{s}, t)$ and $Y(\mathbf{s}, t)$ are random processes of rain gauge and NexRad over locations $\mathbf{s}$ at time t. The power $1/k$ is used to make measurements better conform to a Gaussian distribution. This is important, as it is desirable, if possible, to use likelihood methods to fit covariance model parameters. $I(\cdot)$ denotes an indicator function and $\beta_2(t)$ is a time-dependent threshold. The functions $\beta_0(t)$ and $\beta_1(t)$ are time-dependent additive and multiplicative bias terms, respectively. Let $\boldsymbol{\beta}_t$ denote the parameter vector $[\beta_0(t), \beta_1(t)]^{\mathrm{T}}$ and $\xi[Y(\mathbf{s}, t), \boldsymbol{\beta}_t]$ is a function to account for nonconstant variance. The spatial errors $\boldsymbol{\epsilon}_t \sim F(\mathbf{0}, \boldsymbol{\Sigma}_t)$, where F is a multivariate distribution function.

9.2.2 Model fitting

We give some details on fitting of the model in the previous section. First, for the power of transformation, the Shapiro–Wilk test statistics are computed for the null hypothesis of normality. These are maximized at $\widehat{k} = 3$ over the positive integers $k < 10$, indicating that this value of k is most compatible with a Gaussian distribution.

As discussed, the parameter $\beta_2(t)$ is a time-dependent threshold. The predicted gauge values are set to be equal to zero if $P[Z(\mathbf{s}_i, t) > 0 \,|Y(\mathbf{s}_i, t) > 0] < \frac{1}{2}$. These probabilities are estimated using logistic regression on observed pairs (Z, Y). It is known that rainfall patterns in the warm (April–October) season are relatively different from those in the cold (November–March) season. The estimated parameter function $\widehat{\beta}_2(t)$ is approximately constant within each of these seasons, and through stratifying temporally it is found that $\widehat{\beta}_2 = 0.58$ in the cold season, and $\widehat{\beta}_2 = 5.68$ in the warm season. These thresholds reduce the misclassifications in the 311 gauge measurements from approximately 10.9 percent to 9.7 percent (in the cold season), and from 15.9 percent to 11.3 percent (in the warm season) compared to using $\beta_2 = 0$ as the threshold.

The functions $\beta_0(t)$ and $\beta_1(t)$ are estimated as follows: given that $P[Z(\mathbf{s}_i, t) > 0 \mid Y(\mathbf{s}_i, t) > 0] > \frac{1}{2}$, this means that the rainfall, $Z(\mathbf{s}_i, t)$, will be estimated. The functional parameters $\beta_0(t)$ and $\beta_1(t)$, and the function $\xi[Y(\mathbf{s}, t)]$ can be estimated iteratively using weighted least squares. It turns out, however, that fitting cubic splines to the two functional parameters shows that both are approximately constant within the two seasons. From days with a sufficient number of rainy locations, the parameter estimates (standard errors) are found to be $\widehat{\beta}_0 = -0.314(0.032)$, $\widehat{\beta}_1 = 0.938(0.017)$ in the cold season and $\widehat{\beta}_0 = -0.717(0.058)$, $\widehat{\beta}_1 = 1.029(0.022)$ in the warm season. Note that there is significant additive bias in both seasons, and significant multiplicative bias in the cold season. This suggests that there is a possibility of improvement using this methodology over predicting rain using the NexRad measurements alone.

For the nonconstant variance, given the functions $[\beta_0(t), \beta_1(t)]$, nonconstancy of variance can be assessed by computing the Spearman rank correlation coefficients of the absolute studentized residuals with the fitted values as suggested by Carroll and Ruppert (1998). Spearman's rank correlation test estimates the correlation to be $\widehat{\rho} = 0.020$, with a corresponding two-sided p-value = 0.137. This nonsignificance allows an assumption of $\xi[Y(\mathbf{s}, t), \widehat{\boldsymbol{\beta}}_t] = 1$. These residuals are defined in the discussion of covariance estimation for the errors, which follows now.

The spatial errors $\boldsymbol{\epsilon}_t \sim F(\mathbf{0}, \boldsymbol{\Sigma}_t)$, where F is a multivariate distribution function. We parameterize $\boldsymbol{\Sigma}_t$ from the Matérn class of covariance because, as discussed in Section 3.3, the Matérn covariance function allows for estimation of the smoothness of a random field from the data. It turns out that an alternative parameterization for the Matérn covariance function allows for better parameter estimation, namely:

$$C(\|\mathbf{h}\|) = \frac{\sigma^2}{2^{\eta-1}\Gamma(\eta)} \left(\frac{2\eta^{\frac{1}{2}} \|\mathbf{h}\|}{\nu} \right)^{\eta} \mathcal{K}_\eta \left(\frac{2\eta^{\frac{1}{2}} \|\mathbf{h}\|}{\nu} \right), \tag{9.5}$$

where $\Gamma(\cdot)$ is the gamma function, $\mathcal{K}_\eta$ is the modified Bessel function of the second kind, of order η, $\|\mathbf{h}\|$ is Euclidian distance between two locations, σ^2 is the variance or the sill, and η denotes the smoothness parameter. The larger η is, the smoother the random process. The parameter ν measures how quickly the correlations of the random field decay with distance. In this parameterization, interpretation of ν is largely independent of η, as shown in Handcock and Wallis (1994). This can be seen also from the discussion in Chapter 2 of Stein (1999).

The residuals from the linear regression are now defined, and can be used to find the underlying spatial pattern of the residuals. To estimate the variogram, we use daily residuals since the rainfall patterns differ appreciably from day to day. However, calculating residuals at only the closest NexRad location whose rainfall value is above the threshold for each gauge gives at most 311 residuals at a very limited number of spatial lags. This gives information which is too sparse to capture the spatial dependence pattern. Assume that the nearest eight NexRad locations and the corresponding gauge have the same true rainfall, as discussed by Jayakrishnan *et al.* (2004). Then it is possible to calculate the residuals at the nearest eight NexRad locations for each gauge. This allows for a sufficient number of residuals at a variety of spatial lags, enabling variogram estimation.

The residuals from the mean function enable estimation of spatial dependence. Specifically, they are defined as:

$$\widehat{\epsilon}_{ij}(t) = \frac{1}{\widehat{\xi}[Y_{ij}(t), \widehat{\boldsymbol{\beta}}_t]} \left[Z(\mathbf{s}_i, t)^{1/\widehat{k}} - \widehat{\beta}_0(t) - \widehat{\beta}_1(t) Y_{ij}(t)^{1/\widehat{k}} \right], \quad (9.6)$$

$i = 1, \ldots, 311;\ 0 \leq j \leq 8,\ j$ is integer, where $Y_{ij}(t)$ is the jth nearest NexRad whose rainfall is above the threshold for the ith gauge, and $\widehat{\boldsymbol{\beta}}_t = [\widehat{\beta}_0(t), \widehat{\beta}_1(t)]^{\mathrm{T}}$, and $\widehat{k}$ and $[\widehat{\beta}_0(t), \widehat{\beta}_1(t)]$ are as previously described.

Due to the approximate normality of the residuals, it is appropriate to use maximum likelihood to fit a variogram function to the residuals here. The superiority of likelihood methods for covariance model fitting is recommended in both Stein (1999) and Diggle and Ribeiro (2007), for example. However, moment-based estimators are certainly competitors in this context. The full maximum likelihood method, however, is difficult to implement due to the dimensionality of the problem. A simpler and faster procedure is to use an approximate likelihood over moving windows. This is analogous to using the generalized psuedo-likelihood described in Section 4.2. We center the moving windows on each gauge that is used to calculate the residuals in (9.6), obtaining overlapping moving windows across the whole domain. The results, in principle, then depend on the size of each moving window, but they are relatively stable when dependence is not too strong.

Specifically, the parameters in the variogram model are $\boldsymbol{\theta} = (\tau^2, \sigma^2, \nu, \eta)^{\mathrm{T}}$, which represent nugget and Matérn (sill, range, smoothness). Let K denote the total number of gauges that are involved in (9.6), which is also the total number of moving windows. Let $\mathbf{e}_k$ denote the residual vector in the kth window, $k = 1, 2, \ldots, K$, let $\boldsymbol{\Sigma}_k(\boldsymbol{\theta})$ denote the covariance matrix of $\mathbf{e}_k$, and let $|\cdot|$ denote the determinant of a matrix.

Using the expression for the joint likelihood in Equation 4.1, it holds that the joint loglikelihood, up to a constant that does not depend on $\boldsymbol{\theta}$, is:

$$l(\boldsymbol{\theta}) = -\sum_{k=1}^{K} \left[\log |\boldsymbol{\Sigma}_k(\boldsymbol{\theta})| + \mathbf{e}_k^{\mathrm{T}} \boldsymbol{\Sigma}_k(\boldsymbol{\theta})^{-1} \mathbf{e}_k\right].$$

The chosen variogram parameters are the ones that maximize $l(\boldsymbol{\theta})$.

In 2003, there are 143 days that have a reasonably large rain coverage to analyze the daily data. From the daily analysis across these days we obtain the mean (std) of the sill, range, and smoothness parameters: 1.02 (0.29), 34.13 (13.91), 0.86 (0.19). This shows that there is relatively wide variability in the covariance parameters across days. The medians of the covariance parameters are (1.00, 31.07, 0.84). These covariance functions are typically MSC, but not once differentiable. The estimates for each day are used in the fitted variogram function in order to carry out kriging, as we now detail.

9.2.3 Prediction

Given model (9.4) and the estimated parameters, we now address the prediction of rainfall at unobserved locations on any given day. As usual, let $\mathbf{s}_0$ denote an unobserved location. To predict at $\mathbf{s}_0$ on the tth day, the residual vector $\widehat{\boldsymbol{\epsilon}}_t$ is calculated using the residuals as defined in Equation 9.6. The vector of residuals $\widehat{\boldsymbol{\epsilon}}_t$ is then kriged to location $\mathbf{s}_0$ via ordinary kriging, with the estimated variogram model based on the approximate likelihood method discussed in Section 9.2.2:

$$\widehat{\epsilon}(\mathbf{s}_0, t) = \boldsymbol{\lambda}(\mathbf{s}_0, t)^{\mathrm{T}} \widehat{\boldsymbol{\epsilon}}_t, \tag{9.7}$$

where $\boldsymbol{\lambda}(\mathbf{s}_0, t)$ are the ordinary kriging weights.

Finally, $\widehat{\epsilon}(\mathbf{s}_0, t)$ is plugged into the following formula:

$$\widehat{Z}(\mathbf{s}_0, t) = \left(\left\{ \widehat{\beta}_0(t) + \widehat{\beta}_1(t) \bar{\bar{Y}}(\mathbf{s}_0, t)^{1/\widehat{k}} + \widehat{\xi}\left[\bar{\bar{Y}}(\mathbf{s}_0, t), \widehat{\boldsymbol{\beta}}_t\right] \widehat{\epsilon}(\mathbf{s}_0, t) \right\}^{\widehat{k}} - \widehat{bias} \right) I\left[\bar{\bar{Y}}(\mathbf{s}_0, t) > \widehat{\beta}_2(t)\right] \tag{9.8}$$

where $\bar{\bar{Y}}(\mathbf{s}_0, t)$ denotes the average of the nearest eight NexRad measurements for the desired location on day t, and $\widehat{bias}$ is the estimated bias induced by the power transformation. In this case, where the transformation is the cube root, the true bias can be simply calculated, and estimated unbiasedly by

$$\widehat{bias} = 3[\widehat{\beta}_0(t) + \widehat{\beta}_1(t)W(\mathbf{s}_0,t)^{1/3}]\{\sigma^2(\mathbf{s}_0,t) - 2[\boldsymbol{\lambda}(\mathbf{s}_0,t)^{\mathrm{T}}\boldsymbol{\gamma}_0(\mathbf{s}_0,t)]\},$$

where $\boldsymbol{\lambda}$ is the vector of kriging weights, $\boldsymbol{\gamma}_0(\mathbf{s}_0,t)$ is the vector of variogram values between $\mathbf{s}_0$ and observed locations, and $\sigma^2(\mathbf{s}_0,t)$ denotes the kriging variance. For more general transformations, Taylor series methods can be used to estimate the bias (for example, see Section 2.3.1 for the lognormal situation).

9.2.4 Validation

Next comes the most important practical question: how well does this procedure work? In order to explore this, the model is used to predict the rainfall at each of the 46 stations not used in the model fitting, and then this is compared with the actual station rainfall measurements. Several statistics are used by hydrologists, for instance, Jayakrishnan *et al.* (2004), to compare the model prediction with the gauge rainfalls. Let Z_{kt}, $k = 1,\ldots,46$, $t = 1,\ldots,T$, denote the kth station measurement on the tth day, where T denotes the number of days being used in the comparison. Let $\widehat{Z}_{kt}$ denote the predicted value of Z_{kt}, $\widehat{Z}_{kt} = \widehat{Z}(\mathbf{s}_k,t)$ in Equation 9.8, where $\mathbf{s}_k$ is the location of Z_{kt}. Define $Z_t = \sum_{k=1}^{46} Z_{kt}$, and $\widehat{Z}_t = \sum_{k=1}^{46} \widehat{Z}_{kt}$, the daily total of the station measurements and the predicted daily total, respectively. Let $\bar{Z} = \sum_{t=1}^{T} Z_t/T$ denote the overall mean. To assess accuracy, we calculate (Table 9.1):

i. Total difference in precipitation, $D = \sum_{t=1}^{T}(\widehat{Z}_t - Z_t)$.
ii. Percent estimation bias, $EB = 100D\big/\sum_{t=1}^{T} Z_t$.
iii. Estimation efficiency,

$$EE = 1.0 - \frac{\sum_{t=1}^{T}(\widehat{Z}_t - Z_t)^2}{\sum_{t=1}^{T}(Z_t - \bar{Z})^2}.$$

iv. Sum of square prediction error, $SSPE = \sum_{t=1}^{T}\sum_{k=1}^{46}(\widehat{Z}_{kl} - Z_{kt})^2$.

It is common to use the NexRad measurements themselves as rainfall estimates in hydrological applications. For example, Bedient *et al.* (2000) and Bedient *et al.* (2003) use NexRad measurements in a hydrological

Table 9.1 Comparison of rainfall estimates.

Statistic	Model	NexRad
D (mm)	1344.85	7557.22
EB	7.04	39.58
EE	0.85	0.60
$SSPE$ (mm)	332.54	374.38

The column *Model* gives prediction results from the spatio-temporal model using regression and kriging; the *NexRad* column gives prediction results using the NexRad measurements. *SSPE* is the actual *SSPE*/1000.

model for flood prediction. For this reason, it is of interest to compute the four validation statistics using the nearest NexRad value as the estimate of the rainfall. Let Y_{kt} denote the nearest NexRad measurement for Z_{kt} on the tth day. The corresponding D, EB, EE, and $SSPE$ for the nearest NexRad predictor are computed by replacing $\widehat{Z}$ with Y in criteria (i)–(iv). It is informative to compare the model-based prediction with the nearest NexRad to see what (if any) improvement the regression/kriging method accomplishes.

Comparing the model and NexRad estimates, the model prediction has significantly less total difference in precipitation, lower estimation bias, and higher estimation efficiency. Overall, the results indicate that the model prediction performs substantially better than the NexRad in estimating daily total rainfall.

To further evaluate the regression/kriging method, a comparison of the $SSPE$ from the two methods shows that the model prediction improves by approximately 11% over the nearest NexRad-based prediction. On one particular day, September 11th, the total rainfall amount of the 60 stations was 1917.7 mm. This amount is approximately twice the next largest rainfall amount across the entire year. The model prediction method does not work well on this day. If September 11th is set aside, the prediction improves 16.7% compared to using the nearest NexRad as the predictor. All results suggest that the model-based prediction is less biased and slightly less variable than the closest NexRad to predict the true rainfall.

This example shows that regression modeling and covariance modeling can aid in spatial and spatio-temporal predictions.

9.3 Multivariate covariance functions

We might naively assume that multivariate covariance functions simply add another index to the notation for univariate covariance functions. It turns out, however, that some initially surprising relationships are common here. For example, for univariate covariance functions it is always the case that $|C(0)| \geq |C(k)|$ for all lags k. However, for multivariate covariances we can have $|C_{ij}(0)| < |C_{ij}(k)|$ for variables i and j. In other words, in the multivariate context, the maximal cross-covariance is not necessarily at the origin $k = 0$. Further, $C_{ij}(k) \neq C_{ij}(-k)$ in general, or, equivalently, $C_{ij}(k) \neq C_{ji}(k)$. The covariance is said to be asymmetric when these relationships hold. Symmetry requires $C_{ij}(k) = C_{ij}(-k)$ for all i, j, k.

Wackernagel (2003) demonstrates how asymmetric covariances naturally occur when modeling the gas input and the CO_2 output of a furnace. The fluctuation of CO_2 is delayed with respect to the fluctuation of gas input, due to the chemical reaction time for gas to produce CO_2. Thus, the cross-covariance between these two variables exhibits an asymmetric pattern. This example also demonstrates, again, that the maximal cross-covariance is not necessarily located at the origin. This is in stark contrast to the univariate case, where this particular notion of asymmetry does not exist.

Another situation where asymmetric covariances occur is when the cross-covariance is defined between a random function and its derivative; for instance, the cross-covariance between water head and transmissivity in hydrology, where the latter is the derivative of the former [Dagan (1985)]. In such situations, the cross-covariance is an odd function of lag k. Despite the wide range of asymmetries in applications, there are many cases where the cross-covariance is symmetric or approximately symmetric.

The matrix of covariances $C(k)$, for any specific lag k, is in general neither positive nor negative definite. However, in order that (9.2) be a valid cross-covariance function, the $np \times np$ covariance matrix of the random vector $\{Z(x_1)^{\mathrm{T}}, \cdots, Z(x_n)^{\mathrm{T}}\}^{\mathrm{T}} \in \mathbb{R}^{np}$ must be nonnegative definite for any positive integer n, and locations $x_1, \cdots, x_n$ in $\mathbb{R}^q$. This is completely analogous to the requirement for scalar responses for purely spatial data in Section 3.2, and for the univariate spatio-temporal data structure in Section 6.3.

9.3.1 Variogram function or covariance function?

Recall that, in Chapter 2, several arguments were made for the superiority of the variogram function over the covariance function. In the multivariate setting the situation is more complex, as we now illustrated.

Consider the situation in item (iii), in the introduction to this chapter, multivariate spatial observations. It is not obvious how to define the (cross) variogram function; an apparently natural choice is to define the variogram as:

$$\gamma_{ij}^*(k) := Var[Z_i(x+k) - Z_j(x)],$$

for $1 \leq i, j \leq p$. Another possibility is to define

$$\gamma_{ij}(k) := Cov[Z_i(x+k) - Z_i(x), Z_j(x+k) - Z_j(x)].$$

There is one principal way in which γ^* is superior to γ. Namely, γ requires that all variables be measured at the same spatial locations. For γ^* this is clearly not required. However, letting ρ_{ij} denote the correlation between variables i and j, direct computations show that

$$\gamma_{ij}^*(k) = \frac{1}{2}\left(\sigma_i^2 + \sigma_j^2\right) - \sigma_i\sigma_j\rho_{ij}(k),$$

while

$$\gamma_{ij}(k) = \sigma_i\sigma_j\left\{1 - \frac{1}{2}[\rho_{ij}(k) + \rho_{ij}(-k)]\right\}.$$

We see that the formula for γ^* averages the variances of two different variables which may not even be in the same units or measuring the same type of quantities. Further, the definition of γ^* implicitly assumes that $E[Z_i(u+k)] = E[Z_j(k)]$ for all i, j, u, k. This implies not only stationarity for each variable (as does the covariance function) but also that all variables have the *same* mean. On the other hand, to use γ for the cokriging in Section 9.1 requires that the covariance matrix is symmetric. These observations make neither γ nor γ^* entirely appropriate in the multivariate setting. For these reasons, we consider further only the cross-covariance function, as defined in Equation 9.2.

As in the univariate spatial and space–time settings, it is required that our chosen cross-covariance function be nonnegative definite. As before, a common solution to the nonnegative requirement for covariance functions is to consider parametric models that guarantee it. In recent years, various constructions of valid cross-covariance functions have been proposed. We next present three of the various approaches.

9.3.2 Intrinsic correlation, separable models

The most straightforward model is to assume that the joint variable and spatial correlations can be written as a product of the variable covariances and the spatial covariance, as was proposed by, for instance, Mardia and Goodall (1993). Specifically, this requires that the $p \times p$ covariance matrix $\mathbf{C}(k)$, defined in (9.2), can be written as:

$$\mathbf{C}(k) = \rho(k)\mathbf{T},$$

for all lags $k \in \mathbb{R}^q$, where $\rho(\cdot)$ denotes a scalar correlation function, and T is a positive definite $p \times p$ matrix. If there exists such a correlation function and positive definite matrix T, for all lags, the cross-covariance is said to be separable or that the correlation is intrinsic. The latter term seems preferable given the previously defined use of the term *separable* for univariate space–time covariances.

To see how this assumption influences the properties of a multivariate covariance function, note that under intrinsic correlation it necessarily holds that $C_{ij}(k) = C_{ij}(-k)$ for all i, j, k, and thus intrinsic correlation implies covariance symmetry. Further, this assumption greatly simplifies the modeling of multivariate data. For example, consider situation (iii) in the introduction to this chapter. Specifically, consider bivariate observations ($p = 2$) at $n = 4$ irregularly spaced spatial locations, $\mathbf{s}_1, \cdots, \mathbf{s}_4$. This leads to 8 distinct spatial lags. The number of parameters needed to describe the covariances among the observations $z_1(\mathbf{s}_1), \cdots, z_1(\mathbf{s}_4), z_2(\mathbf{s}_1), \cdots, z_2(\mathbf{s}_4)$ is 27. Under a symmetry assumption, this number reduces slightly to 21, and further reduces to 9 under intrinsic correlation. Further, under the intrinsic correlation assumption, we can often find an adequate spatial covariance model to fit our data, due to the availability of rich classes of spatial covariance models from Chapter 3. For example, if we find an exponential covariance model to be appropriate, then the required number of parameters drops to four (two variable variances, one variable covariance, and one common exponential decay parameter). Such gains in model simplicity become more important when either n or p is large. In general, for p-dimensional responses, at n irregularly spaced locations, the symmetry condition reduces the number of parameters from $p[1 + p + pn(n-1)]/2$ to $[p(p+1)(2-n+n^2)]/4$. Under intrinsic correlation, the number of parameters is further reduced to $[p(p+1)+n(n-1)]/2$. Adding a temporal index to observations adds an additional motivation for the simplification of multivariate covariance functions.

9.3.3 Coregionalization and kernel convolution models

Although intrinsic correlation allows for great simplicity in modeling covariances, it is sometimes not sufficient to describe observed correlations adequately. A more flexible type of model known as the linear model of coregionalization (LMC) is discussed in, for example, Wackernagel (2003) and Gelfand *et al.* (2004). This approach eases the restriction of the assumption of intrinsic correlation, while the covariance still remains in a parsimonious and concise form. Further, this approach can be considered as a generalization of the intrinsic correlation model. The LMC holds that

$$\mathbf{C}(k) = \sum_{g=1}^{r} \rho_g(k)\mathbf{T}_g, \tag{9.9}$$

for some positive integer $r \leq p$, and for all lags $k \in \mathbb{R}^q$, where the ρ_g, $g = 1, \ldots, r$, are scalar correlation functions, and the $\mathbf{T}_g$ are positive definite $p \times p$ matrices. The above intrinsic correlation is the special case of a linear model of coregionalization when $r = 1$. The linear model of coregionalization has been widely employed in many applications; see for example, Gelfand *et al.* (2004). If the data strongly suggest that intrinsic correlation holds, then this model should be assumed. If, however, intrinsic correlation does not hold, then it is desirable to determine if the random field allows for a linear model of coregionalization, with small r. Note that a LMC model implies that symmetry holds and thus is a compromise between intrinsic correlation and symmetry in terms of model complexity.

Another approach to multivariate covariance is through kernel convolution models. Crossvariograms are constructed by explicitly modeling spatial data as moving averages over white noise random processes. Parameters of the moving average functions may be inferred from the variogram, and, with a few additional parameters, crossvariogram models are constructed. Weighted least squares, for example, can then be used to fit the crossvariogram model to the empirical crossvariogram for the data. Some details of this approach in the pure spatial setting are given in Section 3.4. This approach for multivariate observations is given in Ver Hoef and Barry (1998). A related convolution-based approach is given in Majumdar and Gelfand (2007). They build multivariate spatial cross-covariances through a convolution of p separate, stationary one-dimensional covariance functions.

9.4 Testing and assessing intrinsic correlation

Given the parsimoniousness of the intrinsic correlation model, and simplicity of interpretation, it is desirable to fit this model if appropriate. Another benefit from assessing the covariance structure occurs when cokriging is considered. If all variables have been measured at all sample locations and the covariance functions are intrinsically correlated, then cokriging is equivalent to separate kriging for each variable. Although these considerations make an intrinsic correlation model desirable, we should be confident that the phenomena of interest at least approximately satisfy this assumption. Towards this end, note that under intrinsic correlation it holds that

$$\begin{aligned} &Corr[Z_i(x), Z_j(x+k)] \\ &\quad = \rho(k)C_{ij}(k)/\sqrt{\rho(k)C_{ii}(k)\rho(k)C_{jj}(k)} = a_{ij}, \end{aligned}$$

say, which is independent of k. Empirical estimates of this correlation, for different i and j, are known as the codispersion coefficients, and they should be approximately constant as k varies. Through empirical estimation of these correlations, this constancy of the codispersion coefficients can be assessed visually. Another diagnostic is based on the value of crossvariograms between the principal components of variables. This approach is discussed in Wackernagel (2003). In this case, a zero value of such crossvariograms indicates the adequacy of a separable model.

Both approaches, using the codispersion coefficients and viewing crossvariograms of principal components, yield useful diagnostic information. There are, however, no clear guidelines on how to make the judgment of constancy of the codispersion coefficients or proximity to zero of the principle components based on the graphical evidence. For these reasons it is desirable to have more objective methods to assess the appropriateness of intrinsic correlation or LMC models based on the given spatial or spatio-temporal observations.

Specifically, we want to test the hypothesis

$$H_0 : C_{ij}(k) = C_{ij}(-k),$$

that the covariance function is symmetric, and the hypothesis that

$$H_0 : C_{ij}(k_1)/C_{ij}(k_2) = \rho(k_1)/\rho(k_2),$$

that the covariance is intrinsically correlated. Recall that if the former hypothesis is rejected then intrinsic correlation is also rejected. In this case, we then can see if a coregionalization model is more appropriate for some given order $r > 1$. Towards this end, a basic framework is needed to justify a formal testing procedure. We follow the development in Li *et al.* (2008).

In many situations, the observations are taken at a fixed number of locations in a region $S \subset \mathbb{R}^2$, at regular times $T_n = \{1, \cdots, n\}$. In this particular case, to quantify temporal dependence, the mixing coefficient is defined [for example, Ibragimov and Linnik (1971), p. 306 or Doukhan (1994)] as follows:

$$\alpha(u) = \sup_{A,B} \{|P(A \cap B) - P(A)P(B)|, A \in \mathfrak{F}_{-\infty}^{0}, B \in \mathfrak{F}_{u}^{\infty}\},$$

where $\mathfrak{F}_{-\infty}^{0}$ denotes the σ-algebra generated by the past time process until $t = 0$, and $\mathfrak{F}_{u}^{\infty}$ is the σ-algebra generated by the future time process from $t = u$. This temporal mixing coefficient is quite analogous to the spatial mixing coefficient defined in Section 5.6.1. Assume the mixing coefficient $\alpha(u)$ satisfies the following mixing condition:

$$\alpha(u) = O(u^{-\epsilon}), \tag{9.10}$$

for some $\epsilon > 0$. Examples of processes satisfying this condition are given in Chapter 6, Section 6.7.

Let $\mathbf{\Lambda}$ be a set of user-chosen space–time lags and let c denote the cardinality of $\mathbf{\Lambda}$. Define

$$G = \{C_{ij}(\mathbf{h}, u) : (\mathbf{h}, u) \in \mathbf{\Lambda}, i, j = 1, \cdots, p\}$$

to be the length cp^2 vector of cross-covariances at spatio-temporal lags $k = (\mathbf{h}, u)$ in $\mathbf{\Lambda}$. Let $\widehat{C}_{ij,n}(\mathbf{h}, u)$ denote the estimator of $C_{ij}(\mathbf{h}, u)$ based on observations in $S \times T_n$, and let

$$\widehat{G}_n = \{\widehat{C}_{ij,n}(\mathbf{h}, u), (\mathbf{h}, u) \in \mathbf{\Lambda}, i, j = 1, \cdots, p\}.$$

Let $S(\mathbf{h}) = \{s : s \in S, s + h \in S\}$ and let $|S(h)|$ denote the number of elements in $S(\mathbf{h})$.

Further assume the following moment condition:

$$\sup_n E\left\{\left||T_n|^{1/2}[\widehat{C}_{ij,n}(h,u) - C_{ij}(h,u)]\right|^{2+\delta}\right\} \leq C_\delta, \tag{9.11}$$

for some $\delta > 0$, $C_\delta < \infty$ and any $i, j = 1, \cdots, p$, and define the estimator of $C_{ij}(h,u)$ as:

$$\widehat{C}_{ij,n}(h,u) = \frac{1}{|S(h)||T_n|} \sum_{S(h)} \sum_{t=1}^{n-u} Z_i(s,t) Z_j(s+h, t+u).$$

This is the natural empirical estimator of $C_{ij}(h,u)$ for mean-centered observations.

Theorem 9.4.1 *Let $\{Z(s,t) \in \mathbb{R}^p, s \in \mathbb{R}^2, t \in \mathbb{Z}\}$ be a mean zero, strictly stationary spatio-temporal multivariate random field observed in $D_n = S \times T_n$, where $S \subset \mathbb{R}^2$ and $T_n = \{1, \cdots, n\}$. Assume:*

$$\sum_{t \in \mathbb{Z}} |Cov[Z_i(0,0)Z_j(h_1,u_1), Z_l(s,t) \times Z_m(s+h_2, t+u_2)]| < \infty, \tag{9.12}$$

for all $h_1 \in S(h_1)$, $h_2 \in S(h_2)$, $s \in S$, any finite $u_1, u_2 \in \mathbb{Z}$, and $i, j, l, m = 1, \cdots, p$. Then $\mathbf{\Sigma} = \lim_{n\to\infty} |T_n| Cov(\widehat{G}_n, \widehat{G}_n)$ exists. The element of $\mathbf{\Sigma}$ corresponding to the covariance $Cov[\widehat{C}_{ij,n}(h_p, u_p), \widehat{C}_{lm,n}(h_q, u_q)]$ is

$$A_{p,q} \sum_{S(h_p)} \sum_{S(h_q)} \sum_{t \in \mathbb{Z}} Cov[Z_i(s_1, 0)Z_j(s_1 + h_p, u_p), Z_l(s_2, t) Z_m(s_2 + h_q, t + u_q)],$$

where $A_{p,q} := [|S(h_p)||S(h_q)|]^{-1}$. If we further assume that $\mathbf{\Sigma}$ is positive definite and that conditions (9.10) and (9.11) hold, then $|T_n|^{1/2}(\widehat{G}_n - G) \longrightarrow N_{cp^2}(0, \mathbf{\Sigma})$ in distribution, as $n \to \infty$.

The strict stationarity in space S can be relaxed in the circumstances where only location-specific spatial covariances are concerned. For example, given the sparsity of spatial data in Section 9.6, we consider the covariances for each specific pair of stations instead of estimating the covariance at arbitrary lags for the irregularly spaced monitoring stations. Condition (9.12) holds, for example, if $\mathrm{E}[|Z_i(s,t)|^{4+\delta}] < C_\delta$, for each variable $i = 1, \cdots, p$, for some $\delta > 0$, $C_\delta < \infty$, and $\alpha(u) = O(u^{-\epsilon(4+\delta)/\delta})$ for some $\epsilon > 1$ [see Ibragimov and Linnik (1971), Theorem 18.5.3]. No assumption of Gaussianity is required. If, however, Z is Gaussian, then (9.12) reduces to $\sum_{t\in\mathbb{Z}} |C_{il}(\mathbf{h}_1, t)C_{jm}(\mathbf{h}_2, t + u)| < \infty$ for all $\mathbf{h}_1 \in S(\mathbf{h}_1)$, $\mathbf{h}_2 \in S(\mathbf{h}_2)$, any finite $u \in \mathbb{Z}$, and $i, j, l, m = 1, \ldots, p$. Similar bounds could be derived for (9.11), although explicit results for general $\delta > 0$ seem difficult.

The tests to be considered involve ratios or products of elements in $\widehat{G}_n$, and for this reason the following result for smooth functions of covariance estimators is needed. Given a function $f = (f_1, \ldots, f_b)^{\mathrm{T}}$ defined on G such that f is differentiable at G, we have, using the multivariate delta theorem [for example, Mardia *et al.* (1979), p. 52], that

$$|T_n|^{1/2}[f(\widehat{G}_n) - f(G)] \xrightarrow{d} N_b(0, \mathbf{B}^{\mathrm{T}}\mathbf{\Sigma}\mathbf{B}),$$

where $B_{ij} = \partial f_j / \partial G_i$, $i = 1, \ldots, c$, $j = 1, \ldots, b$. Using this result, for a matrix A such that $Af(G) = 0$ under the null hypothesis, it holds that

$$TS = |T_n|[\mathbf{A}f(\widehat{G}_n)]^{\mathrm{T}}(\mathbf{A}\mathbf{B}^{\mathrm{T}}\mathbf{\Sigma}\mathbf{B}\mathbf{A}^{\mathrm{T}})^{-1}[\mathbf{A}f(\widehat{G}_n)] \xrightarrow{d} \chi^2_a, \qquad (9.13)$$

where a is the row rank of the matrix $\mathbf{A}$, and *TS* denotes the test statistic.

9.4.1 Testing procedures for intrinsic correlation and symmetry

In order to assess intrinsic correlation (IC) and symmetry, the proper function, f, and contrast matrix, $\mathbf{A}$, are needed according to the specific hypotheses of interest. To test for IC, the function f is defined to give pairwise products. Clearly f can also be defined as pairwise ratios; nevertheless the product form is more convenient. Under the null hypothesis of intrinsic correlation, we can determine a matrix $\mathbf{A}$ such that $\mathbf{A}f(G) = 0$. For example, if $p = 2$ and $\mathbf{\Lambda} = \{k_1, k_2, k_3, k_4\}$, where $k_i = (\mathbf{h}_i, u_i)$ for $i = 1, 2, 3, 4$, we have

$$G = \{C_{12}(k_1), C_{12}(k_2), C_{12}(k_3), C_{12}(k_4), \rho(k_1), \rho(k_2), \rho(k_3), \rho(k_4)\}^{\mathrm{T}}.$$

Note that we can write $\rho(k)$ rather than $\rho_{11}(k)$, $\rho_{22}(k)$ or $\rho_{12}(k)$, as these correlation functions are all equal under the null hypothesis. Define

$$f(G) = \{C_{12}(k_1)\rho(k_2), C_{12}(k_2)\rho(k_1), C_{12}(k_3)\rho(k_4), C_{12}(k_4)\rho(k_3)\}^{\mathrm{T}},$$

and

$$\mathbf{A} = \begin{pmatrix} 1 & -1 & 0 & 0 \\ 0 & 0 & 1 & -1 \end{pmatrix}.$$

Under the null hypothesis of intrinsic correlation, we have $\mathbf{A}f(G) = 0$. We have given an example of G or $f(G)$, but these choices are clearly not unique. For example, we can extend $f(G)$ and $\mathbf{A}$ to include $C_{12}(k_1)\rho(k_3)$ and $C_{12}(k_3)\rho(k_1)$ in $f(G)$. Empirical evidence suggests that repeated use of the same lags, k_1 and k_3, tends to increase the size of the test. One possible reason for this is the extra dependency between elements in $f(G)$ introduced by using certain testing lags repeatedly. The tests for symmetry properties are analogous to the test for intrinsic correlation, but the symmetry test is simpler as f is the identity function in this case. The correlation $\rho(k)$ appearing in the test statistic is estimated by the empirical average over estimates of $\rho_{ij}(k)$, where $\rho_{ij}(k)$, at lag $k = (\mathbf{h}, u)$, denotes the correlation between the ith and jth variables.

9.4.2 Determining the order of a linear model of coregionalization

Consider a multivariate spatio-temporal random field, $Z(\mathbf{s}, t)$, under a LMC of order r defined by Equation 9.9. Due to the need to minimize the number of necessary covariance parameters, it is desired to determine the smallest r necessary to accurately model $C(k)$, where $k = (\mathbf{h}, u)$. If the hypothesis of intrinsic correlation is not rejected, then the order is $r = 1$.

If the test of intrinsic correlation is rejected, that is, $r > 1$, then we can assess the appropriateness of a higher order of an LMC sequentially. Specifically, we next perform a test of H_0: LMC with $r = 2$. If this hypothesis is rejected, then we move to H_0: LMC with $r = 3$, and continue until there is no significant evidence against the null hypothesis or until r reaches $p - 1$. A test can then be developed based on the fact that the linear model of coregionalization can be constructed hierarchically through univariate conditional random fields. This approach is taken, for example, by Royle and Berliner (1999). This hierarchical approach models dependence of variables through *conditional* means. This allows avoidance of the difficult estimation of the correlation functions, ρ_g, in (9.9), while still taking the variable dependence into account.

As an illustration, consider a linear model of coregionalization with $p = 3$ variables and

$$Z(\mathbf{s},t) = \mathbf{A}W(\mathbf{s},t) = \begin{pmatrix} a_{11} & 0 & 0 \\ a_{21} & a_{22} & 0 \\ a_{31} & a_{32} & a_{33} \end{pmatrix} \begin{pmatrix} W_1(\mathbf{s},t) \\ W_2(\mathbf{s},t) \\ W_3(s,t) \end{pmatrix}. \quad (9.14)$$

The lower triangular form of $\mathbf{A}$ is assumed without loss of generality, by Gelfand *et al.* (2004). Denote the component random field and its correlation as $W_g(s,t)$ and $\rho_g(k)$, $g = 1, 2, 3$, and denote a_g as the gth column of $\mathbf{A}$. Assume, for simplicity, that the $W_g(s,t)$ have zero mean and unit variance. This assumption has no practical effect since the spatial variance term can be absorbed into the matrix $\mathbf{A}$. Assuming $\rho_g(k) = \rho(k)$ results in a separable covariance model, $C(k) = \rho(k)T$, where $T = \mathbf{A}\mathbf{A}^{\mathrm{T}}$ in the given formulation. This corresponds to the intrinsic correlation case when $r = 1$. To obtain a LMC with $r = 2$, there are exactly two distinct functions among the correlations $\rho_g(k)$. Without loss of generality, assume $\rho_2(k) = \rho_3(k)$, and hence $C(k) = \sum_{g=1}^{2} \rho_g(k) T_g$, where $T_1 = a_1 a_1^{\mathrm{T}}$ and $T_2 = a_2 a_2^{\mathrm{T}} + a_3 a_3^{\mathrm{T}}$. If all of the $\rho_g(k)$ are distinct functions, we have $r = 3$ with $C(k) = \sum_{g=1}^{3} \rho_g(k) T_g$, where $T_g = a_g a_g^{\mathrm{T}}$.

The joint distribution of $\mathbf{Z}(s,t) = \{Z_1(\mathbf{s},t), Z_2(\mathbf{s},t), Z_3(\mathbf{s},t)\}^{\mathrm{T}}$ can be factored into the product:

$$\mathrm{P}[Z(\mathbf{s},t)] = \mathrm{P}[Z_1(\mathbf{s},t)]\mathrm{P}[Z_2(\mathbf{s},t)|Z_1(\mathbf{s},t)]\mathrm{P}[Z_3(\mathbf{s},t)|Z_1(\mathbf{s},t), Z_2(\mathbf{s},t)],$$

where $Z_i(\mathbf{s},t)$, $i = 1, 2, 3$, represent the univariate space–time processes corresponding to each variable. Using this formulation, the multivariate $Z(\mathbf{s},t)$ can be generated through univariate conditional random fields as illustrated in Gelfand *et al.* (2004) and Schmidt and Gelfand (2003). Specifically,

$$\begin{aligned} Z_1(\mathbf{s},t) &= a_{11} W_1(\mathbf{s},t), \\ Z_2(\mathbf{s},t)|Z_1(\mathbf{s},t) &= \frac{a_{21}}{a_{11}} Z_1(\mathbf{s},t) + a_{22} W_2(\mathbf{s},t), \text{ and} \\ Z_3(\mathbf{s},t)|Z_1(\mathbf{s},t), Z_2(\mathbf{s},t) &= \left(\frac{a_{31}}{a_{11}} - \frac{a_{21}a_{32}}{a_{11}a_{22}}\right) Z_1(\mathbf{s},t) \\ &\quad + \frac{a_{32}}{a_{22}} Z_2(\mathbf{s},t) + a_{33} W_3(s,t). \end{aligned}$$

The associated unconditional form is exactly (9.14). If $\rho_2(k) = \rho_3(k)$, that is, $r = 2$, then we can express $Z_3(\mathbf{s}, t)$ given $Z_1(\mathbf{s}, t)$ as:

$$Z_3(\mathbf{s}, t)|Z_1(\mathbf{s}, t) = \frac{a_{31}}{a_{11}} Z_1(\mathbf{s}, t) + a_{32} W_2(\mathbf{s}, t) + a_{33} W_3(s, t).$$

Then, letting

$$Z_2^*(\mathbf{s}, t) = Z_2(\mathbf{s}, t) - \frac{a_{21}}{a_{11}} Z_1(\mathbf{s}, t),$$

and

$$Z_3^*(\mathbf{s}, t) = Z_3(\mathbf{s}, t) - \frac{a_{31}}{a_{11}} Z_1(\mathbf{s}, t),$$

we obtain a separable bivariate random field $\mathbf{Z}'(\mathbf{s}, t) = \{\mathbf{Z}_2^*(\mathbf{s}, t), Z_3^*(\mathbf{s}, t)\}^{\mathrm{T}}$, say. Hence, the test of $H_0 : r = 2$ is equivalent to the intrinsic correlation test $H_0 : \mathbf{C}^*(k) = \rho^*(k)\mathbf{T}'$, where $\mathbf{C}^*(k)$, $\rho^*(k)$, and $\mathbf{T}^*$ are the functions defined over $\mathbf{Z}^*(\mathbf{s}, t)$. To estimate a_{21}/a_{11} and a_{31}/a_{11}, first estimate $\mathbf{T}$ by $\widehat{\mathbf{T}} = \widehat{\mathbf{C}}(0)$ and then decompose $\widehat{\mathbf{T}}$ into $\widehat{\mathbf{A}}\widehat{\mathbf{A}}^{\mathrm{T}}$, from which we obtain plug-in estimators for the two ratios a_{21}/a_{11} and a_{31}/a_{11}.

9.4.3 Covariance estimation

To use the large sample distribution in Equation 9.13, estimators of the matrices $\mathbf{B}$, when f is not the identity function in the test, and $\mathbf{\Sigma}$ are required. The matrix $\mathbf{B}$ can be estimated empirically using $\widehat{G}_n$. Although plug-in estimates of $\mathbf{\Sigma}$ are available, it turns out that subsampling tends to better estimate $\mathbf{\Sigma}$. Subsampling has proven to be a successful approach in a wide variety of contexts, as we discuss in Chapter 10. For the specific task of estimating the covariance matrix of empirical covariance estimators, consider the implementation of subsampling with overlapping subblocks in a fixed space domain and an increasing time domain. In the univariate setting, Chapter 6 discusses how the optimal subblock length $l(n)$ for variance estimation of a sample mean computed from an order 1 autoregressive time series based on overlapping subseries is

$$l(n) = \left(\frac{2\widehat{\rho}}{1 - \widehat{\rho}^2}\right)^{2/3} \left(\frac{3n}{2}\right)^{1/3},$$

where $\widehat{\rho}$ denotes the estimated order 1 autoregressive parameter. However, in the multivariate setting, a unique temporal correlation estimator does not typically exist, since the correlation function of each variable may be different. To estimate ρ, a natural approach is to simply average the first-order temporal correlation estimators across all variables.

9.5 Numerical experiments

9.5.1 Symmetry

It is appropriate to examine the size and power of the test of multivariate symmetry for a variety of sample sizes in the space–time context before adopting its use. For each setting considered, one thousand simulated replicate data sets are evaluated. To assess the performance of the test, first consider a situation which distinguishes between symmetric and asymmetric covariance structures. In order to do this, consider a bivariate ($p = 2$) random field. Assume $Z_1(\mathbf{s}, t) = Z_2(\mathbf{s}, t + \Delta_t) + \epsilon(s, t)$, where $Z_2(\mathbf{s}, t)$ is a strictly stationary random field with covariance $C_{22}(\mathbf{h}, u)$ $= \text{cov}[Z_2(\mathbf{s}, t), Z_2(\mathbf{s} + \mathbf{h}, t + u)]$, the error $\epsilon(s, t)$ denotes random noise with mean zero and variance σ^2, and Δ_t denotes a temporal shift parameter. It follows that:

$$C_{11}(\mathbf{h}, u) = Cov[Z_1(\mathbf{s}, t), Z_1(\mathbf{s} + \mathbf{h}, t + u)] = C_{22}(\mathbf{h}, u) + \sigma^2,$$

$$C_{12}(\mathbf{h}, u) = Cov[Z_1(\mathbf{s}, t), Z_2(\mathbf{s} + \mathbf{h}, t + u)] = C_{22}(\mathbf{h}, u - \Delta_t),$$

$$C_{21}(\mathbf{h}, u) = Cov[Z_2(\mathbf{s}, t), Z_1(\mathbf{s} + \mathbf{h}, t + u)] = C_{22}(\mathbf{h}, u + \Delta_t).$$

It is clear that for $\Delta_t = 0$, we have $C_{12}(\mathbf{h}, u) = C_{21}(\mathbf{h}, u)$; while for $\Delta_t \neq 0$, we have $C_{12}(\mathbf{h}, u) \neq C_{21}(\mathbf{h}, u)$. Note that in the latter situation it also holds that $C_{12}(\mathbf{h}, u) \neq C_{12}(\mathbf{h}, -u)$, and $C_{21}(\mathbf{h}, u) \neq C_{21}(\mathbf{h}, -u)$. We can perform three tests with null hypotheses of $H_0^1 : C_{12}(\mathbf{h}, u) = C_{21}(\mathbf{h}, u)$, $H_0^2 : C_{12}(\mathbf{h}, u) = C_{12}(\mathbf{h}, -u)$, and $H_0^3 : C_{21}(\mathbf{h}, u) = C_{21}(\mathbf{h}, -u)$.

In order to illustrate the performance of this test of symmetry, assume that $C_{22}(\mathbf{h}, u)$ is space–time separable and spatially isotropic, $C_{22}(\mathbf{h}, u) = C_{22}(\|\mathbf{h}\|, |u|)$. Testing under nonseparable space–time covariance structures gives qualitatively similar results as those to follow. In the specific setting above, observe that $C_{12}(\mathbf{h}, -u) = C_{21}(\mathbf{h}, u)$ and $C_{21}(\mathbf{h}, -u) = C_{12}(\mathbf{h}, u)$. Hence, in this case, the hypothesis test actually tests $H_0 : C_{12}(\|\mathbf{h}\|, u) = C_{21}(\|\mathbf{h}\|, u)$.

We generate Z_2 via a vector autoregressive model of order 1 for $\{Z_2(\mathbf{s}_1, t), \cdots, Z_2(\mathbf{s}_m, t)\}^{\mathrm{T}}$ with Gaussian random noise and a spatial

Table 9.2 Simulation-estimated sizes for the symmetry and intrinsic correlation tests.

		Sizes for symmetry		Sizes for intrinsic correlation	
		$\lvert T_n \rvert$		$\lvert T_n \rvert$	
Grid size	ρ	200	1000	200	1000
	0.4	0.04	0.03	0.06	0.06
3×3	0.6	0.05	0.04	0.08	0.06
	0.8	0.11	0.05	0.09	0.08
	0.4	0.05	0.03	0.05	0.05
7×7	0.6	0.08	0.04	0.07	0.06
	0.8	0.10	0.06	0.11	0.06

Nominal level is 0.05. Sizes for symmetry are obtained by setting $\eta = 3$ and using contrasts C_1, C_2, C_3, and C_4 defined in Section 9.5.1. Sizes for intrinsic correlation are obtained by setting $\eta_1 = \eta_2 = 3$ and using contrasts C_1 and C_2 defined in Section 9.5.2.

exponential covariance function on a spatial grid of m observations, as described in Section 6.7.4. The variance of the random noise of the vector autoregressive model and σ^2 are both set equal to 1. Let η denote the spatial exponential range parameter, let ρ be the temporal correlation parameter in the vector autoregressive model, and let $|T_n|$ denote the number of temporal replicates. We choose four lags k: $k_1 = (\|h\| = 1, u = 1)$, $k_2 = (\|h\| = 2^{1/2}, u = 1)$, $k_3 = (\|h\| = 2$, $u = 1)$, $k_4 = (\|h\| = 5^{1/2}, u = 1)$, which form four contrasts $C_i = C_{12}(k_i) - C_{21}(k_i)$, $i = 1, 2, 3, 4$.

Fixing $\eta = 3$ and $\Delta_t = 0$ enables empirical estimation of the size of the test. Table 9.2 gives the estimated test sizes across the 1000 experiments for each set of parameter values. Overall, the sizes approach the nominal level of 0.05, as grid size and/or temporal length increase. For a random field of fewer temporal replicates, say $|T_n| = 200$, the strong temporal correlation makes the sizes deviate from nominal. The sizes become more stable, however, as $|T_n|$ increases. Fixing $\eta = 3$ and $\rho = 0.4$, we have examined the powers under various grid sizes (3×3 and 7×7), Δ_t (1, 2, and 3), and $|T_n|$ (200 and 1000) using contrasts C_1, C_2, C_3, and C_4 defined above. Summarizing our findings, the powers increase as grid size and/or temporal length increase. Powers diminish, however, as Δ_t increases, which is to be expected, as k contains only lags with $u = 1$, which cannot capture the strongest cross-correlation when $\Delta_t > 1$. The larger Δ_t, the further $C(k)$ departs from strong cross-correlation. This provides guidance on choosing lags to examine. Specifically, when considering a true physical process, it is appropriate to check that k covers the suspected temporal delays.

9.5.2 Intrinsic correlation

We address the ability of the test to detect the difference between intrinsically correlated and non-intrinsically correlated structures between variables. Towards this goal, consider separately observations from an intrinsic correlation model and from non-intrinsically correlated models.

Consider a bivariate ($p = 2$) random field from a linear model of coregionalization $\mathbf{Z}(\mathbf{s}, t) = \mathbf{A}\mathbf{W}(\mathbf{s}, t)$, where $\mathbf{A}$ is a 2×2 lower triangular matrix such that $\mathbf{A}\mathbf{A}^{\mathrm{T}} = \mathbf{T}$, $\mathbf{W}(\mathbf{s}, t) = \{W_1(\mathbf{s}, t), W_2(\mathbf{s}, t)\}^{\mathrm{T}}$, with $W_1(\mathbf{s}, t)$ and $W_2(\mathbf{s}, t)$ independent vector autoregressive models of order 1 with Gaussian random noise and a spatial exponential covariance function. Each vector autoregressive random field is separable in terms of its space–time covariance, but as noted in Section 9.5.1 this is not necessary. Denote the spatial range parameters of $W_1(\mathbf{s}, t)$ and $W_2(\mathbf{s}, t)$ by η_1 and η_2. We set

$$\mathbf{T} = \begin{pmatrix} 1 & 0.5 \\ 0.5 & 1 \end{pmatrix},$$

and consider the four lags k: $k_1 = (\|h\| = 1, u = 0)$, $k_2 = (\|h\| = 2^{1/2}, u = 0)$, $k_3 = (\|h\| = 2, u = 0)$, $k_4 = (\|h\| = 5^{1/2}, u = 0)$, which form two contrasts:

$$\mathcal{C}_1 = C_{12}(k_1)\rho(k_3) - C_{12}(k_3)\rho(k_1),$$

$$\mathcal{C}_2 = C_{12}(k_2)\rho(k_4) - C_{12}(k_4)\rho(k_2),$$

where $\rho(\cdot)$ represents a univariate space–time correlation function. To study the size of the test, the spatial range parameters are set to be equal: $\eta_1 = \eta_2 = 3$. From 1000 replicated data sets in each setting we find the empirical sizes given in Table 9.2. Overall, the sizes become closer to nominal as the amount of temporal and/or spatial information increases. However, as the temporal correlation increases, the effective sample size is decreased and accordingly the empirical sizes depart more from the nominal level of 0.05.

To evaluate the power in this situation, consider the following three settings using different contrasts:

i. $\eta_1 = 2, \eta_2 = 4$, using contrasts C_1 and C_2;

ii. $\eta_1 = 2, \eta_2 = 4$, using contrasts C_1, C_2, C_3, and C_4, where

$$C_3 = C_{12}\left(k_1'\right)\rho\left(k_3'\right) - C_{12}\left(k_3'\right)\rho\left(k_1'\right),$$
$$C_4 = C_{12}\left(k_2'\right)\rho\left(k_4'\right) - C_{12}\left(k_4'\right)\rho\left(k_2'\right),$$

with $k_1' = (\|h\| = 1, u = 1)$, $k_2' = (\|h\| = 2^{1/2}, u = 1)$, $k_3' = (\|h\| = 2, u = 1)$, $k_4' = (\|h\| = 5^{1/2}, u = 1)$;

iii. Temporal correlation fixed at $\rho = 0.4$, $\eta_1 = 1$, but η_2 varies, using contrasts C_1 and C_2.

The power results from these three situations are given in Table 9.3. The patterns of the results are quite clear. As the amount of temporal data increases, the power increases in all cases. However, as the temporal correlation increases, the empirical powers drop, as is to be expected. In cases where spatial correlation is much stronger than temporal correlation, adding space–time lags does not greatly increase power compared to using

Table 9.3 Empirical powers of the test for intrinsic correlation

		$\|T_n\| = 200$		$\|T_n\| = 1000$		
Grid size	ρ	(a)	(b)	(a)	(b)	η_2
3×3	0.4	0.27	0.19	0.89	0.86	2
	0.6	0.20	0.20	0.71	0.72	3
	0.8	0.15	0.19	0.40	0.45	4
7×7	0.4	0.86	0.82	1.00	1.00	2
	0.6	0.70	0.72	1.00	1.00	3
	0.8	0.43	0.55	0.98	0.99	4

Columns (a) and (b) are powers varying with ρ by setting $\eta_1 = 2, \eta_2 = 4$. (a) uses contrasts C_1 and C_2, while (b) uses contrasts C_1, C_2, C_3, and C_4. The nominal level is 0.05, and C_1, C_2, C_3, and C_4 are defined in Section 9.5.2.

purely spatial lags. If, however, temporal correlation becomes relatively stronger, then including the space–time lags becomes more beneficial. To summarize, using the most highly correlated lags delivers the highest power in testing.

9.5.3 Linear model of coregionalization

Consider a random field from a LMC model: $\mathbf{Z}(\mathbf{s}, t) = \mathbf{A}\mathbf{W}(\mathbf{s}, t)$, as in Equation 9.14, where $\mathbf{A}$ is a lower triangular matrix with $\mathbf{T} = \mathbf{A}\mathbf{A}^{\mathrm{T}}$. The vector

$$\mathbf{W}(\mathbf{s}, t) = \{W_1(\mathbf{s}, t), W_2(\mathbf{s}, t), W_3(\mathbf{s}, t)\}^{\mathrm{T}},$$

where $W_1(\mathbf{s}, t)$,$W_2(\mathbf{s}, t)$, and $W_3(\mathbf{s}, t)$ are independent vector autoregressive models of order 1 with Gaussian random noise and spatial exponential covariance functions. The temporal correlation parameter is ρ. Each vector autoregressive random field is separable in terms of its space–time covariance function. Again, the nonseparable structure does not alter the basic conclusions on the testing performance. The spatial range parameters of $W_1(\mathbf{s}, t)$, $W_2(\mathbf{s}, t)$, and $W_3(\mathbf{s}, t)$ are denoted by η_1, η_2, and η_3. The matrix $\mathbf{T}$ is set as

$$\mathbf{T} = \begin{pmatrix} 1 & 0.5 & 0.5 \\ 0.5 & 1 & 0.5 \\ 0.5 & 0.5 & 1 \end{pmatrix},$$

and again employ the testing lags k_1, k_2, k_3, and k_4, as well as C_1 and C_2, from Section 9.5.2. Then, fixing $\eta_1 = 1$ and $\eta_2 = 2$ while varying η_3, for $H_0 : r = 2$, we obtain empirical sizes and powers under various experimental settings. As usual, 1000 simulation replicates for each situation are considered. Figure 9.1 shows that the empirical sizes are reasonably close to the nominal level of 0.05. Further, powers increase for the longer temporal length and/or the larger grid size. Power increases as η_3 increases. This is natural, as this reflects increasing departure from the null hypothesis.

9.6 A data application to pollutants

We analyze trivariate atmospheric pollution measurements (CO, NO, NO_2 levels) from the California Air Resources Board in 1999. During that year, there were 83 monitoring stations with hourly measurements of CO, NO,

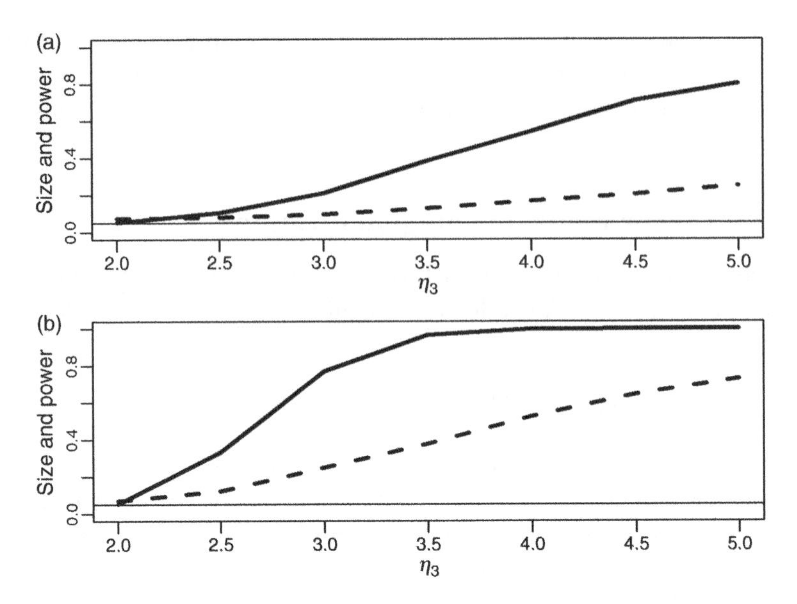

Figure 9.1 LCM experiment analysis. Sizes and powers of the test for the linear model of coregionalization for two spatial grid sizes and temporal lengths. (a) Results for a 3×3 *grid, (b) for a* 7×7 *grid. For both plots, the dotted and solid lines denote the temporal lengths* $|\mathrm{T_n}| = 200$ *and* $|\mathrm{T_n}| = 1000$*, respectively. The solid horizontal line denotes the nominal size of the test (0.05). Spatial decay parameter values are* $\eta_1 = 1$ *and* $\eta_2 = 2$*. The value* $\eta_3 = 2$ *gives the empirical size of the test; larger values of* η_3 *give the power at those values.*

and NO_2. However, not all of them have been recorded continuously in time. Considering only the daily average of the pollutants, we end up with 23 sites which have consecutive daily average measurements of the three above pollutants. Figure 9.2 shows the locations of these 23 monitoring stations. The daily average of the three pollutants over 68 locations on July 16, 1999 was analyzed originally by Schmidt and Gelfand (2003), and later by Majumdar and Gelfand (2007). In their analyses, two different methodologies were presented to build valid multivariate covariance models for the three correlated pollutants, but both suggested different spatial ranges for all three component measurements. The appropriateness of the assumptions of symmetry, intrinsic correlation and a linear model of coregionalization are assessed for these data.

Figure 9.2 Twenty-three monitoring stations in California with consecutive daily measurements of CO, NO, and NO_2. The six line segments denote the 12 stations (6 pairs) used in testing in Section 9.6.

From 12 stations, 6 pairs of stations are formed, as shown in Figure 9.2, to be used for testing. This choice of testing pairs seeks pairs with large correlation, and achieves a wide spatial coverage of monitoring stations. As in Schmidt and Gelfand (2003) and Majumdar and Gelfand (2007), we take the logarithm of the daily averages in order to stabilize the variance. Following Haslett and Raftery (1989), the data are detrended as follows: seasonal effects are estimated by calculating the mean of the logarithm of the daily averages across all 12 stations for each day of the year, and then the resulting time series is regressed on a set of annual harmonics. Subtraction of the estimated seasonal effect from the logarithm of the daily averages for each station yields approximately stationary residuals. These residuals are viewed as the component $\mathbf{Z}(x)$ in the framework of Equation 9.1, and serve as the processes of interest in what follows.

The largest empirical cross-correlations occur at temporal lag $u = 0$ for all variable combinations, so we choose $u = 0$ for all testing lags. Three types of pollutants and 6 pairs of stations generate 18 testing pairs in the symmetry test, and consequently the degrees of freedom for the symmetry test are $a = 18$. We obtain an observed test statistic of $TS = 103.5$ with a corresponding p-value $= 4.97 \times 10^{-14}$. To further understand the structure of cross-covariance of the three pollutants, it is interesting to assess the symmetry for each grouping of two pollutants separately. The tests give a p-value of 2.19×10^{-8} between CO and NO_2, a p-value of 0.00047 between CO and NO, and a p-value of 0.00041 between NO and NO_2. All tests indicate that, overall, the cross-covariance of the three pollutants displays asymmetric properties, while CO and NO_2 depart most from symmetry compared to the other two pairs of pollutants. Overall, it is more appropriate to employ an asymmetric covariance model to respect the asymmetric feature of the cross-covariances observed in the pollution data. In particular, a LMC is not appropriate for these data.

Once the test detects an asymmetric covariance structure, testing the intrinsic correlation and the order of the linear model of coregionalization is formally unnecessary. Still, it can be informative to carry out these tests. Consider three contrasts using the six pairs of stations in the intrinsic correlation test, and a test statistic from nine contrasts. This gives a test with $a = 9$ degrees of freedom. The test statistic is $TS = 129.4$, with an associated p-value $= 9.48 \times 10^{-21}$. Splitting this global intrinsic correlation test into three components gives p-value $= 0.00033$ between CO and NO, p-value $= 2.20 \times 10^{-12}$ between NO and NO_2, and p-value $= 5.27 \times 10^{-12}$ between CO and NO_2. One can also consider a test for a LMC with $r = 2$ also based on the six station pairs. Specifically, a test of the intrinsic correlation of the two residual random fields resulting from the three component random fields can be carried out. Without loss of generality, one residual random field is produced by subtracting CO from NO, and the other one by subtracting CO from NO_2. Symmetry is tested as well as intrinsic correlation for the two residual random fields, and the results show p-value $= 0.027$ for the symmetry test and p-value $= 0.001$ for the intrinsic correlation test. Thus, as expected, all the intrinsic correlation hypotheses and the linear model of coregionalization have been rejected. Comparing the symmetry and intrinsic correlation test in the same scenario, a more significant p-value for intrinsic correlation than for symmetry is observed. This makes sense as it suggests more evidence against the more focused hypothesis of intrinsic correlation.

9.7 Discussion

Modeling multivariate data observed at space–time locations often leads to a complex covariance model involving a large number of spatial, temporal, and variable parameters. The symmetric or intrinsic correlation covariance structure eases this complexity by regulating the interactions between variables and locations. For this reason, they are commonly assumed in data modeling. However, making such assumptions without justification often makes resulting inferences less reliable. In this chapter, we have discussed and developed approaches to testing symmetry and intrinsic correlation based on graphical approaches and on the asymptotic joint normality of cross-covariance estimators and properties of covariances under the hypothesis. Numerical experiments have demonstrated the latter, testing, approach to be accurate (sizes close to nominal), and powerful in the situation where adequate data are available.

Through a univariate conditional construction of multivariate random fields, an approach has been given to find the most parsimonious, yet sufficient model among the family of linear models of coregionalization. Through numerical experiments it has been seen that the empirical sizes and powers behave reasonably. The procedures have been carried out in the case where there are $p = 3$ variables of interest; analogous results can be obtained for a general $p > 3$ with some additional algebraic manipulations. Principally, linear models of coregionalization for nonseparable multivariate space–time covariance functions have been introduced as a tool for dimension reduction in, for example, Grzebyk and Wackernagel (1994). Identifying the smallest number, r, of such separable covariances needed is always desirable. The LMC test is meant to guide the data analyst in model choice and dimension reduction.

Once a strong departure from an assumption is detected, it is more appropriate to turn to a model that respects the data features. For example, Sain and Cressie (2007) proposed a flexible covariance model for multivariate lattice data that allows for asymmetric covariances between different variables at different locations. More research on the construction of flexible parametric cross-covariance models is certainly a worthy future effort.

All tests have been developed in terms of the covariance estimators; all properties are maintained if the user prefers the use of correlation estimators. For example, Gelfand *et al.* (2004) use correlation estimators in place of the covariance estimators considered here. This is based on the invariance of the large sample χ^2 family of distributions for the test statistics. This observation holds as well for the tests of isotropy in Chapter 5, and the tests for symmetry and separability in Chapter 6.

A general comment is applicable to testing procedures in this chapter as well as those in Chapters 5, 6, and 8. It does not seem reasonable to adopt a strict policy of rejecting hypotheses when the associated p-value is smaller than 0.05, say. In general, large data sets often lead to formal rejection of null hypotheses, and moderate departures from the null may be acceptable to allow for more parsimonious models and the associated easier covariance estimation. That being said, when rejection is strongly supported by the data, then assuming the null hypothesis will lead to inappropriate models, and thus inefficient predictions and inferences.

The testing methodology was developed for stationary random fields, yet it is worth noting that the notions of symmetry and intrinsic correlation exist in a far more general context which does not depend on stationarity, or any structure, for the indexing sets. For example, these notions are also well defined for nonstationary random fields as illustrated in, for example, Gneiting *et al.* (2007).

10

Resampling for correlated observations

We have often had occasion to use resampling methods to estimate variances, covariance matrices, and to draw inferences for statistics computed from spatial and spatio-temporal observations. In order to motivate resampling for spatially correlated observations, we first consider a general setup including a variety of data structures including independent and correlated observations. We focus on, first, independent observations, then temporally correlated observations, and finally spatially correlated observations.

In general, the data analyst has a sample of n observations, $\mathbf{Z}_n := (Z_1, Z_2, \ldots, Z_n)$, from some process and we compute a statistic, $s(\mathbf{Z}_n)$, that estimates a parameter of interest, θ, say. In order to draw inferences from $s(\mathbf{Z}_n)$ concerning θ (hypothesis tests or confidence intervals), some knowledge of the distribution of the statistic $s(\mathbf{Z}_n)$ under repeated sampling is necessary. Consider the general situation where, for some sequence of constants, a_n, we have that

$$t_n := a_n[s(\mathbf{Z}_n) - \theta] \tag{10.1}$$

converges in distribution to a random variable with distribution function G. If the distribution function G is fully known, then for any α we have

Spatial Statistics and Spatio-Temporal Data: Covariance Functions and Directional Properties Michael Sherman

both $G(\alpha/2)$ and $G(1-\alpha/2)$, the $\alpha/2$ and $1-\alpha/2$ percentiles of the limiting distribution. Then, basic arithmetic shows that

$$[s(\mathbf{Z}_n) - G(1-\alpha/2)/a_n, s(\mathbf{Z}_n) - G(\alpha/2)/a_n]$$

is a (large sample) confidence interval of confidence level $(1-\alpha)100$ percent.

There are several reasons why the assumption that the distribution function, G, is known is not realistic. First, even in the simplest case where G corresponds to the normal distribution and $a_n = n^{1/2}$, the variance of this limiting normal distribution, σ^2, is in general unknown. Note that in this case $G(\alpha) = \sigma\Phi(\alpha)$, for any $\alpha \in (0, 1)$. There are several difficulties in obtaining σ^2, and thus an estimate of $n\,Var[s(\mathbf{Z}_n)]$, the standardized variance.

i. The statistic $s(\mathbf{Z}_n)$ may be complicated, so that a theoretical calculation of σ^2 may be difficult. There are several examples of this throughout this book. For example, the pseudo-likelihood and generalized pseudo-likelihood estimators of α and β, in Section 4.2, both have variances which are not available in closed form. Estimates of model parameters based on point patterns in Chapter 7 typically have unknown variances, which are not available through a simple formula.
ii. The marginal distribution, F, of each observation, Z_i, is in general unknown. In the case of correlated observations, this distribution is not always easily estimable.
iii. When the observations are spatially correlated, their joint distribution must be accounted for to estimate σ accurately. Recall the serious effects of not accounting for temporal and/or spatial correlation in Chapter 1, Section 1.2.1.

Of course, situations (i)–(iii) could all be present in a particular application. Further, when (ii) holds it is often the case that (iii) is a difficult problem, the reason being that, if the marginal distribution is unknown, it is likely that the more obscure joint distribution is also unknown.

In the situation where the functional form of the distribution G is unknown, or difficult to evaluate, inference becomes even more difficult. To see how this can occur, consider the following two examples:

i. For the sample mean computed from independent and identically distributed observations with finite population variance σ^2, we have by the Central Limit theorem that:

$$n^{1/2}(\overline{X} - \mu) \xrightarrow{D} N(0, \sigma^2).$$

For observations from a stationary time series, it is often the case that

$$n^{1/2}(\overline{Z} - \mu) \xrightarrow{D} N\left[0, \sum_{i=-\infty}^{\infty} Cov(Z_0, Z_i)\right].$$

This large sample variance was derived in Section 1.2.1. The large sample normal distribution holds, for example, with $E[Z_t^{2+\delta}] < \infty$, and $\alpha(t) < Ct^{\delta/(2+\delta)}$ for some constant $\delta > 0$ where $\alpha(\cdot)$ denotes the strong mixing coefficient defined in Section 9.4.

For *heavy tail* populations, that is, those without finite variance in the independent setting, we have $a_n(\overline{X} - \mu) \xrightarrow{D} G$, where $a_n \neq n^{1/2}$ and G comes from the family of stable distributions. In particular, in most cases $G \neq N(0, \sigma^2)$ for any σ^2. Also, if correlation is strong, that is, $\alpha(t)$ does not decay sufficiently fast in time, then the rate of convergence is not $n^{1/2}$ (even if the population F has a finite variance) and the limiting distribution is again not normal.

ii. For the sample variance, consider observations from two binary choice models, $X_i := I[Z_i > 0]$ and $\tilde{X}_i := I[Z_i > 1]$. This models, for example, where the choice to purchase a good is made at two different thresholds. Let the statistic be s^2, the sample variance of the $\mathbf{X}$ and $\tilde{\mathbf{X}}$ variables. It turns out that, for the $\mathbf{X}_n$ data, we have

$$n(s^2 - \sigma^2) \xrightarrow{D} G,$$

where G has a distribution which is skewed and supported on $(-\infty, 0.25)$.

For the $\tilde{\mathbf{X}}_n$ data, however, we have

$$n^{1/2}(s^2 - \sigma^2) \xrightarrow{D} N(0, \omega^2),$$

for some variance ω^2.

Note that in each of these two simple cases ($\overline{X}$ or s^2) we have a different limiting distribution G with different rates of convergence a_n, for the *same* statistic, based on subtle changes in the underlying process generating the observations. This shows that, in general, both the limiting distribution and rate of convergence are unknown. In all of these cases, resampling methods allow us to obtain inferences for parameters when traditional methods are difficult or impossible to implement. On the other hand, in cases where the rate of convergence is known, resampling methods can often give more accurate answers than traditional methods. This is discussed further at the end of Section 10.1.3.

We desire various features of the distribution of the standardized statistic t_n in Equation 10.1, depending on the application. Often of interest are estimates of the moments of t_n (e.g., variance, skewness, bias). We might seek percentiles of the distribution of t_n (or its approximation by percentiles of G) to obtain confidence intervals for θ. We may seek diagnostic information to assess whether G is the normal distribution, and, if not, how it departs from normality. Resampling methods can address each of these goals.

The basic idea in resampling is to generate replicates of the statistic $s(\mathbf{Z}_n)$ from the original observations, and then use the distribution of these replicates to model the true distribution of $s(\mathbf{Z}_n)$. The choice of resampling method depends both on the particular goal (e.g., variance estimation or confidence intervals) and on the joint probability structure generating the observations (e.g., independent, or temporally and/or spatially correlated).

10.1 Independent observations

Although the focus of resampling in this book is on the spatially correlated setting, the motivation for many time series and spatial resampling algorithms comes from algorithms for independent observations. For this reason, we initially discuss resampling algorithms for independent observations all coming from the same distribution.

10.1.1 U-statistics

U-statistics are useful when the parameters of interest can be estimated unbiasedly using only a 'small' number of observations. This methodology was developed by Hoeffding (1948). A more recent mathematical account is in Serfling (1980); a more applied presentation is given in Kowalski and Tu (2008). The setup is as follows: let

$\phi(X_1, \ldots, X_r)$ be a chosen kernel function with $E[\phi(X_1, \ldots, X_r)] = \theta = \phi_0$. The goal is to estimate ϕ_0 based on independent observations $X_1, \ldots, X_n$.

We give several examples of U-statistics.

1. *Population mean*

   Take $r = 1, \phi(x) = x$, then $\phi_0 = E[\phi(X_1)] =: \mu$.

2. *Population variance*

   Take $r = 2$, $\phi(x_1, x_2) = (x_1 - x_2)^2/2$, then $\phi_0 = E[\phi(X_1, X_2)] = \sigma^2$.

3. *Gini's mean difference*

   Take $r = 2$, $\phi(x_1, x_2) = |x_1 - x_2|$, then $\phi_0 = E[\phi(X_1, X_2)] = E(|X_1 - X_2|)$.

4. *Cumulative distribution function (c.d.f.)*

   Take $r = 1$, $\phi(x_1) = I(x_1 \leq y)$, then $\phi_0 = E[\phi(X_1)] = P(X_1 \leq y) =: F(y)$.

The one-sample and two-sample Wilcoxon statistics and the Mann–Whitney test are also examples of U-statistics. Now that we have a goal of estimation, θ, the question becomes how to estimate the parameter of interest. There are (at least) three natural possibilities to obtain replicates of a U-statistic from the observed data:

i. We can consider nonoverlapping series, each of the same length as the kernel. Namely, $\phi(X_1, \ldots, X_r)$, $\phi(X_{r+1}, \ldots, X_{2r}), \ldots, \phi(X_{(k-1)r+1}, \ldots, X_{kr})$, assuming for simplicity that $n = kr$. Then we have k replicates of the kernel ϕ, each based on r observations. A natural estimator is then:

$$\widehat{\phi}_0 := \sum_{i=1}^{k} \phi(X_{(i-1)r+1}, \ldots, X_{ir})/k.$$

These replicates are *subseries*, but this is *not* a U-statistic.

(ii) We can sample *without* replacement, *deterministically* from $X_1, \ldots, X_n$, obtaining $\phi(X_{i_1}, \ldots, X_{i_r})$ where the $i_1, \ldots, i_r$ are distinct indices between 1 and n. Then the estimator based on these replicates is

$$U_n := \sum_c \phi(X_{i_1}, \ldots, X_{i_r}) / \binom{n}{r}.$$

This is a U-statistic.

(iii) We can sample *with* replacement *deterministically* from $X_1, \ldots, X_n$, obtaining $\phi(X_{i_1}, \ldots, X_{i_r})$ where the indices $i_1, \ldots, i_r$ are all between 1 and n. The estimator based on these replicates is then

$$V_n := \sum_c \phi(X_{i_1}, \ldots, X_{i_r}) / n^r.$$

This is a V-statistic or 'Von-Mises Statistic.'

A sufficiency argument shows that the U-statistic in (ii) has smaller variability than the subseries estimator in (i). Note that the estimator with overlapping indices is more efficient than the estimator which retains independent replicates. This increased efficiency by reusing observations in replicates turns out to be the case for temporally or spatially correlated observations as well, as discussed in Sections 10.4.4 and 10.6.1. The V-statistic, while biased, is actually the bootstrap with resample size r, which we consider in Section 10.1.3.

10.1.2 The jackknife

For U-statistics, the order of the kernel ϕ is r with $r < n$. Many statistics cannot be represented as a sum of kernels of lower order. For example, the sample median cannot simply be represented in this way. In this situation we need replicates, but have only n observations, and the statistic is a function of all n observations. The jackknife idea is as follows. Consider the replicates:

$$s_{(i)} = s_{n-1}(X_1, \ldots, X_{i-1}, X_{i+1}, \ldots, X_n),$$

for $i = 1, \ldots, n$, the original statistic with the ith observation removed. Quenouille (1949) and Quenouille (1956) used these jackknife replicates to reduce the bias of biased estimators. The jackknife estimator of variance [Tukey (1958)] is:

$$V_J := \frac{n-1}{n} \sum_{i=1}^{n} (s_{(i)} - \bar{s}_n)^2.$$

Note that this is $n - 1$ times the sample variance (with n denominator) of the replicates. This multiplicative factor accounts for the fact that the replicates are heavily correlated with each other through their sharing of $n - 2$ observations. Why is $n - 1$ the correct inflation factor? Direct computations show that the jackknife estimator of variance of the sample mean is identically equal to s^2/n, where s^2 is the usual sample variance. Clearly this is unbiased for the correct variance, so the jackknife works in this case.

Other examples where the jackknife estimate of variance is useful include differences of means, ratios of means, and the correlation coefficient. Another example, in multivariate analysis, where the jackknife is useful, is to obtain variance estimates for sample eigenvalues and ratios of eigenvalues. Although the jackknife seems quite general, it is only appropriate for certain asymptotically normal statistics. For example, the median is a statistic with a large sample normal distribution [Serfling (1980)], but the jackknife estimate of variance fails badly in this case. To appreciate why, note that when n is even, the jackknife replicates of the median take on only two possible values. This does not mimic the true sampling variability of the median from a continuous distribution. One way around this is to delete d observations in computing each replicate where $d > 1$. This 'delete-d' jackknife is studied in, for example, Shao and Wu (1989).

10.1.3 The bootstrap

The basis of the bootstrap is that the empirical cumulative distribution function,

$$\widehat{F}_n(x) := \frac{\sum_{i=1}^{n} I(X_i \le x)}{n},$$

is a good estimator of the true cumulative distribution function, $F(x) := P(X \le x)$. For example, the estimator is unbiased, mean square convergent, and uniformly almost surely convergent. If we knew $F(x)$ for each x we could compute, in principle, all quantities of interest to us. This suggests that, to estimate any quantity that depends on the unknown distribution F, we can 'plug-in' $\widehat{F}_n$ in place of F.

To consider the bootstrap in general, let $R(X_1, \ldots, X_n, F)$ denote a random quantity dependent on observations and possibly on the (unknown) population cumulative distribution function. The object of interest is typically some feature of the distribution of $R(X_1, \ldots, X_n, F)$,

$H_F[R(X_1, \ldots, X_n, F)]$, say. This formulation allows for the consideration of bias estimation, variance estimation, and distribution function estimation simultaneously. For example, for variance estimation of a statistic, $s_n(X_1, \ldots, X_n)$, we take

$$R(X_1, \ldots, X_n, F) := s_n(X_1, \ldots, X_n)$$

and

$$H_F[R(X_1, \ldots, X_n, F)] := Var_F[R(X_1, \ldots, X_n, F)].$$

To estimate the distribution of standardized or studentized statistics, we take

$$R(X_1, \ldots, X_n, F) = I\{n^{1/2}[s_n(X_1, \ldots, X_n) - \theta(F)]/\sigma(F) \leq x\}$$

and

$$R(X_1, \ldots, X_n, F) = I\{n^{1/2}[s_n(X_1, \ldots, X_n) - \theta(F)]/\widehat{\sigma} \leq x\},$$

respectively, with

$$H_F[R(X_1, \ldots, X_n, F)] := E_F[R(X_1, \ldots, X_n, F)],$$

for any quantile, x, of interest. For bias estimation we take

$$R(X_1, \ldots, X_n, F) = s_n(X_1, \ldots, X_n) - \theta$$

and

$$H_F[R(X_1, \ldots, X_n, F)] := E_F[R(X_1, \ldots, X_n, F)].$$

Again, let $\mathbf{X}_n := (X_1, \ldots, X_n)$. The bootstrap principle [Efron (1979)] to generate replicates to estimate $H_F[R(\mathbf{X}_n, F)]$ is: wherever F is needed, we 'plug in' $\widehat{F}_n$, the empirical distribution function.

Conditional on the data, $\mathbf{X}_n$, the bootstrap replicates of $R(\mathbf{X}_n, F)$ are $R(\mathbf{X}_n^*, \widehat{F}_n)$, where X_i^* are independent and identically distributed as $\widehat{F}_n$, and $H_{\widehat{F}_n}[R(\mathbf{X}_n^*, \widehat{F}_n)]$ is the bootstrap estimator of $H_F[R(\mathbf{X}_n, F)]$. For example, the bootstrap estimate of variance is

$$Var_{\widehat{F}_n}[s_n(\mathbf{X}_n^*)] = E_{\widehat{F}_n}[s_n^2(\mathbf{X}_n^*)] - E^2_{\widehat{F}_n}[s_n(\mathbf{X}_n^*)]$$

$$= \frac{\sum_{b=1}^{n^n} s_n^2\left[\mathbf{X}_n^{*(b)}\right]}{n^n} - \left\{\frac{\sum_{b=1}^{n^n} s_n\left[\mathbf{X}_n^{*(b)}\right]}{n^n}\right\}^2,$$

where the sums are over all n^n possible samples of size n sampled with replacement from the original observations. If the sample size n is not quite small, or this bootstrap estimator cannot be analytically simplified, we estimate the bootstrap estimate of variance by:

$$\frac{\sum_{b=1}^{B} s_n^2\left[\mathbf{X}_n^{*(b)}\right]}{B} - \left\{\frac{\sum_{b=1}^{B} s_n\left[\mathbf{X}_n^{*(b)}\right]}{B}\right\}^2,$$

where the B bootstrap replicates are randomly chosen from the n^n possible replicates. This is the usual Monte Carlo implementation of the bootstrap.

Bootstrap variance estimation is generally more reliable than jackknife variance estimation. In particular, the variance of the sample median can be well estimated using the bootstrap, as shown by Maritz and Jarrett (1978). For the choice of B, more is always better. For expensive calculations based on large data sets and/or difficult-to-compute statistics, two viewpoints on sufficient B are given by Efron and Tibshirani (1993) and Booth and Sarkar (1998). In general, B needs to be larger for distribution function estimation than for variance estimation. Bias estimation often requires larger B than variance estimation, although the balanced bootstrap of Davison *et al.* (1986) can reduce B for statistics which are functions of means.

For distribution function estimation of a standardized or studentized statistic, consider

$$t_n(\mathbf{X}_n) := [s(\mathbf{X}_n) - \theta(F)]/[Var(s_n)]^{1/2},$$

where Var denotes a known or, more commonly, an estimated variance. One way of saying that the bootstrap 'works' is that the bootstrap distribution given the data, $t_n\left(\mathbf{X}_n^*|\mathbf{X}_n\right)$, is 'close to' the distribution of $t_n(\mathbf{X}_n)$. In other words, we want to compare:

$$P[t_n(\mathbf{X}_n) \le x] \tag{10.2}$$

with

$$P_*\left[t_n\left(\mathbf{X}_n^*|\mathbf{X}_n\right) \le x\right], \tag{10.3}$$

where P_* denotes the bootstrap conditional distribution given the data $\mathbf{X}_n$. Note that if $s(\mathbf{X}_n)$ is the sample mean or other 'mean-like' statistic, then we typically have that (10.2) converges to $\Phi(x) = P(Z < x)$. In this case, we require that the limiting bootstrap distribution is also standard normal. Consider $\sup_x |(10.2) - (10.3)|$. We hope that this tends to 0 as $n \to \infty$.

Note that (10.3) is random (it depends on the data $\mathbf{X}_n$), so we hope that the difference between the two distributions becomes close in probability:

$$\sup_x |(10.2) - (10.3)| \xrightarrow{p} 0,$$

or almost surely:

$$\sup_x |(10.2) - (10.3)| \xrightarrow{a.s.} 0.$$

The former condition is usually termed uniform consistency, and the latter uniform strong consistency.

Much work has gone into showing that these types of approximations hold (often at certain rates) for various data structures. When s_n is the sample mean or, more generally, a von Mises functional statistic, Bickel and Freedman (1981) showed that strong uniform consistency holds. For the mean, Singh (1981) showed that this consistency holds even when the left hand side in $\sup_x$ is multiplied by the factor $n^{1/2}$. It is this 'second-order accuracy' that often makes bootstrap confidence intervals like the bootstrap-t interval or the BCA (bias-corrected and accelerated interval) more accurate and more correct ('accurate' and 'correct' inferences are defined in Section 1.2.2). Bretagnolle (1983) showed that, for von Mises functionals, the bootstrap is also strongly uniformly consistent even when the limiting distribution is nonnormal.

For more details on the asymptotic justification of the bootstrap see, for instance, Hall (1992) or Shao and Tu (1995). For more details on applications in a wide variety of settings see, for example, Efron and Tibshirani (1993), Davison and Hinkley (1997), or Manly (2006).

10.2 Other data structures

Do the jackknife and/or bootstrap work for temporally correlated observations? To explore this question, consider the standardized mean:

$$t_n := n^{1/2}(\overline{X}_n - \theta).$$

Under second-order stationarity, we have for all j:

$$C(i) := Cov(X_1, X_{i+1}) = Cov(X_j, X_{i+j}).$$

In this case, we have

$$nVar(\overline{X}_n) \to \sum_{i=-\infty}^{\infty} C(i),$$

if $\left|\sum_{i=-\infty}^{\infty} C(i)\right| < \infty$ using stationarity and Kronecker's lemma as in Section 1.2.1.

For the bootstrap estimate of variance, however, we have

$$nV_B\left(\overline{X}_n^*\right) \xrightarrow{p} C(0) = \sigma^2,$$

as in the independent setting. The bootstrap treats the observations interchangeably, that is, as if independent.

For the jackknife, we've seen that $nV_J(\overline{X}_n) = s_n^2$, the sample variance. Direct calculations show:

$$E\left(s_n^2\right) = C(0) - \frac{2\sum_{i=1}^{n-1}(n-i)C(i)}{n(n-1)} \to C(0) = \sigma^2.$$

We note that the following hold:

i. The jackknife and the bootstrap both estimate $\sigma^2 = Var(X_1)$. They do not estimate the correct variance of the mean which involves (all) covariances $C(i)$.

ii. Both the jackknife estimator of variance (the sample variance) and the bootstrap estimator of variance do estimate the population variance, σ^2, even under temporal correlation!

iii. In order to get correct inferences we need to *account* for the correlation if it exists.

There are (roughly) two ways to account for correlations, which can be characterized as model based or model free.

10.3 Model-based bootstrap

Initially, we consider two types of dependence. In the first, observations depend on covariates (regression), and in the latter, observations are dependent in time or space.

10.3.1 Regression

Consider the general linear model where

$$\mathbf{y} = \mathbf{X}\beta + \epsilon,$$

for responses $\mathbf{y}$, covariates $\mathbf{X}$, parameter vector β, and independent errors ϵ from distribution function F. We typically estimate β by minimizing

$$Q(\mathbf{y} - \mathbf{X}\beta),$$

for some loss function $Q(\cdot)$. Three popular choices of Q are:

i. $Q(\mathbf{e}) = \sum_{i=1}^{n} e_i^2$, minimization gives least squares estimators.

ii. $Q(\mathbf{e}) = \sum_{i=1}^{n} |e_i|$, minimization gives least absolute deviations estimators.

iii. $Q(\mathbf{e}) =$ median of e_i^2, minimization gives least median of squares estimators.

If we use (i) (least squares) and F is the normal distribution, then we have well-known distribution theory to obtain inferences based on $\widehat{\beta}$. Otherwise, the distribution of $\widehat{\beta}$ is not easy to obtain. Specifically, there are no closed form formulas for standard errors in situations (ii) and (iii). The bootstrap can be effective in this situation to obtain estimated standard errors and to estimate the entire distribution of parameter estimates. To illustrate how this can be done, assume we can obtain the estimator of interest, $\widehat{\beta}$. This could be achieved using OLS, ML, MOM (method of moments), PL, or QL (quasi-likelihood) to minimize one of the criteria (i), (ii), (iii), or another minimization criterion. There are two principle bootstrap algorithms to obtain the distribution of parameter estimates. The first one attempts to exploit the (assumed) i.i.d. structure of the errors, ϵ_i.

i. *Algorithm (I)*

   a. Given $\widehat{\beta}$, compute the residuals $e_i := y_i - \mathbf{x}_i^{\mathrm{T}}\widehat{\beta}$, $i = 1, \ldots, n$.
   b. Resample with replacement e_i^*, $i = 1, \ldots, n$ from e_i, $i = 1, \ldots, n$.
   c. Set $y_i^* := \mathbf{x}_i^{\mathrm{T}}\widehat{\beta} + e_i^*$.
   d. Calculate $\widehat{\beta}^{*(1)}$ from $[\mathbf{y}^*, \mathbf{X}]$ in the same manner in which $\widehat{\beta}$ is calculated from the original data, $[\mathbf{y}, \mathbf{X}]$.
   e. Obtain the bootstrap distribution directly or through simulation, $\widehat{\beta}^{*(b)}$, $b = 1, \ldots, B$.

We then use these replicates to estimate bias, variance, and to obtain confidence intervals. One result is the following: when we estimate the variance of the least squares estimator of β, we know that $Var(\widehat{\beta}) = \sigma^2[\mathbf{X}^{\mathrm{T}}\mathbf{X}]^{-1}$. Straightforward calculations show:

$$V_B(\widehat{\beta}) = \widehat{\sigma}^2(\mathbf{X}^{\mathrm{T}}\mathbf{X})^{-1},$$

where $\widehat{\sigma}^2$ is the (slightly biased) MLE of σ^2 in the normal model. If, on the other hand, we use the jackknife by deleting $(\mathbf{x}_i, y_i)$ one at a time, we obtain approximately:

$$V_J(\widehat{\beta}) = (\mathbf{X}^{\mathrm{T}}\mathbf{X})^{-1}\left(\sum_{i=1}^{n} \mathbf{x}_i^{\mathrm{T}}\mathbf{x}_i e_i^2\right)(\mathbf{X}^{\mathrm{T}}\mathbf{X})^{-1},$$

which does not look much like the correct variance under homoskedasticity.

ii. *Algorithm (II)*

Use the usual bootstrap on the data $Z_i := (y_i, \mathbf{x}_i), i = 1, \ldots, n$, to obtain the bootstrap replicates $\widehat{\beta}^{*(b)}, b = 1, \ldots, B$. This is termed the *paired bootstrap*. In the homogeneous case, $\sigma_i^2 = \sigma^2$, *fixed* x is better. If, however, the errors depend on $\mathbf{x}$, then the *paired bootstrap* is better, as was shown by Liu and Singh (1992); see also Wu (1986). Freedman (1984) bootstrapped 'linear dynamical models' including the GLM. There are literally *hundreds* if not *thousands* of papers on using and validating the model-based bootstrap in specific contexts. For example, there is a large amount of work in this direction in survival analysis. Pioneering work in this area was bootstrapping the Kaplan–Meier estimator [Akritas (1986)] and the bootstrap in proportional hazards regression [e.g., Zelterman *et al.* (2004)].

10.3.2 Time series: autoregressive models

Consider observations from the time series model:

$$Z_i = \mu + \sum_{t=1}^{p}(Z_{i-t} - \mu)\beta_t + \epsilon_i.$$

This is an AR(p) model. If the errors are Gaussian, we have standard distribution theory for parameter estimates. Otherwise, distribution theory can be complicated and the bootstrap can be a reliable way to obtain inferences.

We can use a *fixed x* resampling algorithm, analogous to that in the regression setting. Specifically, for a time series of length T:

1. Given $\widehat{\beta}$, compute the residuals $e_i := Z_i - \sum_{t=1}^{p} Z_{i-t}\widehat{\beta}_t$ for $i = p+1, \ldots, T$.

2. Form the 'centered' residuals $e_i := e_i - \overline{e}$.
3. Resample $e_i^*, i = 1, \ldots, T$ from $e_i, i = p+1, \ldots, T$.
4. Set $Z_i^* := \sum_{t=1}^{p} Z_{i-t}^* \widehat{\beta}_t + e_i^*$ for $i = 1, \ldots, T$, where $Z_i^* = Z_{i+p}$ for $i \in \{-(p-1), \ldots, 0\}$.
5. Calculate $\widehat{\beta}^{*(1)}$ from $\mathbf{Z}_n^*$.
6. Obtain the bootstrap distribution from $\widehat{\beta}^{*(b)}, b = 1, \ldots, B$.

Again, there is a huge number of examples of model-based resampling proposals for correlated time series observations. Freedman (1984) showed strong uniform consistency for estimating functions of autoregressive parameters. Bose (1988) in autoregression showed second-order consistency of the bootstrap distribution. Datta (1995) considered explosive AR(p) processes. Bose (1990) considered moving average processes. Kreiss and Frank (1992) considered m-estimators of ARMA(p, q) processes. Ferretti and Romo (1996) proposed a unit-root test for ARMA models. Stoffer and Wall (1991) used the bootstrap to analyze state-space models, while Buhlmann (2002) presented a sieve bootstrap for the AR(p) model with p large, for the analysis of categorical time series observations. These are some of the many applications of model-based resampling using the bootstrap.

10.4 Model-free resampling methods

One main motivation of the bootstrap was to get away from parametric assumptions, like an AR(p) model for a time series. Often it is desirable to account for the dependence in a more model-free way.

For temporal observations, we seek to quantify processes whose correlation decays over time, without having to specify the correlation structure. For example, without assuming an AR(p) or ARMA(p, q) model. Further, we hope to obtain estimates of bias, variance estimates, and estimate distribution functions as in the independent observation setting.

Consider (strictly) stationary observations, $Z_1, Z_2, \ldots$ This means:

i. The Z_is are identically distributed; and
ii. The joint distribution of $\{Z_{i_1}, \ldots, Z_{i_k}\}$ is the same as $\{Z_{i_1+m}, \ldots, Z_{i_k+m}\}$ for any k, locations $i_1, \ldots, i_k$, and any translation m.

This is a strict stationarity assumption as discussed in Section 1.1.

To quantify correlation, consider observations at time s and in the past. For this purpose, define: an event A that depends on observations $(Z_1, Z_2, \ldots, Z_s)$, that is, $A \in \mathcal{F}(Z_1, Z_2, \ldots, Z_s)$. Next consider an event B that depends on observations $(Z_{s+m}, Z_{s+m+1}, \ldots)$, that is, $B \in \mathcal{F}(Z_{s+m}, Z_{s+m+1}, \ldots)$. If m is large, then the events A and B should become approximately independent; that is $|P(A \cap B) - P(A)P(B)|$ should become small. We define the mixing coefficient:

$$\alpha(m) := \sup[|P(A \cap B) - P(A)P(B)|]$$

where the sup (max) is over all sets $A \in \mathcal{F}(Z_1, \ldots, Z_s)$, $B \in \mathcal{F}(Z_{s+m}, \ldots)$. We say the time series is *strong mixing* if $\alpha(m) \to 0$ as $m \to \infty$. In other words, observations separated by a large time lag behave as if independent. This is the temporal analogue to the spatial mixing coefficient defined in Section 5.6.1. This temporal mixing coefficient is as used in Section 9.4.

Examples of strong mixing structures:

i. Independent observations: $\alpha(m) = 0$ for all $m \geq 1$.
ii. M-dependent observations: $\alpha(m) = 0$ for all $m \geq M$.
iii. AR(1) processes with parameter β: satisfy $\alpha(m) \leq m|\beta|^m$. This holds for ϵ normal, double exponential or Cauchy distributed [Gastwirth and Rubin (1975)].

Many of the results that hold for independent observations continue to hold for stationary observations. For example, we have:

i. *Strong law of large numbers*

$$\overline{Z}_n = \frac{\sum_{i=1}^n Z_i}{n} \xrightarrow{a.s.} \mu.$$

The strong law holds under strong mixing with no additional assumptions.

ii. *Central Limit theorem*

$$n^{1/2}(\overline{Z} - \mu) \xrightarrow{D} N\left[0, \sum_{i=-\infty}^{\infty} Cov(Z_0, Z_i)\right],$$

if

$$E|Z_i|^{2+\delta} \quad \text{and} \quad \sum_{m=-\infty}^{\infty} \alpha^{\frac{\delta}{2+\delta}}(m) < \infty,$$

as has been stated previously. For Zs bounded, only $\sum_{m=-\infty}^{\infty} \alpha(m) < \infty$ is required. This result complements the usual Central Limit theorem, where only $E(Z^2) < \infty$ is required, but then the result only holds for independent observations. Here a slightly higher moment is required, and there is a tradeoff between heaviness of the tails of the distribution (quantified by the number of moments) and the strength of correlation (quantified by the mixing condition).

10.4.1 Resampling for stationary dependent observations

Consider the following general situation. Let $h_n(\mathbf{Z}_n)$ be a statistic estimating a target ν such that

$$E[h_n(\mathbf{Z}_n)] \rightarrow \nu.$$

The goal is to estimate the parameter ν under minimal assumptions. Recall the subseries introduced in Section 10.1.1:

$$(X_1, \ldots, X_r), (X_{r+1}, \ldots, X_{2r}), \ldots, (X_{(k-1)r+1}, \ldots, X_{kr}).$$

Switching from X to Z for our correlated observations, note that the subseries:

$$(Z_1, \ldots, Z_r), (Z_{2r+1}, \ldots, Z_{3r}), \ldots$$

are separated by distance r, as are the subseries

$$(Z_{r+1}, \ldots, Z_{2r}), (Z_{3r+1}, \ldots, Z_{4r}), \ldots$$

Now, let $Z_i^r := (Z_{i+1}, \ldots, Z_{i+r})$ denote a group of r consecutive observations starting at the $(i+1)$th observation, and define $h_i^r := h_r\left(Z_i^r\right)$ to be the corresponding replicate. Let

$$\widehat{\nu} = \frac{\sum_{i=0}^{k-1} h_{ir}^r}{k} = \frac{\sum_{[1]} h_{ir}^r + \sum_{[2]} h_{ir}^r}{k}.$$

where $[i]$, $i = 1, 2$, denotes the sum over the two sets of subseries. Within each of the two sets of summands the terms are approximately independent (for large r). Each h_{ir}^r has $E\left(h_{ir}^r\right)$ approaching ν by the definition of ν. If k is large, then $\widehat{\nu}$ will be close to ν as the mean of each of the two sets of replicates targets its expectation. Formally, we have [Carlstein (1986)]:

If:

$$k_n \alpha(r_n) \rightarrow 0$$

and

$$E\left(|h_n|^{1+\delta}\right) < \infty,$$

then

$$\widehat{\nu}_n \xrightarrow{p} \nu.$$

The main application of this result is in variance estimation. To see this, let $s(\mathbf{Z}_n)$ be a statistic and let $t_n = n^{1/2}\{s(\mathbf{Z}_n) - \mathbf{E}[\mathbf{s}(\mathbf{Z}_n)]\}$ and let $\sigma_n^2 := nVar[s(\mathbf{Z}_n)] = Var(t_n)$ be the variance of interest. Let

$$\widehat{\sigma}_n^2 = r_n \frac{\sum_{i=0}^{k_n-1}\left(s_{ir}^r - \bar{s}_n\right)^2}{k_n}, \text{ where}$$

$$\bar{s}_n = \frac{\sum_{i=0}^{k_n-1} s_{ir}^r}{k_n}.$$

Then, the shortcut formula for variance shows that

$$\widehat{\sigma}_n^2 = \sum_{i=0}^{k-1} \frac{\left(t_{ir_n}^{r_n}\right)^2}{k_n} - \left(\sum_{i=0}^{k-1} \frac{t_{ir_n}^{r_n}}{k_n}\right)^2.$$

So, if the mixing condition holds and $E\left(|t_n|^{2+\delta}\right) < \infty$ then

$$\widehat{\sigma}_n^2 \xrightarrow{p} \sigma^2.$$

Thus, for example, whenever, $t_n \xrightarrow{d} N(0, \sigma^2)$, then we have $t_n/\widehat{\sigma}_n \xrightarrow{d} N(0, 1)$. This makes G completely specified, and thus we have large sample confidence intervals for any $s(\mathbf{Z}_n)$ estimating θ satisfying the moment and mixing conditions. Although the consistency results hold for any block length r_n satisfying the above large sample conditions, it is also true that practical good performance of this variance estimator depends on an appropriate choice of block length. We now discuss this important choice.

Choice of subseries length: when $s(\mathbf{Z}_n)$ is the sample mean, extensive arithmetic shows that:

$$\begin{aligned} MSE\left(\widehat{\sigma}_n^2\right) &= Bias^2\left(\widehat{\sigma}_n^2\right) + Var\left(\widehat{\sigma}_n^2\right) \\ &\simeq C_1 r_n^{-2} + C_2 r_n/n, \end{aligned}$$

where C_1 and C_2 are constants depending on the underlying process. Thus we have

$$r_{n,\text{opt}} = \left(\frac{2C_1}{C_2}\right)^{1/3} n^{1/3},$$

so the order of r_n is $n^{1/3}$. For a specific application, we need to know the constant $\frac{2C_1}{C_2}$ in the formula for $r_{n,\text{opt}}$. One option is to assume that the underlying process is an AR(1) process, for example, then arithmetic shows that $\frac{2C_1}{C_2} = \left(\frac{2\rho}{1-\rho^2}\right)^2$. We can then estimate ρ by least squares, plug into the formula for $r_{n,\text{opt}}$ and use this subseries length. This is a semiparametric approach: nonparametric subseries estimation with a parametrically chosen subseries length. A version of this approach was used in Chapter 9.

Note that, as ρ increases, we need the block length, r_n, to also increase. This makes sense, as more correlation demands larger blocks to capture the correlation. One can use other processes than the AR(1) to estimate the constants C_1 and C_2; we have simply given once such choice. More fully nonparametric approaches to the choice of block length r_n are given in, for example, Sherman (1995), Hall and Jing (1996), and Lahiri (2003).

10.4.2 Block bootstrap

Let $s(\mathbf{Z}_n)$ again be the statistic of interest and let $\widehat{F}$ place mass $\frac{1}{n-l+1}$ on each block of observations of length l, $\mathbf{Z}_i^l := (Z_i, \ldots, Z_{i+l-1})$, $i = 1, \ldots, n-l+1$. Here l plays the role that r performed in the subseries approach. Assume that $n = kl$. Resample k blocks (with replacement), each of length l, and obtain:

$$\mathbf{Z}_n^{*(1)} := \left[Z_1^*, \ldots, Z_l^*, Z_{l+1}^*, \ldots, Z_{2l}^*, \ldots, Z_{(k-1)l+1}^*, \ldots, Z_{kl}^*\right].$$

We then compute $s\left[\mathbf{Z}_n^{*(1)}\right]$. This is one block bootstrap replicate. We use the (conditional) distribution of the $s\left[\mathbf{Z}_n^{*(b)}\right]$, $b = 1, \ldots, B$ to model the distribution of $s(\mathbf{Z}_n)$. This is known as the *block bootstrap* or *moving blocks bootstrap*. Künsch (1989) and Liu and Singh (1992) proposed this method independently. Künsch showed that Bickel and Freedman's and Singh's results in the i.i.d. case hold here (first-order accuracy). For more general statistics than the mean, to obtain better performance we consider resampling 'blocks of blocks' as described by Künsch. Also, obtaining second-order accurate inferences is possible here, as shown by Götze and Künsch (1996), but requires careful centering and variance estimation to obtain this higher accuracy. This is discussed in, for example, Davison and Hall (1993). Paparoditis and Politis (2001) consider a tapered block bootstrap to reduce bias. See also Lahiri (2003) for more details on block bootstrapping and implementation.

10.4.3 Block jackknife

In place of resampling blocks of length l, we delete blocks of length l. The replicates are then $s_n^{(j)}$, each based on $n - l$ observations after deleting Z_{j-1}^{l}, $j = 1, \ldots, (n - l + 1)$. The block jackknife estimator of variance is then:

$$\frac{(n-l)^2}{n(n-l+1)}\frac{1}{l}\sum_{j=1}^{n-l+1}\left[s_n^{(j)} - s_n^{(\cdot)}\right]^2.$$

Note that taking $l = 1$ (the ordinary jackknife) gives $(n-1)/n$ times the ordinary jackknife estimator of variance. Künsch (1989) showed that, for the sample mean, this jackknife (and the block bootstrap) estimator of variance has the same (asymptotic) bias as the subseries approach, but has $\frac{2}{3}$ of the variance. This suggests that for long series, *overlapping* blocks perform better than *nonoverlapping* blocks. This is asymptotic; in practice the overlapping blocks can have a larger bias than the subseries. This is particularly so when the correlation is strong relative to series length, as we now illustrate.

10.4.4 A numerical experiment

Consider the sample mean computed from an AR(1) process with parameter ρ. In this case, we know the large sample variance of the standardized mean is, $\sigma^2 = (1-\rho)^{-2}$. Take $\rho = 0.9$ so that $\sigma^2 = 100$. For this statistic, the subseries variance estimator, block bootstrap, and block jackknife all give the same answer when nonoverlapping blocks are used, so we aim to assess how the variance estimators depend on choice of block overlap. We take the series length to be $n = 300$ and subseries length to be $l = 30$. This is asymptotically optimal for nonoverlapping subseries based on an AR(1) model from 10.4.1.

Table 10.1 Comparing overlapping schemes in variance estimation.

Scheme	*Average*$(\widehat{\sigma}^2)$	*Var*$(\widehat{\sigma}^2)$
No overlap	69.5	905
$\frac{1}{2}$ overlap	61.7	727
$\frac{2}{3}$ overlap	55.4	640
All overlap	51.3	608

The performance of four different estimates of variance. The data are from an AR(1) time series with $\rho = 0.9$; the target value of estimation is $\sigma^2 = 100$.

For each of 1000 simulated AR(1) series we use

$$\widehat{\sigma}^2 := l \sum_{j=1}^{k} \frac{\left(\overline{Z}_j^l - \overline{\overline{Z}}\right)^2}{k},$$

where k depends on the overlapping scheme. Specifically, $k = 10, 19, 29, 271$ for no overlap, $\frac{1}{2}$ overlap, $\frac{2}{3}$ overlap, and all overlap, respectively. The first is the subseries variance estimator, the last the originally defined block bootstrap estimator. The middle two are compromise estimators in terms of number of replicates and computation.

The theory of subseries and the block bootstrap states that $E(\widehat{\sigma}^2)$ is asymptotically equal to 68.4 in this case. Note that asymptotic theory says that our estimated standard deviation will be approximately $(68.4/100)^{0.5} = 0.83$ of the truth. This relatively strong correlation for this length of series presents an apparently difficult estimation problem. Table 10.1 gives the average variance estimate and the variance of the variance estimates across 1000 simulated time series of length $n = 300$. We note the following from this experiment:

i. For the nonoverlapping estimator we have that *Average*$(\widehat{\sigma}^2) = 69.5$, compared to 68.4 in theory. This is relatively good agreement. Further, $Bias^2/Var = (69.5 - 100)^2/905 = 1.03$, which shows that there is a good balance between square bias and variance. The large sample best block length of $l = 30$ seems appropriate when $n = 300$ in this model.

ii. *Average*$(\widehat{\sigma}^2)$ should be approximately equal for all four schemes. This is clearly not the case, and we see that the bias becomes more severe as the amount of overlap increases.

iii. The ratio of variances for the subseries and block bootstrap is $Var\left(\widehat{\sigma}^2_{all}\right)/Var\left(\widehat{\sigma}^2_{no}\right) = 608/905 = 0.67 \simeq \frac{2}{3}$, supporting asymptotic theory.

iv. Minimizing MSE determines that the subseries estimator based on no overlap is the winner. When $\rho = 0.8$, however, a further experiment shows that the estimator based on $\frac{2}{3}$ overlap is the winner, while for $\rho = 0.5$, the estimator using all overlapping blocks is the winner. Although all estimators are useful, the conclusion is that the best procedure depends on the underlying strength of correlation. This shows the desirability of initially estimating the strength of correlation before choosing a block length, or using other data-based choices of block length, as suggested in the discussion on subseries.

A hybrid model-based – model-free resampling was proposed by Rajarshi (1990). Assume that the data come from a Markov process; that is, dependence is specified by

$$P(Z_{i+1} < z | Z_i = y),$$

for $i = 1, \ldots, n-1$. The Markov assumption is like an AR(1), so this is more restrictive than the block bootstrap or block jackknife, which only require stationarity and a weak dependence (mixing) condition [of which the AR(1) is a special case]. However, the Markov assumption is not as restrictive as an AR(1), as the dependence of each observation on its immediate previous time observation is not necessarily linear. The idea is to nonparametrically estimate this transition density using kernel estimators. Specifically, given estimates of the marginal and transition densities:

i. Z_1^* is generated from $\widehat{f}(y)$;

ii. Z_i^* is generated from $\widehat{f}(z \mid y)$ for $i = 2, \ldots, n$;

iii. Calculate $s\left(Z_1^*, \ldots, Z_n^*\right)$.

Then we use the distribution of these replicates to model the object of interest. For example, the sample variance of B replicated $s\left(Z_1^*, \ldots, Z_n^*\right)$s estimates the variance of $s(\mathbf{Z}_n)$. Rajarshi showed that the distribution of the standardized mean from this bootstrap converges to the normal distribution at a $n^{1/2}/\log(n)$ rate. This is almost the same rate as the bootstrapped mean, $n^{1/2}$, obtained by Singh (1981) in the independent setting.

Other research considers the case where the standardizing sequence, a_n, that determines the rate of convergence of $s(\mathbf{Z}_n)$ to θ is unknown. Sherman and Carlstein (1996) use subseries to gather diagnostic information on the distribution of s_n. For example: is the distribution normal, skewed, and if skewed in what direction? Politis and Romano (1994) obtain confidence intervals for θ using subseries in the case where the normalizing sequence a_n is assumed to be known, but the limiting distribution G is difficult to evaluate. Bertail *et al.* (1999) consider convergence rate estimation when the rate is known to be of a particular form.

10.5 Spatial resampling

Recall the cancer rates in the eastern United States from Section 4.2. We have assumed that the autologistic model (4.2) holds for which

$$P(Z_i = z_i | Z_j = z_j : j \neq i) = \frac{\exp[z_i(\alpha + \beta \sum_{j \in N(i)} z_j)]}{\exp(\alpha + \beta \sum_{j \in N(i)} z_j) + \exp(-\alpha - \beta \sum_{j \in N(i)} z_j)},$$

where $N(i)$ denotes the four nearest neighbors of county i. Based on the observed data, the MPLEs are $\widehat{\alpha} = -0.322$ and $\widehat{\beta} = 0.106$, suggesting an overall majority of low-rate counties ($\widehat{\alpha} < 0$), and some indication of clumping or positive correlation between neighboring counties ($\widehat{\beta} > 0$). The important inferential question is if $\widehat{\beta} = 0.106$ is significantly different from 0. That is, do high-cancer-rate counties cluster together ($\beta > 0$). In order to answer this question, we require an estimate of the standard error of $\widehat{\beta}$.

Standard errors for MPLEs typically have no closed form solution, so it is a nontrivial problem to estimate them. Motivated by the resampling methods in the time series setting, we consider model-free and model-based resampling approaches in the spatial setting.

10.5.1 Model-based resampling

In an ideal world, we would like to generate multiple data sets from the true model (in addition to the single data set we have) and then we could empirically estimate variability and the distribution of our parameter estimates. A model-based approach to variance estimation generates simulated data sets from the *estimated* model. Then the parameter estimates are computed from each, and then the desired summary of the estimates is computed. This is the same concept as in the independent and time series data settings. In the spatial setting, however, it is less clear how to generate replicate data sets from the estimated model. We discuss how the creation of the simulated data sets can be accomplished using the Gibbs sampler. The Gibbs sampler applies more generally to situations where simulation from joint distributions is difficult, but sampling from corresponding conditional distributions is relatively easy.

To illustrate, in the autologistic model, we have the conditional distribution in Equation 4.2. We want to generate a large number of simulated data sets (B, say) from the estimated model using the estimated parameters from PL. To estimate the variance of the MPL estimates then, we use the sample variance of the B parameter estimates.

To begin, we need a starting point from which to generate the simulated data sets. Consider independent binary variables from $[-1, 1]$ at each location. Now, let $\mathbf{u}^{(0)} = \{u_i^{(0)},\ i = 1, \ldots, |D|\}$ denote this 'initial state.'

We update each observation u_i according to Equation 4.2 and the values of its (four) nearest neighbors. After updating each location we have the pseudo-data $\mathbf{u}^{(1)} = \{u_i^{(1)}\}$, say. Let $\mathbf{u}^{(k)} = \{u_i^{(k)}\}$ denote the data observations after k steps of the Gibbs sampler. We may not want to simply take all the $\mathbf{u}^{(k)}$s as realizations from the model for two reasons. First, we started from a set of independent observations, and thus we typically desire to wait before accepting a given $\mathbf{u}^{(k)}$ as the first replicate data set. Further, it is a fact that at the k and $k+1$ steps, the observations $\{u_i^k\}$ and $\{u_i^{k+1}\}$ may be heavily (temporally) correlated. Our experiment in the time series setting suggests that they may be too heavily correlated to obtain the best results. These considerations lead to defining our Monte Carlo sample to be: $\mathbf{z}^{(b)} = \mathbf{u}^{(d+b\gamma)}$, $b = 1, \ldots, B$, where d and γ are integers. We see that there are three choices of tuning parameters necessary to implement the Gibbs sampler. Namely: (1) the number of Gibbs sampler steps until we accept the $b = 1$ simulated data set (the 'burn in' d); (2) the number of Gibbs sampler steps taken between accepted simulated data sets b and $b + 1$ (the 'spacing' γ); and (3) the total number of simulated data sets (B). The total number of Gibbs sampling steps is $S = d + B\gamma$.

In general, the variance estimates depend on these three choices. Note, in particular, that the choices of γ and B are closely related for fixed S. Increasing the spacing leads to less temporally correlated simulated spatial data sets, but decreases the number of simulated data sets. In this example, the estimates agree reasonably well across all choices of the three tuning parameters. We find that $s.e.(\widehat{\alpha}) \simeq 0.036$ and $s.e.(\widehat{\beta}) \simeq 0.019$ are the averages across a wide variety of Gibbs sampling schemes. Thus, using either the model-based or the model-free estimates of variability, we conclude that the observed geographic clustering is indeed significant. This conclusion is corroborated by the results of a goodness of fit analysis as in Section 4.2.2.

The Gibbs sampling in this setting can be very time consuming. We need to generate n times S observations based on the conditional distribution. In our example we generated $2003 \times (10\,000 + 1000) \simeq 2.0 \times 10^8$ observations from our conditional model. Although this turns out to be sufficient in this example, when correlation is stronger, the amount of simulation necessarily increases. It is also the case that the standard errors are only reliable if the conditional model is correct. This means that the functional form of the dependence model, the number of neighbors in the neighborhood (four) and the function of the four nearest neighbors (the sum) on which each observation depends are all assumed to be correct.

10.5.2 Monte Carlo maximum likelihood

In Section 4.2 we saw that there is no simple method to approximate the joint likelihood of the observations. Several authors, however, have used Monte Carlo methods to approximate the maximum likelihood estimator. Such MCML (Monte Carlo maximum likelihood) stochastic algorithms have been proposed by, for example, Younes (1991), Moyeed and Baddeley (1991), Geyer and Thompson (1992), Seymour and Ji (1996), and Gu and Zhu (2001). To illustrate the flavor of these approaches, we consider the MCML approach of Geyer and Thompson.

They consider

$$d_n(\alpha, \beta) = B^{-1} \sum_{b=1}^{B} \exp\left\{(\alpha - \alpha_0) \sum_{k \in D} z_k^{(b)} + [(\beta - \beta_0)/2] \times \sum_{k \in D} z_k^{(b)} \left[\sum_{\ell \in N(k)} z_\ell^{(b)} \right]\right\}, \quad (10.4)$$

where α_0 and β_0 are initial values for the parameters, and $z_j^{(b)}$, $j = 1, \ldots, |D|$ is the bth realization, $b = 1, \ldots, B$ of data from the autologistic model with parameters α_0, β_0. These realizations can be obtained by an application of the Gibbs sampler with conditional distributions given by (4.2) as described above.

Then

$$\log[L(\alpha, \beta)] + \log[\mathcal{G}(\alpha_0, \beta_0)] = \alpha \sum_{i \in D} Z_i + (\beta/2) \sum_{i \in D} Z_i \left[\sum_{j \in N(i)} Z_j \right] - \log[d(\alpha, \beta)],$$

where $\mathcal{G}(\alpha_0, \beta_0)$ is the denominator of $L(\alpha_0, \beta_0)$ in Equation 4.3, and $d(\alpha, \beta) = \mathcal{G}(\alpha, \beta)/\mathcal{G}(\alpha_0, \beta_0)$. The term $d_n(\alpha, \beta)$ approaches $d(\alpha, \beta)$ as $B \to \infty$. The term $\log[\mathcal{G}(\alpha_0, \beta_0)]$ is taken to be a constant and thus the maximizer of

$$\alpha \sum_{i \in D} Z_i + (\beta/2) \sum\nolimits_{i \in D} Z_i \left(\sum\nolimits_{j \in N(i)} Z_j\right) - \log[d_n(\alpha, \beta)]$$

approaches the MLE as $B \to \infty$. In practice, we find the parameters that maximize

$$\sum_{b=1}^{B} \exp\{(\alpha - \alpha_0) u_{1,b} + [(\beta - \beta_0)/2] u_{2,b}\},$$

where $u_{1,b} = \sum_{i \in D} z_i^{(b)} - \sum_{i \in D} Z_i$ and $u_{2,b} = \sum_{i \in D} z_i^{(b)} \left[\sum_{j \in N(i)} z_j^{(b)}\right] - \sum_{i \in D} Z_i \left[\sum_{j \in N(i)} Z_j\right]$. This yields the same parameter estimates and is more computationally stable.

Although the quantity $d_n(\alpha, \beta) \to d(\alpha, \beta)$ as $B \to \infty$, it may do so very slowly. Good initial values for the parameters are very important for reasonable convergence. Natural candidates are the MPL or one of the MGPL (maximum generalized pseudo-likelihood) estimators. Using the PL estimates we find that $\widehat{\alpha} = -0.303$ and $\widehat{\beta} = 0.117$, with standard errors $s.e.(\widehat{\alpha}) \simeq 0.034$ and $s.e.(\widehat{\beta}) \simeq 0.018$ as given in Section 4.2. These standard errors come from the approximate likelihood from the MCML algorithm. These values are relatively close to the MPL and GPL estimates. There is, however, mild evidence that MCML estimates are more efficient than those of PL due to their slightly smaller standard errors compared to those of the model-based standard errors of the PL estimators found in the previous section.

To evaluate how the different estimation methods compare in a known situation, consider simulated data from the autologistic model with $\alpha = 0$, and $\beta = 0.33$ and $\beta = 0.5$. The value $\beta = 0.33$ indicates very strong correlation, while the value $\beta = 0.5$ is such strong correlation that the spatial field is no longer stationary. We've observed that all methods give similar results under weak dependence (with $\beta \simeq 0.1$), so it is informative to see how they compare under these heavy correlation settings. Table 10.2 gives the average results and standard errors across 100 simulated data sets in each of the two settings. The data sets are simulated from the true model using the Gibbs sampler with spacing of 500 Gibbs steps between accepted data sets.

All methods give approximately unbiased estimators when $\beta = 0.33$. The PL and GPL estimators, however, are seen to be more efficient than the MCML estimators. In the $\beta = 0.5$ setting, there is some suggestion that the GPL method is slightly more efficient than PL. Monte Carlo maximum likelihood, however, is not competitive, even where $S = 200\,000$ spatial fields have been generated. The results show that, although the MCML method gets us to the efficient MLE, a very large number of Gibbs steps may be necessary when the spatial correlation is strong. Model-based resampling methods are very useful, but their adequacy often depends on large amounts of computing, and always depend on the correctness of the model generating the replicates. This raises the desire for a jackknife, bootstrap or subseries-based estimator in the spatial setting that will be appropriate for *any* estimators from *any* correlation structure satisfying basic regularity conditions.

10.6 Model-free spatial resampling

Let D_n, a subset of the two-dimensional integers, denote the domain where equally spaced data are observed. Let $s(D_n) := s(Z_i : i \in D_n)$ be any

Table 10.2 Simulation results from the autologistic (Ising) model.

	$\alpha = 0.0, \beta = 0.33$			
Method	Mean($\widehat{\alpha}$)	S.e.($\widehat{\alpha}$)	Mean($\widehat{\beta}$)	S.e.($\widehat{\beta}$)
PL	0.001	0.010	0.329	0.019
GPL5	0.000	0.009	0.329	0.018
GPL9	0.000	0.011	0.326	0.018
MCML1	−0.001	0.020	0.331	0.034
MCML2	−0.001	0.016	0.331	0.028
MCML3	0.003	0.017	0.328	0.028
	$\alpha = 0.0, \beta = 0.5$			
Method	Mean($\widehat{\alpha}$)	S.e.($\widehat{\alpha}$)	Mean($\widehat{\beta}$)	S.e.($\widehat{\beta}$)
PL	0.001	0.044	0.500	0.030
GPL5	0.002	0.041	0.500	0.029
GPL9	0.003	0.041	0.498	0.029
MCML1	−0.964	8.310	0.647	2.510
MCML2	−0.028	1.710	0.466	0.619
MCML3	−0.002	0.050	0.490	0.042

The number of simulated data sets is 100; the spacing between each is 500 Gibbs steps. 'PL' is the pseudo-likelihood estimate; 'GPL5' is the generalized pseudo-likelihood estimate where the number of sites in the set is $|D| = 5$; 'GPL9' is the generalized pseudo-likelihood estimate with $|D| = 9$; 'MCML1' refers to MCML with $S = 5000$ Gibbs steps, burn in of $d = 100$, and spacing of $\gamma = 2$; 'MCML2' refers to MCML with $S = 20\,000$, $d = 100$, and $\gamma = 10$; 'MCML3' refers to MCML with $S = 200\,000$, $d = 100$, and $\gamma = 20$.

statistic of interest computed from the data in D_n. We desire an estimate of $Var[s(D_n)] := v_n$, say. Let $D^j_{l(n)}$, $j = 1, \ldots, k_n$ denote overlapping subshapes of D_n where $l(n)$ determines the size of each. The method is a natural analogue to the subseries approach for time series observations.

Let $\bar{s}_n := \sum_{j=1}^{k_n} s\left[D^j_{l(n)}\right] / k_n$ denote the mean of the replicates and define

$$\widehat{\nu}_n := \frac{\sum_{j=1}^{k_n} \left|D_{l(n)}^j\right| \left\{s\left[D_{l(n)}^j\right] - \bar{s}_n\right\}^2}{k_n}.$$

This is simply the empirical variance of the standardized replicates of $s(D_n)$. If we let $t(D) := |D|^{1/2}\{s(D) - E[s(D)]\}$ denote the standardized statistic, then we see that $\widehat{\nu}_n$ targets $\nu := \lim_{n\to\infty} |D_n|\nu_n$. Formal justification of the variance estimator is given by the following. Assume:

Equation 5.2 holds

and

$$E[|t(D_n)|^6] < \infty.$$

Then

$$\widehat{\nu}_n \xrightarrow{L_2} \nu,$$

that is, $MSE(\widehat{\nu}_n) \to 0$. See Sherman (1996) for this and results under weaker conditions.

The two conditions, analogous to the time series setting, are that the correlation is not too strong at large spatial lags, and that the statistic of interest has sufficient moments. We illustrate implementation in the cancer data:

The counties in the cancer data set are split up into $k = 10$, partially overlapping subshapes D_j, $j = 1, \ldots, 10$. Each subshape is identical in size ($|D_j| = 141$) and similar in shape to the original data domain shown in Figure 4.4. From the subshapes we calculate the replicate statistics $\widehat{\alpha}^j$ and $\widehat{\beta}^j$, $j = 1, \ldots, 10$. The standardized estimate of Var($\widehat{\beta}$) is:

$$\widehat{\nu}_n = |D_j| \sum_{j=1}^{10} \frac{(\widehat{\beta}^j - \overline{\beta})^2}{10},$$

where $\overline{\beta} = \sum_{j=1}^{10} \widehat{\beta}^j/10$. The 10 replicate statistics are given in Table 10.3.

The average of the 10 replicates yields $\widehat{\alpha} = -0.4368$ and $\widehat{\beta} = 0.0588$. The resulting standardized variances are $\widehat{\nu}_n = 3.887$ (for α) and $\widehat{\nu}_n = 1.305$ (for β). The estimated variances of the parameter estimates are then $\widehat{\nu}_n/2003 = (0.00194, 0.0006165)$.

The square roots of these quantities give the resulting estimates of $s.e.(\widehat{\alpha}) = 0.044$ and $s.e.(\widehat{\beta}) = 0.026$. We saw in Section 4.2 that this indicated that there are more low-rate than high-rate counties, and that county rates are not independent. High-rate counties tend to have

Table 10.3 Replicate parameter estimates for the cancer data.

α	β
−0.2297	0.2526
−0.7693	−0.0174
−0.3702	0.1556
−0.5669	−0.0601
−0.4791	0.1271
−0.4710	0.0085
−0.3804	−0.0247
−0.3204	0.0414
−0.1945	0.1269
−0.5866	−0.0224

The 10 replicate pseudo-likelihood parameter estimates for α and β are computed from subsets of the binary cancer data set.

neighbors that are also high-rate counties. It may seem odd that 4 out of the 10 replicates of $\widehat{\beta}$ are negative, yet we judge $\widehat{\beta}$ to be significantly larger than 0. The replicates, however, vary according to the sample size of $|D_j| = 141$, while $\widehat{\beta}$ varies according to a sample size of $|D_n| = 2003$. Additionally using:

$$\widehat{\nu}_n = |D_j| \sum_{j=1}^{10} \frac{(\widehat{\alpha}^j - \overline{\alpha})(\widehat{\beta}^j - \overline{\beta})}{10},$$

to estimate the standardized covariance between the two parameter estimates, we find a value of $\widehat{\nu}_n = 1.522$, giving $Corr(\widehat{\alpha}, \widehat{\beta}) = (1/2003)[1.522/(0.044 \times 0.026)] = 0.664$. This suggests a moderate positive correlation between the overall rate parameter estimate and the clustering parameter estimate.

The method as presented uses overlapping replicates. Sherman and Carlstein (1994) presented a similar method using nonoverlapping replicates for spatial variance estimation. They also show the large sample normality of the variance estimators in this setting. We will see, however, in Section 10.6.1, that overlapping replicates are typically more efficient than nonoverlapping replicates.

As in the time series setting, determination of a good block size can be a difficult task. For choice of $l(n)$, large sample results show it should be $l(n) = Cn^{1/2}$ for some constant C. For example, for data on an $n \times n$ grid, the subblocks should be $Cn^{1/2} \times Cn^{1/2}$ for some constant C. We note the following:

i. The constant C can chosen using the methods of Hall and Jing (1996) for time series. See Section 10.7 for an implementation in estimating the variance of spatial statistics computed from irregularly spaced locations in the longleaf pine data discussed in Section 5.8.

ii. The subshape size in time series was $l(n) = Cn^{1/3}$; here it is $l(n) = Cn^{1/2}$. Bias is a bigger problem for spatial data, so the blocks have to become relatively larger to diminish this bias.

iii. The use of overlapping blocks was asymptotically $\frac{3}{2}$-times more efficient than nonoverlapping blocks in the time series setting. For data on a square, however, overlapping blocks are $\left(\frac{3}{2}\right)^2 = \frac{9}{4}$-times more efficient. This suggests that we should use overlapping subblocks if we can afford the computations. This result and the comparison for other domain shapes are discussed in Nordman and Lahiri (2004).

iv. Although the pseudo-likelihood estimators can be obtained using standard software for logistic regression, the reported standard errors will be wrong as they erroneously assume independence between terms in the PL. The model-based or model-free resampling estimates are much more reliable.

10.6.1 A spatial numerical experiment

We seek to assess the quality of two different nonparametric variance estimators in the autologistic model. We take $\alpha = 0.0$ and $\beta = 0.1$, and focus on estimation of β. This value of β is close to the observed parameter estimate in the cancer data set. For the target variance, v, we generate 2000 data sets, each from a 500×500 grid. On each we compute the estimator, $\widehat{\beta}$. Due to the lack of a simple closed form solution for the true variance, we compute the sample variance across the 2000 data replicates. The sample variance of the replicates is $v_n = 1.927 \times 10^{-6}$, and we then take $v := |D_n|v_n = 500^2 \times v_n = 0.4817$ as the (standardized) target variance.

Table 10.4 Spatial variance estimation for the pseudo-likelihood estimator.

Scheme	Average($\widehat{\nu}$)	Var($\widehat{\nu}$)
No overlap	0.4973	0.0222
$\frac{3}{4}$ overlap	0.4972	0.0095
True(ν)	0.4817	

A comparison of two overlapping schemes for variance estimation of $\widehat{\beta}$ in the spatial autologistic model. The model parameters are $\alpha = 0$ and $\beta = 0.1$.

For each of 100 data sets we compute the model-free estimate of variance using 100×100 subblocks. This is taking $C \simeq 4.5$ in our large sample recommendation for block size. To see the possible benefits of using overlapping blocks, we consider:

i. 25 nonoverlapping blocks, and
ii. $17^2 = 289$ partially overlapping (with $\frac{3}{4}$ adjacent block overlap) blocks.

The results are shown in Table 10.4. We can see from the table that:

i. Both schemes are approximately unbiased.
ii. The MSEs are 0.02248 and 0.00974 for the two schemes. Almost all mean square error is due to variability, strongly implying that the blocks are too big. Choosing $C = 2$ or possibly even $C = 1$ would likely have been more appropriate to reduce variability in this relatively weak correlation setting.
iii. Recall that in the AR(1) case the overlapping schemes had relatively worse bias than nonoverlapping blocks. That was, however, for a series of length $n = 100$ and AR(1) parameter $\rho = 0.9$. Here, $\beta = 0.1$ is roughly equivalent to $\rho = 0.25$ for an AR(1) time series. This weaker correlation makes all asymptotic results closer to reality.
iv. The ratio of variances, $0.0222/0.0095 = 2.34$, which is very close to the $\frac{9}{4} = 2.25$ predicted by the asymptotic theory.

Both estimators perform well in the relatively large data setting with weak correlation. Overlapping blocks are more computationally demanding but perform better under weak correlation. Again we see, in order to get the best performance it's useful to assess how strong the correlation is before making any decisions on block length and resampling scheme.

10.6.2 Spatial bootstrap

In trying to use a spatial block bootstrap, one difficulty arises. Namely, that block bootstrap replicates in general do not *fit together* properly for general shaped domains. For rectangular grids, however, we can use an analogous approach to that for time series. A semiparametric spatial bootstrap has been explored in Lele (1988), who adapted Rajarshi's nonparametric Markov transition model to the spatial setting. The extension may seem clear, but there is no natural way to generate a bootstrap resample due to the lack of a natural ordering in the spatial setting that exists in the temporal setting.

Consider continuous observations on an equally spaced grid. Assume that the conditional distribution at each location depends only on its four nearest neighbors (as in the autologistic model). For each location i, let $f(z_i|y_i)$ denote this conditional density, where $y_i := \sum_{j \in N(i)} Z_i$. We can then estimate $f(z|y)$ as in Rajarshi's approach. Now, code the grid, and generate a bootstrap sample using the Metropolis algorithm or the Gibb's sampler as described in Section 10.5.1. Lele (1988) shows the asymptotic appropriateness of this for the sample mean.

Hall (1988) divides up the spatial grid in one dimension as in the subseries method of variance estimation. Given the k spatial blocks of width r, he suggests splitting them into two groups of replicates, each with $k/2$ blocks. Within each group, the replicates are separated by distance r and taken to be independent, assuming the process is r-dependent. This is similar to the approach for time series variance estimation in Section 10.4.1. Here, within each of the two sets of replicates, a confidence interval can be computed using any of the large sample percentile, BCA or bootstrap-t methods. This gives two confidence intervals which can be combined using, for instance, Bonferroni's procedure.

10.7 Unequally spaced observations

Until now, we've considered equally spaced observations on a spatial grid. Many (if not most) data sets are not observed on an equally spaced grid.

For example, the locations of longleaf pines trees in the southern USA analysed in Section 5.7.

We consider estimating the variance and distribution of statistics from stationary marked point process data. The observations are now:

$$[(\mathbf{s}_1, Z(\mathbf{s}_1)), (\mathbf{s}_2, Z(\mathbf{s}_2)), \ldots, (\mathbf{s}_n, Z(\mathbf{s}_n))],$$

where $\mathbf{s} \in \mathbf{R}^d$ and $Z(s)$ is the observed mark at location s. The locations $\mathbf{s}$ are assumed to be generated by a stationary point process of the types discussed in Chapter 7.

The continuous parameter random field $\{Z(\mathbf{s}), \mathbf{s} \in \mathbf{R}^d\}$ is assumed to satisfy a weak dependence condition that is quantified in terms of strong mixing coefficients. Let $d(\cdot, \cdot)$ denote sup-distance (i.e., the distance arising from the l_∞ norm) on $\mathbf{R}^d$. We use a particular type of strong mixing coefficients defined by

$$\alpha_Z(k; l) \equiv \sup\{|P(A_1 \cap A_2) - P(A_1)P(A_2)| : A_i \in \mathcal{F}_Z(E_i),$$
$$i = 1, 2, E_2 = E_1 + s\},$$

where $|E_1| = |E_2| \leq l, d(E_1, E_2) \geq k$, and where the supremum is taken over all *compact and convex* sets $E_1 \subset \mathbf{R}^d$, and over all $\mathbf{s} \in \mathbf{R}^d$ such that $d(E_1, E_2) \geq k$; in the above, $\mathcal{F}_Z(E)$ denotes the σ–algebra generated by the random variables $\{Z(\mathbf{s}) : \mathbf{s} \in E\}$, and $|E|$ denotes Lebesgue measure (volume) of set E. The interpretation of this mixing coefficient is as previously given in Chapter 5. We assume that

$$\alpha_Z(l; l^d) \to 0 \text{ as } l \to \infty.$$

We now mimic the approach for subseries and the equally spaced case. For example, to estimate the variance of an arbitrary statistic, $s(\mathbf{Z}_n)$. We assume that

$$t(K_n) := |K_n|^{1/2}\{s(\mathbf{Z}_n) - E[s(\mathbf{Z}_n)]\},$$

is such that

$$Var[t(K_n)] \to \theta \geq 0 \text{ as } n \to \infty.$$

Define the replicates $s(B_n)$ from the rescaled $B_n := K_{cn}$. From the (standardized) variability of the associated replicate statistics, $s(B_n + y)$, we assess the variability of (the standardized) $s(K_n)$ as follows:

$$\widehat{\theta} := \int_{K_n^{1-c}} |B_n|[s(B_n + y) - \overline{s}(B_n)]^2 dy / \left|K_n^{1-c}\right|,$$

where $\bar{s}(B_n) := \int_{K_n^{1-c}} s(B_n + y)dy / \left|K_n^{1-c}\right|$. This subsampling estimator can be approximated by finite sums in applications.

The performance of this estimator depends on the choice of c. This is analogous to the choice of block size in the equally spaced time series and spatial data settings. The optimal choice is $c = Cn^{-1/2}$ for some constant C depending on the underlying spatial process parameters. The effect of this choice is evaluated in the following numerical study where we consider $C = 1$ and $C = 2$.

Under weak dependence of the spatial point process generating the locations, weak dependence of the marks, and moment conditions on the statistic of interest, we have that:

$$\widehat{\theta} \xrightarrow{L_2} \theta.$$

The precise conditions necessary for convergence are given in Politis and Sherman (2001).

To see the behavior of the spatial variance estimator, we give the results of a numerical experiment. Consider the marked point process with locations generated by a stationary Poisson process with intensity λ observed on $K_n := [0, n]^2$. Thus, the expected number of points in K_n is λn^2. For simplicity, take $\lambda = 1.0$. The underlying continuous time process generating the marks (the observations Z) is Gaussian, with isotropic exponential covariance function $Cov[Z(\mathbf{s}), Z(\mathbf{t})] = \exp(-\gamma|t - s|)$, for some $\gamma > 0$. The Gaussian process $Z(t)$ is therefore strong mixing [Rosenblatt (1985)]. Large γ corresponds to approximate independence, while γ near 0 corresponds to strong correlation for $|t - s|$ small. To assess the importance of the strength of correlation on the performance of the variance estimator $\widehat{\theta}$, we consider two different values of γ: $\gamma = 2.0$ and $\gamma = 1.0$. The true variances of the mean are obtained from 10 000 independent simulations from the given setup. The results for variance estimation are given in Table 10.5 based on 1000 independent spatial patterns from the described marked point process.

We see that the theoretical consistency holds in practice. The bias and variance of the variance estimator decrease as the size of the region increases. This holds for either strength of correlation or choice of subblock size. For fixed choice of correlation strength, we see that larger subblocks leads to a decreased bias but to an increase in variability.

To illustrate implementation we assess the width of longleaf pines. As discussed in Chapter 5, Cressie (1993) gives the exact locations and diameters at breast height (cm) of $N = 584$ longleaf pine trees from

Table 10.5 Variance estimation for irregularly spaced observations.

		Mean	Variance	MSE
$\gamma = 2.0$				
$K_n = 10 \times 10$	c = 0.316	1.39	0.35	1.25
	c = 0.632	1.61	0.66	1.19
$K_n = 20 \times 20$	c = 0.223	1.76	0.32	0.65
	c = 0.447	1.89	0.49	0.70
θ		2.34		
$\gamma = 1.0$				
$K_n = 10 \times 10$	c = 0.316	2.09	1.58	17.04
	c = 0.632	3.17	3.24	11.32
$K_n = 20 \times 20$	c = 0.223	3.44	1.01	7.62
	c = 0.447	4.52	2.98	5.20
θ		6.01		

Variance estimation for the sample mean, K_n denotes the sizes of each spatial domain, γ the strength of spatial correlation, c the size of subshapes used. Given are the mean variance estimates and the variance of the variance estimates across 1000 independent experiments. The target variances, θ, are evaluated from 10 000 independent spatial fields.

the Wade Tract, an old-growth forest in Thomas County, Georgia, USA. The domain $K_n = 200\,\text{m} \times 200\,\text{m}$. Using the methods of Chapter 7 we determine that there is significant positive clustering of the locations of the trees. While allowing for this correlation, we assume that the locations are generated by a stationary point process that satisfies our mixing condition (there is some doubt as to the reasonableness of stationarity, as discussed in Chapter 5). One statistic that assesses the size of trees in the forest is the sample mean of the breast height diameters, which is $\overline{x} = 26.9\,\text{cm}$. In order to assess the stability of this estimate, we desire an estimate of the standard deviation of the mean. The sample variance of the heights is $s^2 = 336\,\text{cm}^2$, so a naive estimate of *std*$(\overline{x})$ *assuming independence* is $\widehat{std}_{\text{naive}} = (336/584)^{1/2} = 0.76\,\text{cm}$.

The data-based algorithm of Hall and Jing (1996) is adapted to the spatial setting using the known optimal rate of block size. This algorithm estimates that $\widehat{c}_n \in (0.20, 0.25)$, corresponding to 40 m × 40 m or 50 m × 50 m subshapes. The first value gives $\widehat{\theta}_1 = 315\,000$, while the second gives $\widehat{\theta}_2 = 377\,000$. The former gives an estimated standard deviation of $\widehat{std}_1 = \left(\widehat{\theta}_1/|K_n|\right)^{1/2} = (315\,000)^{1/2}/200 = 2.81$ cm, while the latter gives an estimate of $\widehat{std}_2 = (377\,000)^{1/2}/200 = 3.07$ cm. Both estimates are approximately four-times larger than the naive estimate of standard deviation, which erroneously assumes an underlying structure of independent observations. In particular, confidence intervals assuming normality would be four-times too narrow if independence were assumed, leading to dramatic undercoverage.

Another approach to resampling for unequally spaced observations is to allow for nonstationary locations. One general approach is that of Lahiri and Zhu (2006). This has been discussed in Section 5.7 in an application to a test for spatial isotropy. The method is also appropriate for inferences for regression models under locations coming from a nonconstant intensity.

Bibliography

Akritas, M. (1986). Bootstrapping the Kaplan-Meier estimator, *Journal of the American Statistical Association*, 81, 1032–1037.

Baczkowski, A.J. and Mardia, K.V. (1990). A test of spatial symmetry with general application, *Communications in Statistics: Theory and Methods*, 19, 555–572.

Baddeley, A., Turner, R., Moller, J., and Hazelton, M. (2005). Residual analysis for spatial point processes, *Journal of the Royal Statistical Society, Series B*, 67, 617–666.

Barry, R.D. and Ver Hoef, J.M. (1996). Blackbox kriging: spatial prediction without specifying the variogram, *Journal of Agricultural, Biological, and Environmental Statistics*, 1, 297–322.

Bedient, P.B., Hoblit, B.C., Gladwell, D.C., and Vieux, B.E. (2000). NEXRAD radar for flood prediction in Houston, *Journal of Hydrological Engineering*, 5, 269–277.

Bedient, P.B., Holder, A., Benavides, J.A., and Vieux, B.E. (2003). Radar-based flood warning system applied to tropical storm Allison, *Journal of Hydrological Engineering*, 8, 308–318.

Bertail, P., Politis, D.N., and Romano, J.P. (1999). On subsampling estimators with unknown rate of convergence, *Journal of the American Statistical Association*, 94, 569–579.

Besag, J. (1974). Spatial interaction and the statistical analysis of lattice systems, *Journal of the Royal Statistical Society, Series B*, 36, 192–236.

Besag, J. (1975). Statistical analysis of non-lattice data, *The Statistician*, 24, 179–195.

Bickel, P.J. and Freedman, D.A. (1981). Some asymptotic theory for the bootstrap, *Annals of Statistics*, 9, 1196–1217.

Booth, J.G. and Sarkar, S. (1998). Monte Carlo approximation of bootstrap variances, *American Statistician*, 52, 354–357.

Bose, A. (1988). Edgeworth correction by bootstrap in autoregressions, *Annals of Statistics*, 16, 1709–1722.

Spatial Statistics and Spatio-Temporal Data: Covariance Functions and Directional Properties Michael Sherman

Bose, A. (1990). Bootstrap in moving average models, *Annals of the Institute of Statistical Mathematics*, 42, 753–768.

Boumans, R.M.J. and Sklar, F.H. (1991). A polygon-based spatial model for simulating landscape change, *Landscape Ecology*, 4, 83–97.

Boyd, C.E. (1992). Shrimp pond bottom soil and sediment management. In *Proceedings of the Special Session on Shrimp Farming*, ed. G. Wyban, World Aquaculture Society, Baton Rouge, LA, pp. 166–181.

Bradley, R.C. (1993). Some examples of mixing random fields, *Rocky Mountain Journal of Mathematics*, 23, 495–519.

Bretagnolle, J. (1983). Lois limites du bootstrap de certaines fonctionelles, *Annales de l'Institut Henri Poincare, Section B*, 3, 256–261.

Buhlmann, P. (2002). Sieve bootstrap with variable length Markov chains for stationary categorical time series (with discussion), *Journal of the American Statistical Association*, 97, 443–471.

Burrough, P.A. and McDonnell, R.A. (1998). *Principles of Geographical Information Systems*, Oxford University Press, Oxford.

Cabana, E.M. (1987). Affine processes: a test of isotropy based on level sets, *SIAM Journal of Applied Mathematics*, 47, 886–891.

Calder, C.A. (2007). Dynamic factor process convolution models for multivariate space-time data with application to air quality assessment, *Journal of Environmental and Ecolological Statistics*, 14, 229–247.

Carlstein, E. (1986). The use of subseries values for estimating the variance of a general statistic from a stationary sequence, *Annals of Statistics*, 14, 1171–1179.

Carroll, R.J. and Ruppert, D. (1988). *Transformation and Weighting in Regression*, Chapman and Hall, New York, NY.

Castelloe, J.M. (1998). Issues in reversible jump Markov chain Monte Carlo and composite EM analysis, applied to spatial Poisson cluster processes. Ph.D. Thesis, University of Iowa, Iowa City, IA.

Chiles, J.P. and Delfiner, P. (1999). *Geostatistics; Modeling Spatial Uncertainty*, John Wiley & Sons, Inc., New York, NY.

Christensen, O.F. (2004). Monte Carlo maximum likelihood in model-based geostatistics, *Journal of Computational and Graphical Statistics*, 13, 702–718.

Comets, F. (1992). On consistency of a class of estimators for exponential families of Markov random fields on the lattice, *Annals of Statistics*, 20, 455–468.

Costanza, R., Sklar, F.H., and White, M.L. (1990). Modeling coast landscape dynamics, *Bioscience*, 40, 91–107.

Cox, D.R. and Isham, V. (1988). A simple spatial-temporal model of rainfall, *Proceedings of the Royal Society of London, Series A*, 415, 317–328.

Cressie, N.A.C. (1993). *Statistics for Spatial Data*, John Wiley & Sons, Inc., New York, NY.

Cressie, N. and Hawkins, D.M. (1980). Robust estimation of the variogram, I, *Journal of the International Association for Mathematical Geology*, 12, 115–125.

Cressie, N. and Huang, H.C. (1999). Classes of nonseparable, spatio-temporal stationary covariance functions, *Journal of the American Statistical Association*, 94, 1330–1340.

Cuzick, J. and Edwards, R. (1990). Spatial clustering for inhomogeneous populations, *Journal of the Royal Statistical Society, Series B*, 19, 555–572.

Dagan, G. (1985). Stochastic modeling of groundwater flow by unconditional and conditional probabilities: the inverse problem, *Water Resources Research*, 21, 65–72.

Datta, S. (1995). Limit theory and bootstrap for explosive and partially explosive autoregression, *Stochastic Processes and their Application*, 57, 285–304.

Davison, A.C. and Hall, P. (1993). On studentizing and blocking methods for implementing the bootstrap with dependent data, *Australian Journal of Statistics*, 35, 215–224.

Davison, A.C. and Hinkley, D.V. (1997). *Bootstrap Methods and their Application*, Cambridge University Press, Cambridge.

Davison, A.C., Hinkley, D.V., and Schechtman, E. (1986). Efficient bootstrap simulations, *Biometrika*, 73, 555–566.

de Luna, X. and Genton, M.G. (2005). Predictive spatio-temporal models for spatially sparse environmental data, *Statistica Sinica*, 15, 547–568.

De Oliveira, V. (2006). On optimal point and block prediction in log-gaussian random fields, *Scandinavian Journal of Statistics*, 33, 523–540.

Diciccio, T., Hall, P., and Romano, J. (1991). Empirical likelihood is Bartlett correctable, *Annals of Statistics*, 19, 1053–1061.

Diggle, P.J. (2003). *Statistical Analysis of Spatial Point Patterns*, Oxford University Press, New York, NY.

Diggle, P.J. and Ribeiro, P.J., Jr. (2007). *Model-Based Geostatistics*, Springer, New York, NY.

Donnelly, K. (1978). Simulations to determine the variance and edge effect of total nearest neighbour distance. In *Simulation Methods in Archaeology*, ed. I. Hodder, Cambridge University Press, London, pp. 91–95.

Doukhan, P. (1994). *Mixing: Properties and Examples*, Springer, New York, NY.

Efron, B. (1979). Bootstrap methods: another look at the jackknife, *Annals of Statistics*, 7, 1–26.

Efron, B. and Tibshirani, R.J. (1993). *Introduction to the Bootstrap*, Chapman and Hall, New York, NY.

Ferguson, T. (1996). *A Course in Large Sample Theory*, Chapman and Hall, London, UK.

Ferretti, N. and Romo, J. (1996). Unit root bootstrap tests for AR(1) models, *Biometrika*, 83, 949–960.

Freedman, D.A. (1984). Bootstrapping two-stage least squares estimates in stationary linear models, *Annals of Statistics*, 12, 827–842.

Fuentes, M. (2006). Testing for separability of spatial-temporal covariance functions, *Journal of Statistical Planning and Inference*, 136, 447–466.

Fuentes, M., Chen, L., and Davis, J. (2008). A class of nonseparable and nonstationary spatial temporal covariance functions, *Environmetrics*, 19, 487–507.

Gastwirth, J.L. and Rubin, H. (1975). The asymptotic distribution theory of the empirical CDF for mixing stochastic processes, *Annals of Statistics*, 3, 809–824.

Gelfand, A.E., Schmidt, A.M., Banerjee, S., and Sirmans, C.F. (2004). Nonstationary multivariate process modeling through spatially varying coregionalization, *Test*, 13, 263–312.

Geman, S. and Graffigne, C. (1987). Markov random field image models and their applications to computer vision. In *Proceedings of the International Congress of Mathematicians, Berkeley, California, USA, 1986*, ed. A.M. Gleason, American Mathematical Society, Providence, RI, pp. 1496–1517.

Genton, M.G. (2001). Robustness problems in the analysis of spatial data. In *Spatial Statistics: Methodological Aspects and Applications*, ed. M. Moore, Springer-Verlag, New York, NY, pp. 21–38.

Geyer, C.J. and Thompson, E.A. (1992). Constrained Monte Carlo maximum likelihood for dependent data, *Journal of the Royal Statistical Society, Series B*, 54, 657–699.

Gneiting, T. (2002). Nonseparable, stationary covariance functions for space-time data, *Journal of the American Statistical Association*, 97, 590–600.

Gneiting, T., Genton, M.G., and Guttorp, P. (2007). Geostatistical space-time models, stationarity, separability and full symmetry. In *Statistics of Spatio-Temporal Systems*, eds. B. Finkenstaedt, L. Held, and V. Isham, Chapman and Hall, CRC Press, Boca Raton, FL, pp. 151–175.

Götze, F. and Künsch, H.R. (1996). Second order correctness of the blockwise bootstrap for stationary observations, *Annals of Statistics*, 24, 1914–1933.

Grzebyk, M. and Wackernagel, H. (1994). Challenges in multivariate spatio-temporal modelling. In *Proceedings of the XVIIth International Biometrics Conference, Hamilton, Ontario, Canada, 8–12 August 1994*, IBC '94 Local Organising Committee, McMaster University, Hamilton, Ontario, pp. 19–33.

Gu, M. and Zhu, H. (2001). Maximum likelihood estimation for spatial models by Markov chain Monte Carlo stochastic approximation, *Journal of the Royal Statistical Society, Series B*, 63, 339–355.

Guan, Y. (2008). On consistent nonparametric intensity estimation for inhomogeneous spatial point processes, *Journal of the American Statistical Association*, 103, 1238–1247.

Guan, Y. and Sherman, M. (2007). On least squares fitting for stationary spatial point processes, *Journal of the Royal Statistical Society, Series B*, 69, 31–49.

Guan, Y., Sherman, M., and Calvin, J.A. (2004). A nonparametric test for spatial isotropy using subsampling, *Journal of the American Statistical Association*, 99, 810–821.

Gupta, V.K. and Waymire, E. (1987). On Taylor's hypothesis and dissipation in rainfall, *Journal of Geophysical Research*, 92, 9657–9660.

Guyon, X. and Künsch, H.R. (1992). Asymptotic comparison of estimators in the Ising model. In *Stochastic Models, Statistical Methods, and Algorithms in Image Analysis*, ed. P. Barone, Lecture Notes in Statistics, 74, Springer, Berlin, pp. 177–198.

Haas, T.C. (1990). Lognormal and moving window methods of estimating acid deposition, *Journal of the American Statistical Association*, 85, 950–963.

Haase, P. (2001). Can isotropy vs. anisotropy in the spatial association of plant species reveal physical vs. biotic facilitation? *Journal of Vegetation Science*, 12, 127–136.

Hall, P. (1988). On confidence intervals for spatial parameters estimated from nonreplicated data, *Biometrics*, 44, 271–277.

Hall, P. (1992). *The Bootstrap and Edgeworth Expansion*, Springer Verlag, New York, NY.

Hall, P. and Jing, B.Y. (1996). On sample re-use methods for dependent data, *Journal of the Royal Statistical Society, Series B*, 58, 727–738.

Hall, P. and Patil, P. (1994). Properties of nonparametric estimators of autocovariance for stationary random fields, *Probability Theory and Related Fields*, 99, 399–423.

Hall, P., Fisher, N.I., and Hoffmann, B. (1994). On the nonparametric estimation of covariance functions, *Annals of Statistics*, 22, 2115–2134.

Handcock, M.S. and Wallis, J.R. (1994). An approach to statistical spatial-temporal modelling of meteorological fields (with discussion), *Journal of the American Statistical Association*, 89, 368–390.

Hartley, H.O. (1949). Tests of significance in harmonic analysis, *Biometrika*, 36, 194–201.

Haslett, J. and Raftery, A.E. (1989). Space-time modelling with long-memory dependence: assessing Ireland's wind power resource, *Applied Statistics*, 38, 1–50.

Heagerty, P. and Lele, S. (1998). A composite likelihood approach to binary spatial data, *Journal of the American Statistical Association*, 93, 1099–1111.

Higdon, D. (1998). A process-convolution approach to modeling temperatures in the North Atlantic Ocean, *Journal of Environmental and Ecolological Statistics*, 5, 173–190.

Hoeffding, W. (1948). A class of statistics with asymptotically normal distributions, *Annals of Mathematical Statistics*, 19, 293–325.

Howe, H.F. and Primack, R.B. (1975). Directional seed dispersal by birds of the tree, *Casearia nitida (Flacourtiaceae), Biotropica*, 7, 278–283.

Huang, F. and Ogata, Y. (2002). Generalized pseudo-likelihood estimates for Markov random fields on lattice, *Annals of the Institute of Statistical Mathematics*, 54, 1–18.

Huang, C., Hsing, T., and Cressie, N. (2008). Nonparametric estimation of variogram and its spectrum, Technical Report 08-05, Department of Statistics, Indiana University.

Ibragimov, I.A. and Linnik, U.V. (1971). *Independent and Stationary Sequences of Random Variables*, Wolters-Noordhoff, Groningen, The Netherlands.

Illian, J., Penttinen, A., Stoyan, H., and Stoyan, D. (2008). *Spatial Analysis and Modelling of Spatial Point Patterns*, John Wiley & Sons, Ltd, Chichester.

Isaaks, E.H. and Srivastava, R.M. (1989). *An Introduction to Applied Geostatistics*, Oxford University Press, Oxford.

Ising, E. (1925). Beitrag zur Theorie des Ferromagnetismus, *Zeitschrift Physik*, 31, 253–258.

Jayakrishnan, R., Srinivasan, R., and Arnold, J. (2004). Comparison of raingage and WSR-88D Stage III precipitation data over the Texas-Gulf basin, *Journal of Hydrology*, 292, 135–152.

Jona-Lasinio, G. (2001). Modeling and exploring multivariate spatial variation: a test procedure for isotropy of multivariate spatial data, *Journal of Multivariate Analysis*, 77, 295–317.

Jones, R.H. and Zhang, Y. (1997). Models for continuous stationary space-time processes. In *Modelling Longitudinal and Spatially Correlated Data*, eds. T. Gregoire, D. Brillinger, P. Diggle, *et al.* Springer, New York, NY, pp. 289–298.

Jun, M. and Stein, M.L. (2007). An approach to producing space-time covariance functions on spheres, *Technometrics*, 49, 468–479.

Kadlec, R.H. and Hammer, D.E. (1988). Modelling nutrient behavior in wetlands, *Ecological Modelling*, 40, 37–66.

Kaiser, M.S. and Cressie, N. (1997). Modeling Poisson variables with positive spatial dependence, *Statistics and Probability Letters*, 33, 423–432.

Karr, A.F. (1986). Inference for stationary random fields given Poisson samples, *Advances in Applied Probability*, 18, 406–422.

Kessell, S.R. (1977). Gradient modeling: a new approach to fire modeling and resource management. In *Ecosystem Modeling in Theory and Practice*, eds. C.A.S. Hall and J.W.J. Day, John Wiley & Sons, Inc., New York, NY, pp. 576–605.

Kitamura, Y. (1997). Empirical likelihood with weakly dependent processes, *Annals of Statistics*, 25, 2084–2102.

Kowalski, J. and Tu, X.M. (2008). *Modern Applied U-Statistics*, John Wiley & Sons, Inc., Hoboken, NJ.

Kreiss, J.P. and Frank, J. (1992). Bootstrapping stationary autoregressive moving-average models, *Journal of Time Series Analysis*, 13, 297–317.

Künsch, H. (1989). The Jackknife and the bootstrap for general stationary observations, *Annals of Statistics*, 17, 1217–1241.

Lahiri, S.N. (2003). *Resampling Methods for Dependent Data*, Springer, New York, NY.

Lahiri, S.N. and Zhu, J. (2006). Resampling methods for spatial regression models under a class of stochastic designs, *Annals of Statistics*, 34, 1774–1813.

Lahiri, S.N., Kaiser, M.S., Cressie, N., and Hsu, N.J. (1999). Prediction of spatial cumulative distribution functions using subsampling (with discussion), *Journal of the American Statistical Association*, 94, 86–110.

Lee, H., Higdon, D., Calder, C., and Holloman, C. (2005). Efficient models for correlated data via convolutions of intrinsic processes, *Statistical Modelling*, 5, 53–74.

Lee, Y.D. and Lahiri, S.N. (2002). Least squares variogram fitting by spatial subsampling, *Journal of the Royal Statistical Society, Series B*, 64, 837–854.

Lele, S. (1988). Nonparametric bootstrap for spatial processes, Technical Report 671, Department of Biostatistics, Johns Hopkins University.

Li, B., Genton, M.G., and Sherman, M. (2007). A nonparametric assessment of properties of space-time covariance functions, *Journal of the American Statistical Association*, 102, 736–744.

Li, B., Genton, M.G., and Sherman, M. (2008). Testing the covariance structure of multivariate random fields, *Biometrika*, 95, 813–829.

Li, B., Murthi, A., Bowman, K., *et al.* (2009). Statistical tests of Taylor's hypothesis: an application to precipitation fields, *Journal of Hydrometeorology*, 10, 254–265.

Lim, J., Wang, S., and Sherman, M. (2007). An adjustment for edge effects using an augmented neighborhood model in the spatial auto-logistic model, *Computational Statistics and Data Analysis*, 8, 3679–3688.

Liu, R. and Singh, K. (1992). Moving blocks jackknife and bootstrap capture weak dependence. In *Exploring the Limits of Bootstrap*, eds. R. Lepage and L. Billard, John Wiley and Sons, Inc., New York, NY.

Lu, H. and Zimmerman D.L. (2001). Testing for isotropy and other directional symmetry properties of spatial correlation, *preprint*.

Lu, H. and Zimmerman, D.L. (2005). Testing for directional symmetry in spatial dependence using the periodogram, *Journal of Statistical Planning and Inference*, 129, 369–385.

Maguire, L.A. and Porter, J.W. (1977). A spatial model of growth and competition strategies in coral communities, *Ecological Modelling*, 3, 249–271.

Majumdar, A. and Gelfand, A.E. (2007). Multivariate spatial modeling using convolved covariance functions, *Mathematical Geology*, 39, 225–245.

Manly, B.F.J. (2006). *Randomization, Bootstrap and Monte Carlo Methods in Biology*, Chapman and Hall, Boca Raton, FL.

Mardia, K.V. and Goodall, C.R. (1993). Spatial-temporal analysis of multivariate environmental monitoring data. In *Multivariate Environmental Statistics*, eds. G.P. Patel, C.R. Rao, North-Holland Series in Statistics and Probability, 6, North-Holland, Amsterdam, pp. 347–386.

Mardia, K.V., Kent, J.T., and Bibby, J.M. (1979). *Multivariate Analysis*, Academic Press, London.

Maritz, J.S. and Jarrett, R.G. (1978). A note on estimating the variance of the sample median, *Journal of the American Statistical Association*, 73, 194–196.

Martin, R.J. (1979). A subclass of lattice processes applied to a problem in planar sampling, *Biometrika*, 66, 209–217.

Mercer, W.B. and Hall, A.D. (1911). The experimental error of field trials, *Journal of Agricultural Science (Cambridge)*, 4, 107–132.

Mitchell, M.W., Genton, M.G., and Gumpertz, M.L. (2005). Testing for separability of space-time covariances, *Environmetrics*, 16, 819–831.

Mitchell, M.W., Genton, M.G., and Gumpertz, M.L. (2006). A likelihood ratio test for separability of covariances, *Journal of Multivariate Analysis*, 97, 1025–1043.

Molina, A. and Feito, F.R. (2002). A method for testing anisotropy and quantifying its direction in digital images, *Computers and Graphics*, 26, 771–784.

Moller, J. and Waagepetersen, R.P. (2004). *Statistical Inference and Simulation for Spatial Point Processes*, Chapman Hall, Boca Raton, FL.

Moyeed, R.A. and Baddeley, A.J. (1991). Stochastic approximation of the MLE for a spatial point pattern, *Scandinavian Journal of Statistics*, 18, 39–50.

Neyman, J. and Scott, E.L. (1958). Statistical approach to problems of cosmology, *Journal of the Royal Statistical Society, Series B*, 20, 1–43.

Nordman, D.J. and Caragea, P.C. (2008). Point and interval estimation of variogram models using spatial empirical likelihood, *Journal of the American Statistical Association*, 103, 350–361.

Nordman, D.J. and Lahiri, S.N. (2004). On optimal spatial subsample size for variance estimation, *Annals of Statistics*, 32, 1981–2027.

Ohser, J. and Stoyan, D. (1981). On the second-order and orientation analysis of planar stationary point processes, *Biometrical Journal*, 23, 523–533.

Owen, A. (1990). Empirical likelihood ratio confidence regions, *Annals of Statistics*, 18, 90–120.

Paparoditis, E. and Politis, D.N. (2001). Tapered block bootstrap, *Biometrika*, 88, 1105–1119.

Phillips, P.L. and Barnes, P.W. (2002). Spatial asymmetry in tree-shrub clusters in a subtropical savanna, *American Midland Naturalist*, 149, 59–70.

Politis, D.N. and Romano, J.P. (1994). Large sample confidence regions based on subsamples under minimal assumptions, *Annals of Statistics*, 22, 2031–2050.

Politis, D.N. and Sherman, M. (2001). Moment estimation for statistics from marked point processes, *Journal of the Royal Statistical Society, Series B*, 63, 261–275.

Politis, D.N., Paparoditis, E., and Romano, J.P. (1998). Large sample inference for irregularly spaced dependent observations based on subsampling, *Sankhya, Series A*, 60, 274–292.

Possolo, A. (1991). Subsampling a random field. In *Spatial Statistics and Imaging*, ed. A. Possolo, IMS Lecture Notes – Monograph Series, Volume 20, Institute of Mathematical Statistics, Hayward, CA, pp. 286–294.

Priestley, M.B. (1981). *Spectral Analysis and the Time Series*, Academic Press, London.

Pukelsheim, F. (1994). The three sigma rule, *The American Statistician*, 48, 88–91.

Quenouille, M. (1949). Approximate tests of correlation in time series, *Journal of the Royal Statistical Society, Series B*, 11, 18–84.

Quenouille, M. (1956). Notes on bias in estimation, *Biometrika*, 43, 353–360.

Rajarshi, M.B. (1990). Bootstrap in Markov sequences based on estimates of transition density, *Annals of the Institute of Statistical Mathematics*, 42, 253–268.

Riggan, W.B., Creason, J.P., Nelson, W.C., *et al.* (1987). *U.S. Cancer Mortality Rates and Trends*, 1950–1979, Volume IV, Maps, US Environmental Protection Agency, Washington, DC.

Rosenberg, M.S. (2004). Wavelet analysis for detecting anisotropy in point patterns, *Journal of Vegetation Science*, 15, 277–284.

Rosenblatt, M. (1956). A central limit theorem and a strong mixing condition, *Proceedings of the National Academy of Science, U.S.A.*, 42, 43–47.

Rosenblatt, M. (1985). *Stationary Sequences and Random Fields*, Birkhauser, Boston, MA.

Royle, J.A. and Berliner, L.M. (1999). A hierarchical approach to multivariate spatial modeling and prediction, *Journal of Agricultural, Biological, and Environmental Statistics*, 4, 29–56.

Sain, S.R. and Cressie, N. (2007). A spatial model for multivariate lattice data, *Journal of Econometrics*, 140, 226–259.

Scaccia, L. and Martin, R.J. (2002). Testing for simplification in spatial models. In *COMPSTAT 2002, Proceedings in Computational Statistics*, eds. W. Härdle and B. Rönz, Physica, Heidelberg, pp. 581–586.

Scaccia, L. and Martin, R.J. (2005). Testing axial symmetry and separability of lattice processes, *Journal of Statistical Planning and Inference*, 131, 19–39.

Schenk, H.J. and Mahall, B.E. (2002). Positive and negative interactions contribute to a north-south-patterned association between two desert shrub species, *Oecologia*, 132, 402–410.

Schlather, M., Ribeiro, P.J. and Diggle, P.J. (2004). Detecting dependence between marks and locations of marked point processes, *Journal of the Royal Statistical Society, Series B*, 66, 79–93.

Schmidt, A.M. and Gelfand, A.E. (2003). A Bayesian coregionalization approach for multivariate pollutant data, *Journal of Geophysical Research Atmospheres*, 108, 8783.

Serfling, R.J. (1980). *Approximation Theorems of Mathematical Statistics*, John Wiley & Sons, Inc., New York, NY.

Seymour, L. and Ji, C. (1996). Approximate Bayes model selection procedures for Gibbs-Markov random fields, *Journal of Statistical Planning and Inference*, 51, 75–97.

Shao, J. and Tu, D. (1995). *The Jacknife and Bootstrap*, Springer, New York, NY.

Shao, J. and Wu, C.F.J. (1989). A general theory of jackknife variance estimation, *Annals of Statistics*, 20, 1571–1593.

Sherman, M. (1995). On batch means in the simulation and statistics communities. In *Proceedings of the 1995 Winter Simulation Conference*, eds. C. Alexopoulos, K. Wang, W.R. Lilegdon, and D. Goldsman, IEEE Press, Piscataway, NJ, pp. 297–302.

Sherman, M. (1996). Variance estimation for statistics computed from spatial lattice data, *Journal of the Royal Statistical Society, Series B*, 58, 509–523.

Sherman, M. and Carlstein, E. (1994). Nonparametric estimation of the moments of a general statistic computed from spatial data, *Journal of the American Statistical Association*, 89, 496–500.

Sherman, M. and Carlstein, E. (1996). Replicate histograms, *Journal of the American Statistical Association*, 91, 566–576.

Sherman, M., Speed, M., and Speed, M. (1997). Analysis of tidal data via the blockwise bootstrap, *Journal of Applied Statistics*, 25, 333–340.

Show, I.T., Jr. (1979). Plankton community and physical environment simulation for the Gulf of Mexico region. In *Proceedings of the 1979 Summer Computer Simulation Conference*. Society for Computer Simulation, San Diego, CA, pp. 432–439.

Singh, K. (1981). On the asymptotic accuracy of Efron's bootstrap, *Annals of Statistics*, 9, 1187–1195.

Smith, P.T. (1996). Physical and chemical characteristics of sediments from prawn farms and mangrove habitat on the Clarence river, Australia, *Aquaculture*, 146, 47–83.

Staelens, J., Nachtergale, L., Luyssaert, S., and Lust, N. (2003). A model of wind-influenced leaf litterfall in a mixed hardwood forest, *Canadian Journal of Forest Research*, 33, 201–209.

Stein, M.L. (1999). *Interpolation of Spatial Data*, Springer, New York, NY.

Stein, M.L. (2002). The screening effect in kriging, *Annals of Statistics*, 30, 298–323.

Stein, M.L. (2005). Space-time covariance functions, *Journal of the American Statistical Association*, 100, 310–321.

Stoffer, D.S. and Wall, K.D. (1991). Bootstrapping state-space models: Gaussian maximum likelihood estimation and Kalman filter, *Journal of the American Statistical Association*, 86, 1024–1033.

Stoyan, D. (1991). Describing the anisotropy of marked planar point processes, *Statistics*, 22, 449–462.

Subba Rao, S. (2008). Statistical analysis of a spatio-temporal model with location dependent parameters and a test for spatial stationarity, *Journal of Time Series Analysis*, 29, 673–694.

Taylor, G.I. (1938). The spectrum of turbulence, *Proceedings of the Royal Society of London, Series A*, 164, 476–490.

Tukey, J. (1958). Bias and confidence in not-quite large samples (abstract), *Annals of Mathematical Statistics*, 29, 614.

Upton, G.J.G. and Fingleton, B. (1985). *Spatial Data Analysis by Example. Volume 1, Point Pattern and Quantitative Data*. John Wiley & Sons, Ltd, Chichester.

Ver Hoef, J. and Barry, R. (1998). Constructing and fitting models for cokriging and multivariable spatial prediction, *Journal of Statistical Planning and Inference*, 69, 275–294.

Wackernagel, H. (2003). *Multivariate Geostatistics: An Introduction with Applications*, 2nd edition, Springer-Verlag, Berlin.

Whittle, P. (1963). Stochastic processes in several dimensions, *Bulletin of the International Statistical Institute*, 40, 974–994.

Wikle, C.K. and Cressie, N. (1999). A dimension reduction approach to space-time Kalman filtering, *Biometrika*, 86, 815–829.

Wu, C.F.J. (1986). Jackknife, bootstrap, and other resampling methods in regression analysis (with discussion), *Annals of Statistics*, 14, 1261–1350.

Younes, L. (1991). Maximum likelihood estimation for Gibbs fields. In *Spatial Statistics and Imaging*, ed. A. Possolo, IMS Lecture Notes – Monograph Series, Volume 20, Institute of Mathematical Statistics, Hayward, CA, pp. 403–426.

Zelterman, D., Le, C.T. and Louis, T.A. (2004). Bootstrap techniques for proportional hazards models with censored observations, *Statistics and Computing*, 6, 191–199.

Zhang, H. and Zimmerman, D.L. (2005). Towards reconciling two asymptotic frameworks in spatial statistics, *Biometrika*, 92, 921–936.

Zimmerman, D. (1993). Another look at anisotropy in geostatistics, *Mathematical Geology*, 25, 453–470.

Zimmerman, D.L. and Zimmerman, M.B. (1991). A comparison of spatial semivariogram estimators and corresponding ordinary kriging predictors, *Technometrics*, 33, 77–91.

Index

WILEY SERIES IN PROBABILITY AND STATISTICS

Established by WALTER A. SHEWHART and SAMUEL S. WILKS

The Wiley Series in Probability and Statistics is well established and authoritative. It covers many topics of current research interest in both pure and applied statistics and probability theory. Written by leading statisticians and institutions, the titles span both state-of-the-art developments in the field and classical methods.

Reflecting the wide range of current research in statistics, the series encompasses applied, methodological and theoretical statistics, ranging from applications and new techniques made possible by advances in computerized practice to rigorous treatment of theoretical approaches.

This series provides essential and invaluable reading for all statisticians, whether in academia, industry, government, or research.

ABRAHAM and LEDOLTER · Statistical Methods for Forecasting
AGRESTI · Analysis of Ordinal Categorical Data
AGRESTI · An Introduction to Categorical Data Analysis
AGRESTI · Categorical Data Analysis, Second Edition
ALTMAN, GILL and McDONALD · Numerical Issues in Statistical Computing for the Social Scientist
AMARATUNGA and CABRERA · Exploration and Analysis of DNA Microarray and Protein Array Data
ANDĚL · Mathematics of Chance
ANDERSON · An Introduction to Multivariate Statistical Analysis, Third Edition
*ANDERSON · The Statistical Analysis of Time Series
ANDERSON, AUQUIER, HAUCK, OAKES, VANDAELE and WEISBERG · Statistical Methods for Comparative Studies
ANDERSON and LOYNES · The Teaching of Practical Statistics
ARMITAGE and DAVID (editors) · Advances in Biometry
ARNOLD, BALAKRISHNAN and NAGARAJA · Records
*ARTHANARI and DODGE · Mathematical Programming in Statistics
*BAILEY · The Elements of Stochastic Processes with Applications to the Natural Sciences
BALAKRISHNAN and KOUTRAS · Runs and Scans with Applications

*Now available in a lower priced paperback edition in the Wiley Classics Library.

BALAKRISHNAN and NG · Precedence-Type Tests and Applications
BARNETT · Comparative Statistical Inference, Third Edition
BARNETT · Environmental Statistics: Methods & Applications
BARNETT and LEWIS · Outliers in Statistical Data, Third Edition
BARTOSZYNSKI and NIEWIADOMSKA-BUGAJ · Probability and Statistical Inference
BASILEVSKY · Statistical Factor Analysis and Related Methods: Theory and Applications
BASU and RIGDON · Statistical Methods for the Reliability of Repairable Systems
BATES and WATTS · Nonlinear Regression Analysis and Its Applications
BECHHOFER, SANTNER and GOLDSMAN · Design and Analysis of Experiments for Statistical Selection, Screening and Multiple Comparisons
BEIRLANT, GOEGEBEUR, SEGERS, TEUGELS and DE WAAL · Statistics of Extremes: Theory and Applications
BELSLEY · Conditioning Diagnostics: Collinearity and Weak Data in Regression
BELSLEY, KUH and WELSCH · Regression Diagnostics: Identifying Influential Data and Sources of Collinearity
BENDAT and PIERSOL · Random Data: Analysis and Measurement Procedures, Third Edition
BERNARDO and SMITH · Bayesian Theory
BERRY, CHALONER and GEWEKE · Bayesian Analysis in Statistics and Econometrics: Essays in Honor of Arnold Zellner
BHAT and MILLER · Elements of Applied Stochastic Processes, Third Edition
BHATTACHARYA and JOHNSON · Statistical Concepts and Methods
BHATTACHARYA and WAYMIRE · Stochastic Processes with Applications
BIEMER, GROVES, LYBERG, MATHIOWETZ and SUDMAN · Measurement Errors in Surveys
BILLINGSLEY · Convergence of Probability Measures, Second Edition
BILLINGSLEY · Probability and Measure, Third Edition
BIRKES and DODGE · Alternative Methods of Regression
BISWAS, DATTA, FINE and SEGAL · Statistical Advances in the Biomedical Sciences: Clinical Trials, Epidemiology, Survival Analysis, and Bioinformatics
BLISCHKE and MURTHY (editors) · Case Studies in Reliability and Maintenance
BLISCHKE and MURTHY · Reliability: Modeling, Prediction and Optimization
BLOOMFIELD · Fourier Analysis of Time Series: An Introduction, Second Edition
BOLLEN · Structural Equations with Latent Variables
BOLLEN and CURRAN · Latent Curve Models: A Structural Equation Perspective
BOROVKOV · Ergodicity and Stability of Stochastic Processes
BOSQ and BLANKE · Inference and Prediction in Large Dimensions
BOULEAU · Numerical Methods for Stochastic Processes
BOX · Bayesian Inference in Statistical Analysis
BOX · R. A. Fisher, the Life of a Scientist
BOX and DRAPER · Empirical Model-Building and Response Surfaces
*BOX and DRAPER · Evolutionary Operation: A Statistical Method for Process Improvement
BOX · Improving Almost Anything *Revised Edition*

*Now available in a lower priced paperback edition in the Wiley Classics Library.

BOX, HUNTER and HUNTER · Statistics for Experimenters: An Introduction to Design, Data Analysis and Model Building
BOX, HUNTER and HUNTER · Statistics for Experimenters: Design, Innovation and Discovery, Second Edition
BOX and LUCEÑO · Statistical Control by Monitoring and Feedback Adjustment
BRANDIMARTE · Numerical Methods in Finance: A MATLAB-Based Introduction
BROWN and HOLLANDER · Statistics: A Biomedical Introduction
BRUNNER, DOMHOF and LANGER · Nonparametric Analysis of Longitudinal Data in Factorial Experiments
BUCKLEW · Large Deviation Techniques in Decision, Simulation and Estimation
CAIROLI and DALANG · Sequential Stochastic Optimization
CASTILLO, HADI, BALAKRISHNAN and SARABIA · Extreme Value and Related Models with Applications in Engineering and Science
CHAN · Time Series: Applications to Finance
CHARALAMBIDES · Combinatorial Methods in Discrete Distributions
CHATTERJEE and HADI · Regression Analysis by Example, Fourth Edition
CHATTERJEE and HADI · Sensitivity Analysis in Linear Regression
CHERNICK · Bootstrap Methods: A Practitioner's Guide
CHERNICK and FRIIS · Introductory Biostatistics for the Health Sciences
CHILÉS and DELFINER · Geostatistics: Modeling Spatial Uncertainty
CHOW and LIU · Design and Analysis of Clinical Trials: Concepts and Methodologies, Second Edition
CLARKE · Linear Models: The Theory and Application of Analysis of Variance
CLARKE and DISNEY · Probability and Random Processes: A First Course with Applications, Second Edition
*COCHRAN and COX · Experimental Designs, Second Edition
CONGDON · Applied Bayesian Modelling
CONGDON · Bayesian Models for Categorical Data
CONGDON · Bayesian Statistical Modelling, Second Edition
CONOVER · Practical Nonparametric Statistics, Second Edition
COOK · Regression Graphics
COOK and WEISBERG · An Introduction to Regression Graphics
COOK and WEISBERG · Applied Regression Including Computing and Graphics
CORNELL · Experiments with Mixtures, Designs, Models and the Analysis of Mixture Data, Third Edition
COVER and THOMAS · Elements of Information Theory
COX · A Handbook of Introductory Statistical Methods
*COX · Planning of Experiments
CRESSIE · Statistics for Spatial Data, Revised Edition
CSÖRGÖ and HORVÁTH · Limit Theorems in Change Point Analysis
DANIEL · Applications of Statistics to Industrial Experimentation
DANIEL · Biostatistics: A Foundation for Analysis in the Health Sciences, Sixth Edition
*DANIEL · Fitting Equations to Data: Computer Analysis of Multifactor Data, Second Edition
DASU and JOHNSON · Exploratory Data Mining and Data Cleaning

*Now available in a lower priced paperback edition in the Wiley Classics Library.

DAVID and NAGARAJA · Order Statistics, Third Edition
*DEGROOT, FIENBERG and KADANE · Statistics and the Law
DEL CASTILLO · Statistical Process Adjustment for Quality Control
DEMARIS · Regression with Social Data: Modeling Continuous and Limited Response Variables
DEMIDENKO · Mixed Models: Theory and Applications
DENISON, HOLMES, MALLICK and SMITH · Bayesian Methods for Nonlinear Classification and Regression
DETTE and STUDDEN · The Theory of Canonical Moments with Applications in Statistics, Probability and Analysis
DEY and MUKERJEE · Fractional Factorial Plans
DILLON and GOLDSTEIN · Multivariate Analysis: Methods and Applications
DODGE · Alternative Methods of Regression
*DODGE and ROMIG · Sampling Inspection Tables, Second Edition
*DOOB · Stochastic Processes
DOWDY, WEARDEN and CHILKO · Statistics for Research, Third Edition
DRAPER and SMITH · Applied Regression Analysis, Third Edition
DRYDEN and MARDIA · Statistical Shape Analysis
DUDEWICZ and MISHRA · Modern Mathematical Statistics
DUNN and CLARK · Applied Statistics: Analysis of Variance and Regression, Second Edition
DUNN and CLARK · Basic Statistics: A Primer for the Biomedical Sciences, Third Edition
DUPUIS and ELLIS · A Weak Convergence Approach to the Theory of Large Deviations
EDLER and KITSOS (editors) · Recent Advances in Quantitative Methods in Cancer and Human Health Risk Assessment
*ELANDT-JOHNSON and JOHNSON · Survival Models and Data Analysis
ENDERS · Applied Econometric Time Series
ETHIER and KURTZ · Markov Processes: Characterization and Convergence
EVANS, HASTINGS and PEACOCK · Statistical Distribution, Third Edition
FELLER · An Introduction to Probability Theory and Its Applications, Volume I, Third Edition, Revised; Volume II, Second Edition
FISHER and VAN BELLE · Biostatistics: A Methodology for the Health Sciences
FITZMAURICE, LAIRD and WARE · Applied Longitudinal Analysis
*FLEISS · The Design and Analysis of Clinical Experiments
FLEISS · Statistical Methods for Rates and Proportions, Second Edition
FLEMING and HARRINGTON · Counting Processes and Survival Analysis
FULLER · Introduction to Statistical Time Series, Second Edition
FULLER · Measurement Error Models
GALLANT · Nonlinear Statistical Models.
GEISSER · Modes of Parametric Statistical Inference
GELMAN and MENG (editors) · Applied Bayesian Modeling and Casual Inference from Incomplete-data Perspectives
GEWEKE · Contemporary Bayesian Econometrics and Statistics
GHOSH, MUKHOPADHYAY and SEN · Sequential Estimation

*Now available in a lower priced paperback edition in the Wiley Classics Library.

GIESBRECHT and GUMPERTZ · Planning, Construction and Statistical Analysis of Comparative Experiments
GIFI · Nonlinear Multivariate Analysis
GIVENS and HOETING · Computational Statistics
GLASSERMAN and YAO · Monotone Structure in Discrete-Event Systems
GNANADESIKAN · Methods for Statistical Data Analysis of Multivariate Observations, Second Edition
GOLDSTEIN and WOLFF · Bayes Linear Statistics, Theory & Methods
GOLDSTEIN and LEWIS · Assessment: Problems, Development and Statistical Issues
GREENWOOD and NIKULIN · A Guide to Chi-Squared Testing
GROSS, SHORTLE, THOMPSON and HARRIS · Fundamentals of Queueing Theory, Fourth Edition
GROSS, SHORTLE, THOMPSON and HARRIS · Solutions Manual to Accompany Fundamentals of Queueing Theory, Fourth Edition
*HAHN and SHAPIRO · Statistical Models in Engineering
HAHN and MEEKER · Statistical Intervals: A Guide for Practitioners
HALD · A History of Probability and Statistics and their Applications Before 1750
HALD · A History of Mathematical Statistics from 1750 to 1930
HAMPEL · Robust Statistics: The Approach Based on Influence Functions
HANNAN and DEISTLER · The Statistical Theory of Linear Systems
HARTUNG, KNAPP and SINHA · Statistical Meta-Analysis with Applications
HEIBERGER · Computation for the Analysis of Designed Experiments
HEDAYAT and SINHA · Design and Inference in Finite Population Sampling
HEDEKER and GIBBONS · Longitudinal Data Analysis
HELLER · MACSYMA for Statisticians
HERITIER, CANTONI, COPT and VICTORIA-FESER · Robust Methods in Biostatistics
HINKELMANN and KEMPTHORNE · Design and Analysis of Experiments, Volume 1: Introduction to Experimental Design
HINKELMANN and KEMPTHORNE · Design and analysis of experiments, Volume 2: Advanced Experimental Design
HOAGLIN, MOSTELLER and TUKEY · Exploratory Approach to Analysis of Variance
HOAGLIN, MOSTELLER and TUKEY · Exploring Data Tables, Trends and Shapes
*HOAGLIN, MOSTELLER and TUKEY · Understanding Robust and Exploratory Data Analysis
HOCHBERG and TAMHANE · Multiple Comparison Procedures
HOCKING · Methods and Applications of Linear Models: Regression and the Analysis of Variance, Second Edition
HOEL · Introduction to Mathematical Statistics, Fifth Edition
HOGG and KLUGMAN · Loss Distributions
HOLLANDER and WOLFE · Nonparametric Statistical Methods, Second Edition
HOSMER and LEMESHOW · Applied Logistic Regression, Second Edition
HOSMER and LEMESHOW · Applied Survival Analysis: Regression Modeling of Time to Event Data
HUBER · Robust Statistics
HUBERTY · Applied Discriminant Analysis

*Now available in a lower priced paperback edition in the Wiley Classics Library.

HUBERY and OLEJNKI · Applied MANOVA and Discriminant Analysis, 2nd Edition
HUNT and KENNEDY · Financial Derivatives in Theory and Practice, Revised Edition
HURD and MIAMEE · Periodically Correlated Random Sequences: Spectral Theory and Practice
HUSKOVA, BERAN and DUPAC · Collected Works of Jaroslav Hajek – with Commentary
HUZURBAZAR · Flowgraph Models for Multistate Time-to-Event Data
IMAN and CONOVER · A Modern Approach to Statistics
JACKMAN · Bayesian Analysis for the Social Sciences
JACKSON · A User's Guide to Principle Components
JOHN · Statistical Methods in Engineering and Quality Assurance
JOHNSON · Multivariate Statistical Simulation
JOHNSON and BALAKRISHNAN · Advances in the Theory and Practice of Statistics: A Volume in Honor of Samuel Kotz
JOHNSON and BHATTACHARYYA · Statistics: Principles and Methods, Fifth Edition
JOHNSON and KOTZ · Distributions in Statistics
JOHNSON and KOTZ (editors) · Leading Personalities in Statistical Sciences: From the Seventeenth Century to the Present
JOHNSON, KOTZ and BALAKRISHNAN · Continuous Univariate Distributions, Volume 1, Second Edition
JOHNSON, KOTZ and BALAKRISHNAN · Continuous Univariate Distributions, Volume 2, Second Edition
JOHNSON, KOTZ and BALAKRISHNAN · Discrete Multivariate Distributions
JOHNSON, KOTZ and KEMP · Univariate Discrete Distributions, Second Edition
JUDGE, GRIFFITHS, HILL, LU TKEPOHL and LEE · The Theory and Practice of Econometrics, Second Edition
JUREČKOVÁ and SEN · Robust Statistical Procedures: Asymptotics and Interrelations
JUREK and MASON · Operator-Limit Distributions in Probability Theory
KADANE · Bayesian Methods and Ethics in a Clinical Trial Design
KADANE and SCHUM · A Probabilistic Analysis of the Sacco and Vanzetti Evidence
KALBFLEISCH and PRENTICE · The Statistical Analysis of Failure Time Data, Second Edition
KARIYA and KURATA · Generalized Least Squares
KASS and VOS · Geometrical Foundations of Asymptotic Inference
KAUFMAN and ROUSSEEUW · Finding Groups in Data: An Introduction to Cluster Analysis
KEDEM and FOKIANOS · Regression Models for Time Series Analysis
KENDALL, BARDEN, CARNE and LE · Shape and Shape Theory
KHURI · Advanced Calculus with Applications in Statistics, Second Edition
KHURI, MATHEW and SINHA · Statistical Tests for Mixed Linear Models
*KISH · Statistical Design for Research
KLEIBER and KOTZ · Statistical Size Distributions in Economics and Actuarial Sciences
KLUGMAN, PANJER and WILLMOT · Loss Models: From Data to Decisions
KLUGMAN, PANJER and WILLMOT · Solutions Manual to Accompany Loss Models: From Data to Decisions

*Now available in a lower priced paperback edition in the Wiley – Interscience Paperback Series.

KOSKI and NOBLE · Bayesian Networks: An Introduction
KOTZ, BALAKRISHNAN and JOHNSON · Continuous Multivariate Distributions, Volume 1, Second Edition
KOTZ and JOHNSON (editors) · Encyclopedia of Statistical Sciences: Volumes 1 to 9 with Index
KOTZ and JOHNSON (editors) · Encyclopedia of Statistical Sciences: Supplement Volume
KOTZ, READ and BANKS (editors) · Encyclopedia of Statistical Sciences: Update Volume 1
KOTZ, READ and BANKS (editors) · Encyclopedia of Statistical Sciences: Update Volume 2
KOVALENKO, KUZNETZOV and PEGG · Mathematical Theory of Reliability of Time-Dependent Systems with Practical Applications
KOWALSI and TU · Modern Applied U-Statistics
KROONENBERG · Applied Multiway Data Analysis
KULINSKAYA, MORGENTHALER and STAUDTE · Meta Analysis: A Guide to Calibrating and Combining Statistical Evidence
KUROWICKA and COOKE · Uncertainty Analysis with High Dimensional Dependence Modelling
KVAM and VIDAKOVIC · Nonparametric Statistics with Applications to Science and Engineering
LACHIN · Biostatistical Methods: The Assessment of Relative Risks
LAD · Operational Subjective Statistical Methods: A Mathematical, Philosophical and Historical Introduction
LAMPERTI · Probability: A Survey of the Mathematical Theory, Second Edition
LANGE, RYAN, BILLARD, BRILLINGER, CONQUEST and GREENHOUSE · Case Studies in Biometry
LARSON · Introduction to Probability Theory and Statistical Inference, Third Edition
LAWLESS · Statistical Models and Methods for Lifetime Data, Second Edition
LAWSON · Statistical Methods in Spatial Epidemiology, Second Edition
LE · Applied Categorical Data Analysis
LE · Applied Survival Analysis
LEE · Structural Equation Modelling: A Bayesian Approach
LEE and WANG · Statistical Methods for Survival Data Analysis, Third Edition
LEPAGE and BILLARD · Exploring the Limits of Bootstrap
LEYLAND and GOLDSTEIN (editors) · Multilevel Modelling of Health Statistics
LIAO · Statistical Group Comparison
LINDVALL · Lectures on the Coupling Method
LIN · Introductory Stochastic Analysis for Finance and Insurance
LINHART and ZUCCHINI · Model Selection
LITTLE and RUBIN · Statistical Analysis with Missing Data, Second Edition
LLOYD · The Statistical Analysis of Categorical Data
LOWEN and TEICH · Fractal-Based Point Processes
MAGNUS and NEUDECKER · Matrix Differential Calculus with Applications in Statistics and Econometrics, Revised Edition
MALLER and ZHOU · Survival Analysis with Long Term Survivors
MALLOWS · Design, Data and Analysis by Some Friends of Cuthbert Daniel

MANN, SCHAFER and SINGPURWALLA · Methods for Statistical Analysis of Reliability and Life Data

MANTON, WOODBURY and TOLLEY · Statistical Applications Using Fuzzy Sets

MARCHETTE · Random Graphs for Statistical Pattern Recognition

MARKOVICH · Nonparametric Analysis of Univariate Heavy-Tailed Data: Research and Practice

MARDIA and JUPP · Directional Statistics

MARKOVICH · Nonparametric Analysis of Univariate Heavy-Tailed Data: Research and Practice

MARONNA, MARTIN and YOHAI · Robust Statistics: Theory and Methods

MASON, GUNST and HESS · Statistical Design and Analysis of Experiments with Applications to Engineering and Science, Second Edition

MCCULLOCH and SERLE · Generalized, Linear and Mixed Models

MCFADDEN · Management of Data in Clinical Trials

MCLACHLAN · Discriminant Analysis and Statistical Pattern Recognition

MCLACHLAN, DO and AMBROISE · Analyzing Microarray Gene Expression Data

MCLACHLAN and KRISHNAN · The EM Algorithm and Extensions

MCLACHLAN and PEEL · Finite Mixture Models

MCNEIL · Epidemiological Research Methods

MEEKER and ESCOBAR · Statistical Methods for Reliability Data

MEERSCHAERT and SCHEFFLER · Limit Distributions for Sums of Independent Random Vectors: Heavy Tails in Theory and Practice

MICKEY, DUNN and CLARK · Applied Statistics: Analysis of Variance and Regression, Third Edition

*MILLER · Survival Analysis, Second Edition

MONTGOMERY, JENNINGS and KULAHCI · Introduction to Time Series Analysis and Forecasting Solutions Set

MONTGOMERY, PECK and VINING · Introduction to Linear Regression Analysis, Fourth Edition

MORGENTHALER and TUKEY · Configural Polysampling: A Route to Practical Robustness

MUIRHEAD · Aspects of Multivariate Statistical Theory

MULLER and STEWART · Linear Model Theory: Univariate, Multivariate and Mixed Models

MURRAY · X-STAT 2.0 Statistical Experimentation, Design Data Analysis and Nonlinear Optimization

MURTHY, XIE and JIANG · Weibull Models

MYERS and MONTGOMERY · Response Surface Methodology: Process and Product Optimization Using Designed Experiments, Second Edition

MYERS, MONTGOMERY and VINING · Generalized Linear Models. With Applications in Engineering and the Sciences

†NELSON · Accelerated Testing, Statistical Models, Test Plans and Data Analysis

†NELSON · Applied Life Data Analysis

NEWMAN · Biostatistical Methods in Epidemiology

*Now available in a lower priced paperback edition in the Wiley Classics Library.

†Now available in a lower priced paperback edition in the Wiley – Interscience Paperback Series.

OCHI · Applied Probability and Stochastic Processes in Engineering and Physical Sciences
OKABE, BOOTS, SUGIHARA and CHIU · Spatial Tesselations: Concepts and Applications of Voronoi Diagrams, Second Edition
OLIVER and SMITH · Influence Diagrams, Belief Nets and Decision Analysis
PALTA · Quantitative Methods in Population Health: Extentions of Ordinary Regression
PANJER · Operational Risks: Modeling Analytics
PANKRATZ · Forecasting with Dynamic Regression Models
PANKRATZ · Forecasting with Univariate Box-Jenkins Models: Concepts and Cases
PARDOUX · Markov Processes and Applications: Algorithms, Networks, Genome and Finance
PARMIGIANI and INOUE · Decision Theory: Principles and Approaches
*PARZEN · Modern Probability Theory and Its Applications
PEÑA, TIAO and TSAY · A Course in Time Series Analysis
PESARIN and SALMASO · Permutation Tests for Complex Data: Theory, Applications and Software
PIANTADOSI · Clinical Trials: A Methodologic Perspective
PORT · Theoretical Probability for Applications
POURAHMADI · Foundations of Time Series Analysis and Prediction Theory
POWELL · Approximate Dynamic Programming: Solving the Curses of Dimensionality
PRESS · Bayesian Statistics: Principles, Models and Applications
PRESS · Subjective and Objective Bayesian Statistics, Second Edition
PRESS and TANUR · The Subjectivity of Scientists and the Bayesian Approach
PUKELSHEIM · Optimal Experimental Design
PURI, VILAPLANA and WERTZ · New Perspectives in Theoretical and Applied Statistics
PUTERMAN · Markov Decision Processes: Discrete Stochastic Dynamic Programming
QIU · Image Processing and Jump Regression Analysis
RAO · Linear Statistical Inference and its Applications, Second Edition
RAUSAND and HØYLAND · System Reliability Theory: Models, Statistical Methods and Applications, Second Edition
RENCHER · Linear Models in Statistics
RENCHER · Methods of Multivariate Analysis, Second Edition
RENCHER · Multivariate Statistical Inference with Applications
RIPLEY · Spatial Statistics
RIPLEY · Stochastic Simulation
ROBINSON · Practical Strategies for Experimenting
ROHATGI and SALEH · An Introduction to Probability and Statistics, Second Edition
ROLSKI, SCHMIDLI, SCHMIDT and TEUGELS · Stochastic Processes for Insurance and Finance
ROSENBERGER and LACHIN · Randomization in Clinical Trials: Theory and Practice
ROSS · Introduction to Probability and Statistics for Engineers and Scientists

*Now available in a lower priced paperback edition in the Wiley Classics Library.

ROSSI, ALLENBY and MCCULLOCH · Bayesian Statistics and Marketing
ROUSSEEUW and LEROY · Robust Regression and Outline Detection
ROYSTON and SAUERBREI · Multivariable Model - Building: A Pragmatic Approach to Regression Anaylsis based on Fractional Polynomials for Modelling Continuous Variables
RUBIN · Multiple Imputation for Nonresponse in Surveys
RUBINSTEIN · Simulation and the Monte Carlo Method, Second Edition
RUBINSTEIN and MELAMED · Modern Simulation and Modeling
RYAN · Modern Engineering Statistics
RYAN · Modern Experimental Design
RYAN · Modern Regression Methods
RYAN · Statistical Methods for Quality Improvement, Second Edition
SALEH · Theory of Preliminary Test and Stein-Type Estimation with Applications
SALTELLI, CHAN and SCOTT (editors) · Sensitivity Analysis
*SCHEFFE · The Analysis of Variance
SCHIMEK · Smoothing and Regression: Approaches, Computation and Application
SCHOTT · Matrix Analysis for Statistics
SCHOUTENS · Levy Processes in Finance: Pricing Financial Derivatives
SCHUSS · Theory and Applications of Stochastic Differential Equations
SCOTT · Multivariate Density Estimation: Theory, Practice and Visualization
*SEARLE · Linear Models
SEARLE · Linear Models for Unbalanced Data
SEARLE · Matrix Algebra Useful for Statistics
SEARLE and WILLETT · Matrix Algebra for Applied Economics
SEBER · Multivariate Observations
SEBER and LEE · Linear Regression Analysis, Second Edition
SEBER and WILD · Nonlinear Regression
SENNOTT · Stochastic Dynamic Programming and the Control of Queueing Systems
*SERFLING · Approximation Theorems of Mathematical Statistics
SHAFER and VOVK · Probability and Finance: Its Only a Game!
SILVAPULLE and SEN · Constrained Statistical Inference: Inequality, Order and Shape Restrictions
SINGPURWALLA · Reliability and Risk: A Bayesian Perspective
SMALL and MCLEISH · Hilbert Space Methods in Probability and Statistical Inference
SRIVASTAVA · Methods of Multivariate Statistics
STAPLETON · Linear Statistical Models
STAUDTE and SHEATHER · Robust Estimation and Testing
STOYAN, KENDALL and MECKE · Stochastic Geometry and Its Applications, Second Edition
STOYAN and STOYAN · Fractals, Random and Point Fields: Methods of Geometrical Statistics
STREET and BURGESS · The Construction of Optimal Stated Choice Experiments: Theory and Methods
STYAN · The Collected Papers of T. W. Anderson: 1943–1985

*Now available in a lower priced paperback edition in the Wiley Classics Library.

SUTTON, ABRAMS, JONES, SHELDON and SONG · Methods for Meta-Analysis in Medical Research
TAKEZAWA · Introduction to Nonparametric Regression
TAMHANE · Statistical Analysis of Designed Experiments: Theory and Applications
TANAKA · Time Series Analysis: Nonstationary and Noninvertible Distribution Theory
THOMPSON · Empirical Model Building
THOMPSON · Sampling, Second Edition
THOMPSON · Simulation: A Modeler's Approach
THOMPSON and SEBER · Adaptive Sampling
THOMPSON, WILLIAMS and FINDLAY · Models for Investors in Real World Markets
TIAO, BISGAARD, HILL, PEÑA and STIGLER (editors) · Box on Quality and Discovery: with Design, Control and Robustness
TIERNEY · LISP-STAT: An Object-Oriented Environment for Statistical Computing and Dynamic Graphics
TSAY · Analysis of Financial Time Series
UPTON and FINGLETON · Spatial Data Analysis by Example, Volume II: Categorical and Directional Data
VAN BELLE · Statistical Rules of Thumb
VAN BELLE, FISHER, HEAGERTY and LUMLEY · Biostatistics: A Methodology for the Health Sciences, Second Edition
VESTRUP · The Theory of Measures and Integration
VIDAKOVIC · Statistical Modeling by Wavelets
VINOD and REAGLE · Preparing for the Worst: Incorporating Downside Risk in Stock Market Investments
WALLER and GOTWAY · Applied Spatial Statistics for Public Health Data
WEERAHANDI · Generalized Inference in Repeated Measures: Exact Methods in MANOVA and Mixed Models
WEISBERG · Applied Linear Regression, Second Edition
WELSH · Aspects of Statistical Inference
WESTFALL and YOUNG · Resampling-Based Multiple Testing: Examples and Methods for p-Value Adjustment
WHITTAKER · Graphical Models in Applied Multivariate Statistics
WINKER · Optimization Heuristics in Economics: Applications of Threshold Accepting
WONNACOTT and WONNACOTT · Econometrics, Second Edition
WOODING · Planning Pharmaceutical Clinical Trials: Basic Statistical Principles
WOODWORTH · Biostatistics: A Bayesian Introduction
WOOLSON and CLARKE · Statistical Methods for the Analysis of Biomedical Data, Second Edition
WU and HAMADA · Experiments: Planning, Analysis and Parameter Design Optimization
WU and ZHANG · Nonparametric Regression Methods for Longitudinal Data Analysis: Mixed-Effects Modeling Approaches
YANG · The Construction Theory of Denumerable Markov Processes

YOUNG, VALERO-MORA and FRIENDLY · Visual Statistics: Seeing Data with Dynamic Interactive Graphics
ZACKS · Stage-Wise Adaptive Designs
*ZELLNER · An Introduction to Bayesian Inference in Econometrics
ZELTERMAN · Discrete Distributions: Applications in the Health Sciences
ZHOU, OBUCHOWSKI and McCLISH · Statistical Methods in Diagnostic Medicine

*Now available in a lower priced paperback edition in the Wiley Classics Library.

Printed and bound by CPI Group (UK) Ltd, Croydon, CR0 4YY

27/10/2024

14580284-0001